LOEB CLASSICAL LIBRARY
FOUNDED BY JAMES LOEB 1911

EDITED BY
JEFFREY HENDERSON

ARISTOTLE
IX

LCL 437

ARISTOTLE

HISTORY OF ANIMALS
BOOKS I–III

WITH AN ENGLISH TRANSLATION BY

A. L. PECK

HARVARD UNIVERSITY PRESS
CAMBRIDGE, MASSACHUSETTS
LONDON, ENGLAND

Copyright © 1965 by the President and Fellows
of Harvard College
All rights reserved

First published 1965

LOEB CLASSICAL LIBRARY® is a registered trademark
of the President and Fellows of Harvard College

ISBN 978-0-674-99481-2

*Printed on acid-free paper and bound by
The Maple-Vail Book Manufacturing Group*

CONTENTS

INTRODUCTION v

    The Manuscripts   xxxiv
    Printed Editions   xxxviii
    Translations   xlii
    Students of Aristotle's Zoology   xlv
    Authenticity   lv
    Date of the Treatise   lx
    This Edition   lxi

NOTES ON TERMINOLOGY lxiv

TABLES xcii

SIGLA ci

HISTORY OF ANIMALS

    Book I   2
    Book II   74
    Book III   148

ADDITIONAL NOTES 237

INTRODUCTION

HISTORIA—information obtained through investigation : this is how Aristotle describes the fundamental requirement in zoology ; this is what we must have before us to begin with : the ascertained facts about each kind of animal (*H.A.* 491 a 12). If we ask more precisely what this involves, Aristotle's answer is clear : it is that we must inform ourselves of the differences which actually exist among the various animals. The emphasis is laid upon the actual facts. This work is first and foremost to be a factual survey. Until we have got this *historia*, we are not in a position to see what we are dealing with, or how to deal with it. Once we have got it, we can go on to the second stage, which is to attempt to find out the " causes " of these observed and recorded differences.

This is Aristotle's own account of the purpose of the treatise, and not surprisingly it accords more closely with what we find in it than some alternative accounts which have been put forward. Many attempts have been made to represent it as an essay in taxonomy, and these attempts have continued although as long ago as 1855 J. B. Meyer [a] showed that this view was mistaken. Most of the differentiae, he pointed out, which scholars had regarded as characteristics distinguishing genera, were not so used by Aristotle himself : the group-names were not meant to indicate systematic divisions, but were intended by him as descriptions. Not only had earlier attempts to work out a systematic classification been widely divergent in their results ; none of them had been successful. Indeed, Meyer's own attempt to work out a scheme cannot be regarded as satisfactory. At

[a] Jürgen Bona Meyer. *Aristoteles Thierkunde*, Berlin, 1855.

ARISTOTLE

the other extreme we have Whewell's view,[a] quoted by Meyer, that there is no system at all. The error underlying all these interpretations, as Mr. D. M. Balme has shown in a very valuable study,[b] is the assumption that Aristotle " put systematics first in zoology, and morphology first in systematics." Aristotle himself frequently points out how impossible it is to produce a neat hierarchical system on the basis of the obvious physical differences, because such divisions cut across each other ; and in fact the differences which he lists and records relate not only to the physical parts of animals, though he agrees that these are of the first importance, but also to their manner of life, their activities, and their dispositions (see *H.A.* 487 a 12 ff., 491 a 15 ff.). His purpose is not to construct a system, but to collect data for ascertaining the causes of the observed phenomena ; and this is to be done by looking to see whether certain characteristics are regularly found in combination : this is how the clues to the causes will be brought to light. No arbitrary and premature selection of this or that

---

[a] William Whewell, *History of the Inductive Sciences from the Earliest to the Present Times*, London, 1837. Meyer also quotes the views of several other writers. See too P. Louis, *op. cit. infra* (p. xxiii). To which may be added Jaeger's opinion (*Aristotle*, Eng. tr., p. 374) : "Aristotle's writings always present a disparate picture if we examine their systematic structure in detail. In this respect the *H.A.* is the same as the *Metaphysics* or the *Politics*. Outlines of a systematic arrangement, often introduced only during the subsequent labour of welding the parts together, are carried only half through or remain entirely unfulfilled. To produce an external architectonic was not the original idea of this builder and therefore none can be ' reconstructed.' "

[b] D. M. Balme, *Aristotle's use of differentiae in Zoology*, in *Aristote et les problèmes de méthode*, Louvain, 1962, p. 205.

HISTORIA ANIMALIUM

differentia will be relevant for this purpose ; a proper and thorough study of *all* observable differentiae must be made ; and that, as we have seen, is what Aristotle professes to be undertaking in this treatise. He does, it is true, distinguish certain " main groups " (μέγιστα γένη) of animals, but the purpose of this is not to provide a starting-point for systematic division and subdivision ; it is for convenience in reviewing the various observable differences.[a] Some of these " main groups" have popular names (such as Birds and Fishes), some have not (τὰ μαλάκια, for instance, is not, strictly speaking, a name but a descriptive adjective). Some animals cut across the divisions ; some (such as man) do not fall under any " main group," and so on.

There is, then, no need to look for any thorough and exhaustive classification of animals in the *H.A.*, for we shall find none. Nevertheless, in some parts of Aristotle's zoological works, as well as in the non-zoological works, we find statements about division and classification. It is perhaps unnecessary here to discuss fully what Aristotle says in his non-zoological works, though the outlines will be relevant. The procedure as stated in the *Organon* and *Metaphysics* is this : the highest genus is divided by means of differentiae into subaltern genera, and each of these is then divided and subdivided, until the ultimate species is reached. In making these divisions it is, of course, important to make sure that the determining characteristics are points of essential and not merely quali-

---

[a] It has in fact been claimed that these groups are pre-Aristotelian (see below, p. li, Rudolf Burckhardt, *Das koische Tiersystem*). Indeed, it is obvious that some are as old as the Greek language.

ARISTOTLE

tative difference, and the differentiae must be taken in the right order. This method originated in the Academic method of *diairesis*, and through the Stoics, Porphyry and the Greek commentators came down to Linnaeus and to modern systematics. In some, though not all, of Aristotle's logical treatises, dichotomy appears to be accepted; but in the first book of the *Parts of Animals* (642 b 10 ff.) dichotomy is criticized on the ground that it splits up natural groups (γένη): *e.g.*, the dichotomy land-animals/water-animals splits up the γένος Birds. The right procedure would be to divide a group not into two, and again into two, but into its various natural divisions. Accordingly, in Book I of the *Parts of Animals* we often (though not invariably) find γένος and εἶδος are practically equivalent to "genus" and "species," *i.e.*, they are used to distinguish a larger group from its subdivisions. This practice, however, is hardly ever followed by Aristotle in the main part of his zoological works. There are indications of it here and there, but for the most part he does not appear to envisage any continuous series of divisions such as would be required by the principle just described. At the upper end of the scale we find, as I have already said, certain "main groups" of animals (such as Bird and Fish), and at the lower end the commonly accepted and named types (such as dog, eagle, etc.), and subdivisions of these (what we might now describe as species, varieties, and breeds); but normally the intermediate stages are missing; and it is these intermediate stages, chiefly, which commentators have been concerned to supply, and which they have supplied with such lack of unanimity. The question therefore arises, Why, if Aristotle believed in the

HISTORIA ANIMALIUM

value of division of the sort described, is there so little of it in his zoological works?

I think there are at least three things which can usefully be said in reply to this question. The first is that in the passages of the non-zoological works referring to " division " where zoological terms are introduced (such as πεζόν, ἔνυδρον, footed, footless, two-footed, winged, wingless, and the like), it is clear that Aristotle's concern is not primarily or directly with animals but with logical problems, and the zoological terms are used as convenient ones for illustration; they may well have come to be accepted as standard for this purpose. In any case, it is obvious that no full-scale or exhaustive scheme for the division or classification of animals is being propounded, although, as we noticed in the preceding paragraph, Aristotle sometimes indicates how a particular principle of division would work out if followed through. But not only is the treatment in these passages non-zoological, it is also non-physical, in the sense that no attempt is made to correlate the characteristics mentioned with the physical composition of animals; or, to put it another way, no attempt is made to diagnose the *cause* of these characteristics, not even the " material " cause.[a] When we come to the zoological works, we are in another atmosphere. At the beginning of *H.A.* Aristotle gives a conspectus of the

---

[a] To say nothing of the " formal " and " final " causes. These, and the " efficient " cause, are fully dealt with in *P.A.* (for which Aristotle's title is " The Causes of the Parts of Animals ") and in *G.A.*; and it is inconceivable that Aristotle would have been satisfied (as most of his commentators are) with a " classification " which confined itself to the material cause (τὸ ἐξ ἀνάγκης) and took no account of the final cause (τὸ βέλτιον).

ARISTOTLE

sorts of " differences " exhibited by animals which will have to be recorded and examined (including, of course, those used by way of illustration in the non-zoological treatises), and he then tells us immediately what the purpose of this examination is : it is (not to produce a scheme for the " division " of animals, but) to ascertain the *causes* of the observed characteristics (491 a 10). This operation does in fact incidentally provide a means of " classification " on the level of the material cause, if we need one (*G.A.* 732 a, b), but this is not the primary purpose of the undertaking. As we shall see, on this level the " cause " is to be found in the physical composition of the animals : this part of the problem, then, is to be dealt with by referring to the fundamentals of Aristotle's physics. Many observed characteristics have something to tell us about an animal's physical composition : among them are its posture (*P.A.* 686 a 24 ff.), the number of feet it has (*ibid.* 686 b 27), its habitat (see Notes, §§ 39 ff.), its sex (*G.A.* 765 b 7 ff.), its method of reproduction (*G.A.* 732 b 15 ff.), and so on. But not all of these are useful as means for making " divisions " (see, *e.g.*, *G.A.* 732 b 27) ; in any case, they are all symptoms ; and the " cause " is to be looked for beyond them, *viz.*, in the physical composition of the animal (*G.A. ibid.* and ff.). Some characteristics are more illuminating and informative than others, and offer the possibility of obtaining a longer and more significant series of differences—above all, the methods of reproduction. It is true that Aristotle does not always find it easy to correlate convincingly all the observed phenomena by reference to the fundamental physical " cause "—not as easy as we find it to say that he was foredoomed to failure by

HISTORIA ANIMALIUM

the inadequacy of his theoretical and practical equipment. But he himself is in no doubt about the nature of the " cause."

The second consideration relates to the *Historia animalium* itself. If we are right in supposing that the purpose of the treatise is preliminary, *viz.*, to collect together as many records of differentiae as possible as a basis for discovering the causes of observed phenomena in animals, then it constitutes an earlier stage of the whole process than classification, to which it will be a prelude. As Aristotle himself says, we must find out what the differences are before we can see what it is we are dealing with and how to deal with it. It may be that the undertaking of such a collection was the result of his dissatisfaction with the method of Division, and that the need for it is adumbrated by his statement at *Politics* 1290 b 25 ff. that there can be many different combinations of the varieties of those parts which are indispensable to animals (*e.g.*, some of the sense-organs, the digestive organs, the mouth, the organs of locomotion), and that when all these combinations have been determined we shall have arrived at the various kinds of animals. We have also his assertion in the first Book of the *Parts of Animals* (643 b 12 ff.) that the various groups of animals, such as Birds and Fishes, will be marked off from one another by *many* differentiae (not by one differentia), and that this is a result which can never be achieved by the method of dichotomous Division.

The third observation relates to the *Generation of Animals*. And here we find something which may reasonably be taken as the final outcome of Aristotle's view, already quoted, that groupings must be

based upon natural divisions. In the *G.A.* we find Aristotle working with the same " main groups " as he mentions elsewhere ; but he introduces another basis of grouping, *viz.*, the various methods of reproduction. And differences in the methods of reproduction are in their turn due to the different degrees of " perfection " of the animals concerned, as we shall see later. In the course of this discussion (*G.A.* 732 b 15 ff.), Aristotle explicitly points out that it is impossible to divide animals on the basis of an obvious anatomical character, in this case the organs of locomotion, for there is a great deal of overlapping between the various groups ($\gamma\acute{\epsilon}\nu\eta$). Thus

Not all bipeds are viviparous—birds are oviparous—
    nor are they all oviparous—man is viviparous ;
Not all quadrupeds are oviparous—horse, ox, etc., are viviparous—
    nor are they all viviparous—lizards, crocodiles, etc., are oviparous ;
Some footless animals are viviparous—*e.g.*, vipers, and the Selachia—
    some are oviparous—Fisnes, and serpents other than vipers ;
Many footed animals are oviparous—the quadrupeds named above, lizards, etc.—
    many are viviparous—the quadrupeds named above, horse, etc. ;
Some footed animals are internally viviparous—such as man—
Some footless animals are internally viviparous—such as the whale and the dolphin.

Organs of locomotion, therefore, do not provide a satisfactory basis for making divisions. A better basis is that of degrees of " perfection " : those

HISTORIA ANIMALIUM

animals are viviparous which partake of a purer
" principle "; in other words, no animal is viviparous
unless it respires. Such are those animals which are
" hotter and more fluid in their nature, and are not
earthy " (*G.A.* 732 b 32); and the test of natural
heat in an animal is the presence of a lung—a lung
well supplied with blood. And the more " perfect "
an animal is, the more perfect is the state of the offspring it produces. Here we have briefly and quite
clearly stated one principle on which Aristotle thinks
it worth while to " divide " animals. We could not
ask for any more " natural " basis than the method
of reproduction: Aristotle's requirement as stated
above is completely met on that score; and furthermore, as he himself points out, we can see how fine
and orderly a succession of stages Nature has produced
—a *continuous* succession (ὡς εὖ καὶ ἐφεξῆς τὴν γένεσιν
ἀποδίδωσιν ἡ φύσις, 733 a 34). And what is more,
as will appear later, we can name the reasons for it.
This is a fundamental basis for division, and it cuts
across not only obvious anatomical differences such
as the organs of locomotion, but even across the
division between blooded and bloodless animals,
which, judging from the extensive use which Aristotle makes of it, we might otherwise suppose to be
a clear and unimpeachable distinction.

It will be best to consider the scheme first according
to the degrees of perfection attained by the offspring
when they leave the parent, and then to consider the
reasons assigned by Aristotle, and the animals concerned.

Degrees of perfection of the offspring (*G.A.* 732 a 25 ff.).

1. *Vivipara.* The young are produced " perfect," *i.e.*,

they have the same appearance and structure as the adult, but are smaller in size.

2. *Ovovivipara.* As in (1), the young animal when it emerges is " perfect," but before that a " perfect " egg is produced within the parent. This class therefore exhibits a combination of (1) preceding and (3) following. (See also 754 a 24.)

3. *Ovipara* (i). The product here is not a young animal but an egg. The egg, however, is " perfect," in the sense that it does not increase in size after deposition.

4. *Ovipara* (ii). The product here is an " imperfect " egg, *i.e.*, it has not completed its growth when laid, but has to do so after emerging from the parent. However, both here and in (3) the young when it emerges from the egg is " perfect."

5. *Larvipara.* The product here is not an egg but a larva, which according to Aristotle is a stage previous to an egg : the larva develops into an " egg-like object," and it is not until this has been formed that the young can emerge from it in a " perfect " state.

It is interesting to note that in another passage (*G.A.* 758 a 30 ff.) Aristotle extends the domain of eggs and larvae even to the first of the above-mentioned classes. It seems, he says, that in a way all animals produce a larva to begin with, for the fetation in its most imperfect state is something of this sort. In all the Vivipara, and in all the Ovipara which produce a " perfect " egg, the fetation in its earliest stage is still undifferentiated and growing, and this is precisely what a larva is. And in the internally viviparous animals the fetation, after it has been constituted, in a way becomes " egg-like " : its fluid con-

HISTORIA ANIMALIUM

tent becomes enclosed in a fine membrane—like an egg with the shell taken off. (In other words, it is rather like the soft—though not " imperfect "—egg produced internally by the Ovovivipara.) [a]

We have here, then, an absolutely clearly defined scale, the stages of which are determined by the degree of " perfection " to which the parent is able to bring the offspring before parting with it. If we ask what is the " cause " of these differences between the various animals, again Aristotle's answer is absolutely clear: the cause lies in the physical composition of the animals concerned, *i.e.*, in their κρᾶσις [b]; and the factors involved here are heat and cold, fluidity and solidity (or " earthiness ").[c]

We therefore now pass on to Aristotle's characterization of these groups on the basis of these four fundamental constituents. (This should be read in conjunction with the closely parallel series of phenomena involved in the characterization of animals as πεζά and ἔνυδρα; see Notes, §§ 39-60.)

Characterization according to hot, cold, solid, fluid (G.A. 732 a 25 ff., 732 b 28 ff.).

1. The *internally viviparous* animals are " hotter, more fluid, and not earthy "; they have a lung well

[a] At 733 a 25 ff. Aristotle points out that, although all Insects are bloodless, and although that is why they produce larvae, we cannot simply equate " bloodless " and " larviparous ": there is " overlapping " between larviparous animals and those which produce imperfect eggs (*i.e.*, the scaly fishes, the Crustacea and the Cephalopods), because the eggs of the latter are larva-like (they grow after deposition) and the larvae of the former become egg-like as they develop. And Crustacea, Cephalopods, and Insects are " bloodless," whereas the scaly fishes are " blooded."

[b] See Notes, §§ 33 ff.   [c] See Notes, § 15.

supplied with blood, and soft. They are thus able to perform all three stages within themselves, the first two of which seem to be more or less " telescoped "—at any rate, Aristotle does not reckon them as distinct stages; but see above, p. xiv. The animal, owing to its natural heat (and fluidity), has no trouble about effectively managing these two preliminary stages.

To this group belong Man, horse, ox, etc., dog, and all the hairy animals, " those quadrupeds which bend their hind legs inwards," and the Cetacea.

This group therefore includes two-footed, four-footed, and footless animals, and also cuts across the land-animals/water-animals division.

2. The *ovoviviparous* animals, which produce first a " perfect " egg, and then " perfect " offspring, internally, are " less hot " (*i.e.*, " colder "), though still they are " more fluid "; they are oviparous because they are cold, they are viviparous because they are fluid—fluid matter being particularly conducive to life. Because they are not solid (or earthy), as is shown by their possessing neither feathers nor horny scales—both signs of a solid and earthy constitution—their eggs are soft (though the reason given at 718 b 37 is that they are *cold* creatures; see above); hence the eggs are produced internally, for safety's sake.

To this group belong the Selachian fishes and vipers. This division therefore cuts into the group of Fishes and into the group of Serpents.

3. The first class of *Ovipara* are, like the internally viviparous animals, " hotter " (as is shown by their possessing a lung), but they are more solid, and it is this latter circumstance which reduces them

to their oviparous status: they cannot produce living offspring, but only an egg; still, it is a " perfect " egg.

To this group belong the Birds, the horny-scaled animals, and serpents (other than the viper).

This group therefore includes two-footed, four-footed, and footless animals.

4. The *Ovipara* of the second class are colder as well as being solid; hence they cannot produce a " perfect " egg. Nevertheless, their solidity enables them to provide it with a hard covering, which it needs for protection owing to its imperfect condition.[a]

The scaly fishes come into this group: their scales are a sign of their earthiness or solidity; and so do the Crustacea, which are " earthy ". Cephalopods too produce " imperfect " eggs, though they protect them not with a hard covering, but by means of a sticky substance which they exude over them, similar to the texture of their own bodies. This group therefore not only cuts across the footed/footless division, but also across the blooded/bloodless division.

5. The *Larvipara* are the coldest of all, and therefore produce something even more imperfect than an egg, *viz.*, a larva. This is true of some Insects:

[a] The reason given at *G.A.* 751 a 26 for the " imperfect " state of fishes' eggs is that the group of fishes is prolific, and it is impossible for a very large number of eggs to attain perfection inside the parent; there just is not room. At 755 a 31 the reason why they are prolific is that the majority of their eggs get destroyed, and Nature endeavours to make good the loss by sheer weight of numbers. Aristotle notes one exception among the fishes: the " fishing-frog " (*Lophius piscatorius*), wrongly included by him among the Selachian fishes, lays a " perfect " egg externally (754 a 26, 755 a 9).

they copulate, produce a larva, which develops into an " egg," and a creature of the same kind as its parents is produced from it. But it is not true of *all* Insects ; see (6) following.

Examples are: locusts, cicadas, spiders, wasps, ants.

6. Although all Insects arise from larvae, not all larvae arise from the copulation of parent insects : some arise spontaneously out of putrefying matter. Insects which result from such larvae may copulate and generate, but the larvae they produce (*konides*) are asexual, and generate nothing further.

Examples are : fleas, lice, flies, bugs.

Some Insects, however, which arise from spontaneously-produced larvae do not even copulate. Examples are : gnats, and other such Insects. Insects, therefore, are divided into two (or perhaps three) classes.

Here Aristotle inserts an important warning (*G.A.* 758 b 10). Some larvae are hard ; this must not lead us to mistake them for eggs. It would be wrong to suppose that eggs are hard and larvae soft, or that whatever is round in shape is an egg, or that larvae move about and eggs do not. The difference between a larva and an egg is that in the larva the young creature is formed out of the whole of the contents and not out of part of them (" the pupa feeds upon itself "), whereas in the egg the young creature is formed from part of the contents, the remainder serving as nourishment for it (" the embryo feeds on the yolk "). The same difference is pointed out at 732 a 29 ff., and at *H.A.* 489 b 6 ff.

All larvae, whether spontaneously produced or not, follow the same sequence of stages : (1) larva, (2) pupa, (3) imago.

7. The *Testacea* seem to stand at an even lower level than the Insects. Compared with animals, they resemble plants; compared with plants, they resemble animals. None of them copulates, except the snail, and it has not been established that snails are in fact produced by means of copulation (762 a 33). Some Testacea arise from a spontaneous composition, as some Insects do. Some, such as the whelks and the purpura, produce " honeycombs," or a quasi-seminal fluid, similar in substance to that from which they were constituted to begin with, which gives rise to " sideshoots " (Aristotle compares young mussels with the sideshoots of onions), though these kinds often arise spontaneously too (761 a 14 ff.) ; elsewhere (763 a 26) he says that all the Testacea arise spontaneously. The growth of Testacea is similar to that of larvae (763 a 9).

The Testacea thus fall into two divisions (or three, if " honeycombing " is substantially different from " sideshooting ").

It should be noted that when spontaneous generation is said to take place out of putrefying matter (and this must apply to Insects as well as to Testacea), what is meant is that the putrefaction is what remains over after the process has occurred, the process itself being one of *concoction*, and concoction of course is effected by heat, in this case by the " soul-heat " present in the *pneuma*, *pneuma* being present in water (where some spontaneously-generated animals originate), and water being present in earth (where others originate) (762 a 10 ff.).

It is also interesting to note that a spontaneously-generated animal's position on the scale of " honour "

ARISTOTLE

(τιμιώτερον etc.) depends on the situation where it is produced and the physical substances available (762 a 24 ff.).

Meyer (pp. 97-100 ; see also p. 459) rejects the view that Aristotle intended to put forward methods of reproduction as a satisfactory basis for division—a view which had been favoured earlier by Tiedemann and by Ehrenberg [a]—on the grounds (a) that it splits up a kindred group such as Fishes, or Insects ; and (b) that Aristotle's observation (at *G.A.* 758 a ; see above, p. xiv) that in a way all embryos, even those of Vivipara, begin as a kind of larva and pass through the egg stage, and that the larvae of Larvipara also pass through the egg stage, deprives such a method of any real precision. In making this rejection Meyer seems to me to overlook two important considerations. (1) This is the only method of division which can give a *continuous* series of stages : an advantage which is pointed out by Aristotle himself (*G.A.* 733 a 34 ff.), though I cannot find that his remark to this effect is quoted by Meyer in this connexion. In fact, it is just before he makes this remark that Aristotle mentions that Larvipara and imperfect-egg-producing Ovipara " overlap " : the larvae of the former become egglike as they develop, the eggs of the latter are larva-like in that they grow after deposition. At *P.A.* 681 a 12 Aristotle points out that Nature passes in a continuous gradation (μεταβαίνει συνεχῶς) from lifeless things to animals, and on the

---

[a] Friedrich Tiedemann, *Zoologie*, vol. i, 1808 ; Christian Gottfried Ehrenberg, *Über die Formbeständigkeit und den Entwicklungskreis der organischen Formen*, 1852. Tiedemann (1781–1861) was professor of anatomy and zoology at Heidelberg ; Ehrenberg (1795–1876) was a zoologist, geologist and botanist.

way there are living things which are not actually animals; the result is that the one set is so close to the other that the difference between them is very small indeed, and in some cases it is not clear whether some particular creature (*e.g.*, Ascidians) should be put with the animals or the plants. Actually, the test is whether they have any kind of αἴσθησις; or, to express it in another way, whether they have any fleshy part, which is the seat of αἴσθησις; but it is not always easy to determine this. Furthermore (2), as I have already shown, the methods of reproduction are by Aristotle correlated, by means of the degrees of " perfection " of the various animals, with degrees of heat and cold. None of the methods of grouping based on other criteria—certainly not the popular method of grouping into Birds, Fishes, etc.—gives the possibility of so complete a scheme; and the fact that it is based upon difference of degrees of heat and cold means that Aristotle has a unified frame of reference, under which can be brought various sorts of differences, including even the physical differences between the sexes (*G.A.* 765 b 15 ff.).

In attempting to find a satisfactory approach to the problem of classification, we must not be deflected even by what appear to be Aristotle's own views as expressed in some parts of his works, *e.g.*, the importance of not splitting up a kindred group such as Fishes (*P.A.* 642 b 15), which has already been mentioned. For even this group, as Aristotle is frequently obliged to point out, is one which includes the Selachia, creatures which differ in some notable respects from most fishes; and although (*P.A.* 643 b 10) popular usage has on the whole been right in recognizing and naming the general group Fishes, and although

ARISTOTLE

it is often convenient to use such a group, there is nothing sacrosanct about it; and we must be prepared to go down until we find what is, in Aristotle's scheme of things, the fundamental and most "natural" basis for division. The question we must ask is, What, after an animal's "form," is the most important thing that determines its character? And once we have asked and answered this question, there will be no doubt whatever about how to classify animals as "higher" and "lower" (or, as Aristotle would say, as more and less "honourable"). Of course, various factors come into play; but throughout it is Nature, the servant of "form," using her instrument σύμφυτον πνεῦμα,[a] itself a pre-eminent and extraordinary kind of θερμόν, and ably assisted by the ordinary "hot substance" (which Aristotle tells us is—as "cold substance" also is—*poiētic* (*Meteor.* IV. 390 b 2 ff.) and able to fashion the "uniform parts" such as flesh, bone, sinews, etc.)—it is Nature which is working upon the passive stuffs, and particularly upon "the fluid," which is most conducive to life (ζωτικόν, *G.A.* 733 a 11; *cf.* ζωτικώτερον τοῦ ξηροῦ, 761 a 27). To some extent, Nature is hampered as well as helped by the materials she has to work with and to work upon; indeed, it is a common occurrence for "residues" to be produced; and though many of these can be made use of by Nature, some have to be rejected by the organism as useless. Sometimes the "movements" proper to "form" cannot gain full control, and a deformity or a monstrosity is produced (even the female is, according to Aristotle, a sort of deformity, characterized by deficiency of

---

[a] It is not possible here to discuss this important subject. For a full account, see *G.A.* (Loeb ed.), Appendix B.

heat); and it is when we get sufficient heat acting upon satisfactory fluid matter, when both are in the right proportion (see *G.A.* 743 a 28, 767 a 15 ff., 772 a 17 ff.) that the highest perfection is attained: this is reflected in the high degree of perfection attained at birth by the young of the " hotter and more fluid " animals.

In tracing back the differences in degrees of " perfection," as well as differences of sex, to " the hot," Aristotle is basing his explanation on one of the most fundamental physical substances recognized by him [a];

[a] It is not until the very end of his article " Remarques sur la classification des animaux chez Aristote " in *Autour d'Aristote*, 1955, p. 304, that P. Louis mentions the method of division according to modes of reproduction, based on natural heat, and he seems wholly unaware of the paramount importance of this criterion. The same failure is apparent in L. Robin's account of Aristotle's biology (*Aristote*, Paris, 1944, pp. 179-181, where in addition he makes the mistake of saying that according to Aristotle the octopus was spontaneously generated, and misspells both μαλάκια and μαλακόστρακα). The same failure is to be seen also in Dr. G. E. R. Lloyd's useful article, " The development of Aristotle's theory of the classification of animals " (*Phronesis*, 6 (1961), pp. 59 ff.), for although he recognizes that the division according to degrees of " perfection " and methods of reproduction in *G.A.* is based on the same criterion of natural heat as the (earlier) " stratification " according to posture and number of locomotive organs in *P.A.* (686 a 26 ff.), he does not recognize the overriding significance of the criterion, and indeed strangely describes its use in *G.A.* as " new." This results from his assumption that Aristotle is preoccupied with finding a satisfactory anatomical criterion for classification—and indeed *one* such criterion only—and is trying one after another. Dr. Lloyd's attention is thus so far diverted from the " cause " to the symptom that he supposes Aristotle's citation of another set of symptoms in *G.A.* implies that he has " rejected " the set which he cited in *P.A.* (though Aristotle did not offer it as a unique scheme of classification,

ARISTOTLE

and it is when the "movements" proper to the "form" concerned, superimposed upon and using the "movements" proper to "hot substance," can satisfactorily control the "movements" proper to "fluid substance," that we get the most "perfect" animal.

Anyone who has followed Aristotle's theory to this point, where he brings the material cause into conjunction with the other causes, can have little doubt what he considers to be the fundamental source of "difference" between animals. His view becomes even more explicit when we find him using the same basic principle of "movements" in his discussion of heredity; and here there is much that is pertinent to our subject. In his discussion of heredity (*G.A.* 767 a 36 ff.) Aristotle is speaking of human beings, but it is obvious that the principle involved is not to be confined to them. Whenever, he says, the offspring fails to resemble its parents, we really have a sort of monstrosity, for then "Nature has in a way strayed ἐκ τοῦ γένους." The first beginning of this "straying" is seen when a female offspring is produced instead of a male (although Aristotle recognizes that the production of females is "necessary" for Nature). If the seminal residue of the female parent is well concocted (*i.e.*, has been properly acted upon by "the hot"), then the "movements" supplied to it by the male parent will fashion it after the

---

or even as one at all). The meaning of the symptoms πεζόν and ἔνυδρον, and the important discussion of their interpretation at *H.A.* VIII. 2 is not dealt with at all by Dr. Lloyd (he mentions the *H.A.* passage only in a footnote *en passant*), nor is the relevant discussion in *De Respiratione*, although in these too the same criterion is involved. See Notes, §§ 29 ff., and *cf.* Introd. pp. xv ff.

male's own shape, *i.e.*, will produce a male offspring. We should note, however, that the " movement " which the male possesses and supplies is not a simple one. It contains (*a*) the " movements " proper to him as an individual ; (*b*) those proper to him as a male ; (*c*) those proper to him as a human being ; and (*d*) those proper to him as an animal. Further, under (*a*) we must allow also for other " movements " which are present not *actually* but *potentially* only, *viz.*, those of previous generations ; and here are involved both those of the father's side and those of the mother's side. In this contest, as it were, of " movements," unless all goes with perfect success and a male offspring resembling its father is engendered, there are two main possibilities of shortcoming : the movements may either (1) " depart from type," *i.e.*, change over to the opposite sex ; or (2) they may " relapse " into those of some earlier ancestor. All this is worked out in great detail by Aristotle ; and it is applied by him not merely to the physique of the offspring as a whole, but even to its various parts (768 b 1 ff.). If the " relapsing " proceeds far enough, the offspring engendered bears no resemblance to any of its family or kindred, but is just a human being : there is no family καθ' ἕκαστον about it, but only the καθόλου. Still further " relapsing " results in the offspring no longer having the appearance of a human being at all, but merely that of an animal (ζῷον) ; and this is what we call a monstrosity. Indeed, two modes of shortcoming are concerned here : the " movements " from the male " relapse," and the material provided by the female does not get " mastered "; we then have left τὸ καθόλου μάλιστα, just " τὸ ζῷον."

xxvii

ARISTOTLE

The imperfection of monstrosities may also be due to more obvious causes, for instance, in the case of animals which are prolific, such as sheep and goats, but especially in fissipede animals such as the dog (most puppies are born blind), the wolf and the mouse (770 a 35 ff., 771 a 22 ff); but the terminology of the explanation is identical: there are so many young being produced simultaneously that they hamper the "movements which effect generation" (τὰς κινήσεις τὰς γεννητικάς). Elsewhere further examples are given (774 b 5 ff.) of prolific Vivipara which produce imperfect young, and also of prolific birds, whose young are born blind, though in this case Aristotle does not explicitly mention the "movements."

It is obvious that Aristotle's discussion of heredity does not help a great deal towards working out a taxonomic system. The example chosen by Aristotle, *viz.*, Man, is one which, as we have seen, is not included in any of the "main groups" which he mentions elsewhere; and here, therefore, naturally, we find no intermediate stage between "man" and "animal." When the movements "relapse" they pass from "human being" to "merely animal." We might perhaps wish that Aristotle had chosen "horse" as his example, and then we should have seen what stages he inserted between "horse" and "animal" —whether he would have put in "lophouron," or more stages of that sort, or stages of a different sort altogether—or none. But since αἰσθήσει ζῷον ὥρισται, at which end ought we to begin? Would not a monstrosity, which could be described as "merely an animal," be at the bottom of the scale? And is not a creature such as an Ascidian (which is only just, or

not quite, an animal) also at the bottom of the scale ? Or are they at the bottom of different scales ? We may be reminded of Empedocles' " whole-natured," undifferentiated forms, which appear at two different stages in the cycle. Nevertheless, the discussion on heredity gives us a very valuable insight into Aristotle's way of thinking, and it seems to suggest that anything of the nature of dichotomy, or any similar taxonomic scheme, is foreign to his maturest attitude.

Another striking feature which is to be noticed in the discussion on heredity is that it leaps over without any difficulty the supposed barrier between the *infima species* and the individual member of it : this is something that taxonomy cannot do : even the sexes of an *infima species* cannot be dealt with by it. But Aristotle's theory of " movements " passes *continuously* right through from " animal " to " individual " ; and must we not here recognize that in this respect it is in full accord with his well-known attitude towards the τόδε τι, which is the most fully actual of all οὐσίαι ? (*Cf. G.A.* 767 b 36.)

I am inclined to think that if we take into account Aristotle's close association of proper " movements " with every " form," and his frequent comparisons of Nature to craftsmen of various kinds, we may get a truer view of his attitude towards the classification of animals. The activities of Nature are of course not in all respects parallel to those of an artist or a craftsman, but in one important respect they are identical, in that τὸ οὗ ἕνεκα is paramount ; and this implies the actualizing, in matter, of the " form " by means of the " movements " appropriate to it. There is, it would seem, no obvious reason why we should try to arrange the objects made, *e.g.*, by the carpenter

in any taxonomic scheme : what, for instance, would be the relation of a table to a chair or to a cupboard ? No doubt, as we have seen, Aristotle recognizes a definite *scala naturae* ; but the rungs of this ladder are not the stages of a taxonomic scheme, and there is no evidence that Aristotle felt they should be. Once we have reached the rung of externally viviparous animals, there is nothing to choose between man, quadrupeds, and Cetacea on that particular score (just as there is nothing to choose between table, chair, and cupboard) ; and although man no doubt is τιμιώτερον because his " form " includes " rational soul," there appears to be no criterion which would justify us in putting quadrupeds "higher" than Cetacea or vice versa, or horses above elephants. All we can say is that the body given to this or that animal is the body which its " form " requires : man has hands because he is the most intelligent animal (*P.A.* 687 a 9) ; in other words, the " form " Man involves the ability to perform certain actions for which a body with hands is required. Why there are " forms " of inferior animals at all (if such we may call them) is a question Aristotle does not explicitly raise, though he remarks that all the other animals exist for the sake of man (*Pol.* 1256 b 16 ff.) ; he is not normally interested in such a problem. It may be that he thought of them (or some of them) as indispensable adjuncts to human life, just as slaves were : in that sense they would be " part of the nature of things." What interests him is to trace and to record how the οὗ ἕνεκα is attained in each kind of animal, and to point out how regularly Nature succeeds in embodying the various " forms " in matter in successive generations ; and no doubt each

HISTORIA ANIMALIUM

kind of animal, however "lowly," has its part to play in ensuring that γένεσις is continually going on (*G. & C.* 336 b 34), in other words, that the sublunary part of the cosmos comes as near as it can to the individual eternity exhibited by the things of the upper cosmos. The kind, though not the individual, is eternal. And although Aristotle tells us (*G. & C.* 336 b 30) why there cannot be individual eternity in the sublunary world, he does not offer to explain why, if ἔμψυχα are better than ἄψυχα (*G.A.* 731 b 29), there are any ἄψυχα at all. Perhaps his answer would be on similar lines to the suggestion I made just now: that ἄψυχα are indispensable: they provide the material for the bodies of ἔμψυχα. We may note that here again Aristotle's principle of "movements" is able to leap over a barrier—the barrier between living and lifeless things, for what distinguishes lifeless matter from living is not that the former is destitute of "movements." Just as it is the "movements" proper to the "form" Dog which maintain the canine body as such (*G.A.* 737 a 18, etc.) and distinguish it from other living bodies such as elephant, so it is the "movements" proper to the "form" hot which make hot substance what it is and distinguish it from other kinds of substance.

The question why there are ἄψυχα may then be partly answered in the way I suggested; but in another way it is comparable to the question (which arises directly out of Aristotle's own remarks), Why is it that as we go down the *scala naturae* we see a progressive weakening of the "movements" proper to "form" (until ultimately—in lifeless matter—the "form" is no longer soul at all), *e.g.*, in those insects where the natural "movements" are so weak that

the female has to insert part of itself into the male in order to ensure generation (*G.A.* 730 b 25 ff.), and in plants where the male and female are permanently united, presumably because otherwise the " movements " would be too weak to effect generation at all ?

I think another part of the answer to this question may be that somehow distance is the reason : this is a factor which Aristotle cites more than once as having considerable importance. The reason why the things of the sublunary world cannot have individual eternity (*i.e.*, " being " in the full sense of that word) is that they are too far away from the ἀρχή (which is here, no doubt, the Unmoved Mover ; *G. & C.* 336 b 30). The reason why the eyes of the embryo are so late in being formed is that a very strong movement is needed to move parts which are so far away from the ἀρχή (in this context, the heart) and so much subjected to cold (*G.A.* 744 b 3 ff.). And at *G.A.* 761 a 20-b 16, Aristotle presents us with a general correlation of three divisions of living things and three scenes of habitat :

Land-animals (πεζά),	which live in the air,
Testacea,	which live in the water,
Plants,	which live in the earth.

It will be noted that the further any habitat is from the circumference of the cosmos, the less " perfect " are the creatures which inhabit it : the most " perfect " creatures are those which breathe (*G.A.* 732 b 28 ff.) : they are also by their nature " hotter and more fluid and not earthy " ; and Testacea differ from Plants in proportion as water (where most of them live) and fluid matter are better able to support

HISTORIA ANIMALIUM

life (ζωτικώτερα) than earth and solid matter (*G.A.* 761 a 27). That distance is an important factor in this connexion is shown by Aristotle's concluding remark (761 b 15) : " the more and less and *the nearer and further* make a surprisingly great difference." It is true that Aristotle finds himself in a difficulty when it comes to trying to fit fire into the scheme of habitats ; but that need not concern us here.[a]

Nevertheless, I do not think that Aristotle would have claimed even for a scheme of classification according to methods of generation (*i.e.*, according to degrees of " perfection," or of " heat," or of strength of " movements," whichever way we choose of expressing it), even though it is based on the most fundamental principles of his physics, that it can be made wholly tidy. Just as he is prepared, where the observed phenomena appear to demand it, to allow an anomaly—*e.g.*, the " divine " fifth element in physics, " rational soul " functioning through no bodily organ, the " unmoved mover " which is pure form, matter moving itself with the movements which normally form would have to supply (in spontaneous generation ; though even here it is " heat " which is responsible : κίνησις καὶ θερμότης, *G.A.* 743 a 35), all of which are odd and do not fit into an otherwise tidy and self-consistent pattern—so he is prepared to accept bees as an anomaly : their method of reproduction is not fully understood, but it is certainly odd (περισσόν, *G.A.* 760 a 5, 761 a 4), and this is in keeping with the obvious fact that they possess some " divine " ingredient. But how, we may ask, can an *insect* be capable of possessing any such thing ?[b] Perhaps, after

[a] See Notes, § 59.
[b] See, however, Notes, § 34.

xxxiii

all, this is not a much more difficult question than if we were to ask, How can the highest activity of which a human being is capable be a " divine " rather than a distinctively human one (ἐφ' ὅσον ἐνδέχεται ἀθανατίζειν) ?

As Aristotle remarks (*G.A.* 769 b 3) at the end of his discussion of heredity, it is not easy, by stating one single mode of cause, to explain the causes of all the instances of resemblance and lack of it ; and I do not see why he should have held a substantially different opinion about the classification of animals.

THE MANUSCRIPTS

The chief MSS. THE manuscripts on which the text of *H.A.* is based were thoroughly investigated by Dittmeyer in preparation for his edition of 1907, and the information given below is almost wholly derived from his account. Only the more important MSS. are mentioned here. Those most frequently cited by him and by Bekker are C^a, A^a, P, and D^a, and E^a for parts of Books VI and VIII and Book IX. The MSS. are discussed also by G. Rudberg, *Textstudien zur Tiergeschichte des Aristoteles*, Uppsala, 1908.

Dittmeyer distinguishes two main groups of manuscripts, of which the first, which is the better, contains the following : A^a, C^a, and their copies (including cod. Rhen., which he says is copied from A^a). The second group contains P, D^a, E^a, m, Ambrosianus, etc.

I have added at the end of each notice the serial number of the MS. in Wartelle (André Wartelle, *Inventaire des manuscrits grecs d'Aristote*, Paris, 1963).

HISTORIA ANIMALIUM

C^a — Laurentianus LXXXVII. 4. 14th century. Written by Joanicius. Contains Books I-IX in the following order (which is the order normally found in the mss.): I-VI, VIII, IX, VII. Collated by Bandini against Du Val's 1619 edition for Camus; this was apparently the first time it had been used for establishing the text. Collated also by Bekker and Dittmeyer. [583]

A^a — Marcianus gr. Z 208. Formerly the property of Cardinal Bessarion (d. 1472). 12th or 13th century. Contains Books I-IX. Four later hands of correctors or annotators are recognizable. This ms. and C^a were probably copied from one and the same original, C^a being a somewhat more careful copy. Some parts of it were used by Camus. Collated by Bekker and Dittmeyer. [2109]

Rhen. — Parisinus supp. gr. 212. 15th-16th century. Agrees closely with the first hand of A^a. It was at one time owned by Beatus Rhenanus (1485-1547), a German scholar, whence its name Codex Rhenani. It was used by Schneider, but it has little of its own to contribute. [1582]

P — Vaticanus gr. 1339. 14th-15th century. Contains *H.A.* Books I-IX and several other zoological and physical works of Aristotle. The collation by Foggini against Sylburg's edition was used by Camus. Also collated by Bekker, and the first Book and other passages by Dittmeyer. [1782]

D^a — Vaticanus gr. 262. 14th century. The only ancient ms. which contains Book X. It has

ARISTOTLE

	some good readings not found in C^a or A^a. Collated by Bekker and Dittmeyer. [1709]
m	Parisinus gr. 1921. 14th century. From the same source as P. Used by Camus. [1407]
Ambr.	Ambrosianus 46 (I. 56 sup.) 15th century. Includes Book X in a later hand. A collation by Branca was used by Camus. [948]
E^a	Vaticanus gr. 506. A late MS., from the same source as P, according to Dittmeyer, though Wartelle's *Inventaire* dates it 13th century. Partly collated by Bekker. [1745]
Σ	In conjunction with this list should be noted Michael Scot's Latin version, made in the early thirteenth century from an Arabic version, very probably from that of Ibn al-Batriq, which was itself made in the early ninth century (see p. xli). Ibn al-Batriq's Greek manuscript was therefore at least three centuries older than our earliest existing Greek manuscript (A^a). In disputed passages it is often possible to infer with absolute certainty from Scot's version the reading of the Greek MS. which Ibn al-Batriq had in front of him (examples in the present volume are at 491 b 14 and 26, 496 b 15, 507 a 18, 508 b 12, 512 b 10, 513 a 22), and at 520 b 4 Scot's rendering preserves an obviously correct reading ($\pi i o v$) which is not found in any existing Greek MS. At 487 b 5 ff., a passage which is undoubtedly corrupt, Scot's version suggests a means of restoring the text, which I have endeavoured to make use of. The symbol Σ can therefore represent an older and better Greek text than any now directly

HISTORIA ANIMALIUM

available to us,[a] and it is much to be wished that a full collation of the text of *H.A.* with the Arabic version could be made.

The MS. Laurentianus LXXXVII. 1 [Wartelle 580], of the fifteenth century, is of interest in that it contains the nine Books, apparently transcribed from D[a], but in the order established by Gaza, *viz.*, I-IX as in most printed texts (see next paragraph).

The order of the Books as given in most printed editions of the *H.A.* is not that of the Greek manuscripts, in which the Book now numbered seventh comes after that now numbered ninth. The removal of this Book to its present place is due to Theodore Gaza, whose translation of the first nine Books was published in 1476, and the transference was adopted by the Greek *editio princeps* of Aldus in 1497, since when it has been followed in almost all printed editions. Gaza's reason for making the change was this. At 539 a 6, early in Book V, Aristotle remarks that whereas in treating of the parts of animals he began with Man, in treating of reproduction he will leave Man until the end (or until the last), because the subject is a most troublesome one to deal with (νῦν δὲ περὶ τούτου τελευταῖον λεκτέον, διὰ τὸ πλείστην ἔχειν πραγματείαν). This remark, thought Gaza, was

Order of the Books.

---

[a] My experience of Michael Scot's translation is independently confirmed by G. Rudberg, " Die Tiergeschichte des Michael Scotus," *Eranos*, 9 (1909), pp. 92 ff., who, after an examination of Book I of *H.A.*, concludes that the Greek MS. used by the Arabic translator was written in uncials, earlier than A.D. 800 ; that its text was good, and antedated many of the variants found in our present Greek MSS. ; and that it contained scholia, some of which have persisted into Michael's translation, and in some cases have displaced the original text.

wrongly taken by Apellicon, the owner of Aristotle's manuscripts in the first century B.C., to mean that the subject would be deferred to the end of the whole treatise; whereas what Aristotle really meant was that it would come at the end of the section on reproduction, *i.e.*, after Books V and VI; so that is where Gaza placed it. Since it has followed Book VI in all printed editions (except that of Aubert and Wimmer) including the Berlin edition, by the pages of which Aristotle's works are cited, it seems most convenient to allow it to remain there, and not to restore it to the place it occupies in the manuscripts,

Printed Editions

The following list is not exhaustive, but it includes the most important editions of *H.A.*

1. The Aldine edition. *Editio princeps.* *H.A.* is contained in vol. ii, Venice, 1497.
2. Aristotelis Opera de Animalibus. One volume. Per haeredes Juntae, Florence, 1527. "Omnia ex exemplaribus N. Leonici Thomaei diligenter emendata."
3. Edited by Simon Grynaeus. Part of the Collected works. Basel, 1531. Reissued 1539 and 1550.
4. Edited by J. B. Camotius (Giovanni Battista Camozzi). Part of the Collected works of Aristotle and Theophrastus. *H.A.* is contained in vol. iii, Venice, 1553.
5. Edited by Friedrich Sylburg. Part of the Collected works. Frankfurt, 1587. Based on the Basel edition; shows influence of the 1527 Juntine edition and of Camotius, and of Gaza's translation.

HISTORIA ANIMALIUM

6. Edited by Isaac Casaubon. Part of the two-volume edition of the Collected works. Lyons, 1590. Reissued Geneva, 1605. The Greek text is that of Sylburg. Contains Gaza's translation of Books I-IX and J. C. Scaliger's of Book X (see Translations, p. xlii).
7. Edited by Julius Pacius. Part of a two-volume edition of the Collected works. Lyons, 1597. Reissued Geneva, 1607. The Greek text is that of Casaubon. Includes Gaza's translation of Bks. I-IX and Scaliger's of Bk. X.
8. Edited by Julius Caesar Scaliger. Historia de Animalibus, Toulouse, 1619 (*i.e.*, 61 years after Scaliger's death; brought out by Philippe Jacques de Moussac). Contains Books I-X, a commentary, and a new Latin translation by Scaliger.
9. Edited by Guillaume Du Val. Part of a two-volume edition of the Collected works. Paris, 1619. Contains Casaubon's Greek text, incorporating proposals by Turnebus and Pacius, and a Latin translation. Reissued 1629, 1638, 1639, 1654, 1690.
10. Edited by Armand-Gaston Camus. Histoire des Animaux d'Aristote. Two volumes, Paris, 1783. With a French translation. Omits Book X. Records readings of Bernard Canisianus in a copy of the 1527 Juntine edition, deriving mainly from the MSS. C^a and A^a. Camus takes account of the Latin versions of Michael Scot and William of Moerbeke. This edition marks the beginning of the modern era of criticism and interpretation.
11. Edited by Johann Gottlob Schneider. De animalibus historiae libri X. Four volumes, Leipzig, 1811. Dedicated to Cuvier. Vol. i contains J. C.

Scaliger's translation of Books I-IX, revised by Schneider ; the translation of Book X (which is " manifestly spurious ") is given in Albertus Magnus' " version," which in fact is a commentary-expansion of Michael Scot's translation ; vol. ii contains the Greek text. Vols. iii and iv contain the commentary and the " curae posteriores " on Books I-X. Schneider's revision of Scaliger's translation of Books I-IX is printed in the Berlin edition of 1831.

12. Edited by August Immanuel Bekker. Part of the Collected works. *H.A.* is contained in vol. i, pp. 486-638, including Book X. Berlin, 1831. References to the works of Aristotle are now given by the pages, columns, and lines of this edition. (An octavo edition of *H.A.* was published, Berlin, 1829.)

13. Edited by Carolus Hermannus Weise. One-volume edition of Aristotle's works. Leipzig, 1843. The text of *H.A.* is taken from Schneider ; no *apparatus criticus*.

14. Edited by Ulco Cats Bussemaker. Part of the Collected works. A. F. Didot, Paris, 1854. With a Latin translation. *H.A.* is contained in vol. iii. The text is fundamentally Bekker's, but introduces some better readings and occasional conjectures.

15. Edited by Nicolaos S. Piccolos.[a] Ἀριστοτέλους περὶ ζῴων ἱστορίας βιβλία θ′, ἐφ' οἷς καὶ δέκατον τὸ νόθον. Firmin Didot frères, fils et cie, Paris, 1863. Vol. i contains the Greek text of Books I-X, of which the tenth is " spurious." The second

---

[a] This is his own spelling of his name when he writes it in Roman letters.

volume was never published. The text is substantially Bekker's, but P. makes many interesting suggestions for emendation. His edition marks a considerable advance in the criticism of the text.
16. Edited by Hermann Rudolf Aubert and Christian Friedrich Heinrich Wimmer. Aristotles Thierkunde. Two volumes, Leipzig, 1866. With introduction, *apparatus criticus*, German translation, and commentary, etc. (but no *apparatus*, translation, or commentary for Book X).
17. Edited by Leonardus Dittmeyer. One volume, Teubner, Leipzig, 1907. Greek text of Books I-X, with introduction and *apparatus criticus*. Dittmeyer made a fresh collation of the whole of the MSS. C^a, A^a and D^a, and of parts of other MSS., and also offers a considerable number of suggestions for emending the text.

The stages in the history of *H.A.* scholarship may be roughly summarized as follows :

(1) The *editio princeps*, which formed the essential basis of the text until

(2) Camus' edition of 1783. Camus obtained collations of several MSS. (see p. xxxiii), and took account of the two early Latin translations. His edition marks the beginning of the modern era of the study of *H.A.*

(3) Schneider's commentary was a remarkable performance, the fruit of 30 years' work, based on a wide knowledge of Greek literature and of zoology. He also took account of the Codex Rhenani, and introduced some improvements into the text.

ARISTOTLE

(4) Bekker's edition soon established itself as the accepted text, and it was not until Piccolos that any considerable advance was made in this direction.

(5) Aubert and Wimmer, the one a zoologist and the other a classical scholar,[a] appear not to have made any further research into the MSS., although they offer a number of suggested emendations; but they were the first editors to question the authenticity of further large parts of the treatise other than Book X. In this they were followed by Dittmeyer, who accepted their conclusions. Dittmeyer also made fresh collations of some of the MSS., and himself offers further emendations.

(6) D'Arcy Thompson combined the knowledge of an expert and original zoologist with the training of a classical scholar, and his work is an outstanding achievement in both respects. He offers a number of conjectures for the text, in addition to generous annotations.

Early Translations

Early Latin translations. I GAVE a detailed account in my edition of *P.A.* (pp. 39 ff.) of the early history of the translation of Aristotle's zoological works,[b] and it is unnecessary to repeat it here. A few notes on three early Latin translators may however be useful.

Michael Scot. Michael Scot, who among his other accomplishments was astrologer to Frederick II, king of Sicily,

---

[a] Aubert (1826–1892) was a teacher of physiology at the University of Breslau, Wimmer (1803–1868) was head of the Friedrich-Gymnasium there.

[b] For a full discussion of this subject see S. D. Wingate, *The Medieval Latin Versions*, London, 1931.

HISTORIA ANIMALIUM

at his court at Palermo, had begun work at Toledo before 1217 on his translations of Aristotle. The *De animalibus* (a translation of Aristotle's zoological works in 19 books, *i.e.*, it includes *H.A.* X) is one of his earliest, and was certainly finished before 1217. This translation was made from the Arabic; and it is probable that Michael used the Arabic version made in the early years of the ninth century by the physician Ibn al-Batriq, who was the leader of the school of translators at Bagdad under the Caliphate of Harun-al-Raschid. There is in the British Museum a thirteenth-century MS. (Add. 7511) of an Arabic translation of the zoological works, and a comparison of passages in Michael's translation with this version shows a close correspondence. The fact that Michael's preface to the whole work also corresponds exactly with the preface in the Arabic MS. seems to leave no doubt that this was the Arabic version used by him. Michael's translation was the basis of Albertus Magnus' treatise *De animalibus*. The existence of fifteenth-century manuscripts of Michael's translation indicates that it was not wholly superseded by the later translation made direct from the Greek.

William of Moerbeke, a small town south of Ghent on the borders of Flanders and Brabant, was born about 1215. He was confessor to Popes Clement IV and Gregory X, was archbishop of Corinth, and acted as Greek secretary at the council of Lyons in 1274. At the request of St. Thomas Aquinas, who was a pupil of Albertus Magnus, William undertook to make new translations direct from the Greek. The earliest dated translation of his, made at Thebes, is that of the *P.A.*; the date 1260 occurs in a fifteenth-century MS. of it at Florence, and this MS. contains also *H.A.*,

William of Moerbeke.

ARISTOTLE

G.A., and *De progressu animalium*. William's translation of *H.A.* was used by St. Thomas in the *Summa* (before 1264). The translation is discussed at length by G. Rudberg,[a] who prints the whole of the first Book.

Theodore Gaza. Theodore Gaza was born at Thessalonica about 1400, and in 1430 fled as a fugitive from the Turks to Italy. In 1447 he was professor of Greek at Ferrara, and in 1450 he was invited to Rome by the pope to make Latin versions of Aristotle and of other Greek authors. The most important of his translations, dedicated to Pope Sixtus IV, is the *De animalibus* in 18 books (*viz.*, nine of *H.A.*, four of *P.A.*, and five of *G.A.*). This translation was printed in Venice in 1476, and more editions followed before the end of the century. It was the standard version of the zoological treatises throughout the Renaissance period and beyond.

Later Translations (without Text)

1. Aristotelis liber qui decimus historiarum inscribitur, nunc primum latinus factus a Julio Caesare Scaligero. Latin translation of Book X only, with full commentary. Published by his son, Silvanus Caesar Scaliger, Lyons, 1584.

2. Thomas Taylor. English translation of Books I-IX, and of the treatise on Physiognomy. From Camus' text. London, 1809. The tenth Book is " not sufficiently conformable either to the style or the doctrine of Aristotle."

3. Friedrich Strack. German translation of Books I-IX, with notes and an index of animals. Frankfurt am Main, 1816. The tenth Book is rejected as spurious.

[a] *Textstudien zur Tiergeschichte des Aristoteles*, Uppsala, 1908.

HISTORIA ANIMALIUM

4. Philipp Hedwig Külb. German translation of Books I-X, with footnotes. Stuttgart, 1856/7.
5. Richard Cresswell. English translation of Books I-X, with index of animals. London, 1862. Reprinted 1897, 1907. From Schneider's text. C. includes Book X, although he does not regard it as genuine.
6. Anton Karsch. German translation of the " Naturgeschichte der Thiere. Zehn Bücher." Vol. i, containing Books I-III, Stuttgart, 1866 ; vol. ii, containing Books IV and V, Stuttgart, n.d. ; vol. iii, containing Books VI-VIII, Stuttgart, n.d. The work appears not to have been completed.
7. J. Barthélemy-Saint-Hilaire. French translation of Books I-IX in three volumes, with introduction and copious notes. Paris, 1883. The tenth Book is " apocryphe."
8. D'Arcy Wentworth Thompson. English translation of Books I-IX, with very full and valuable notes and suggestions for textual emendation. Oxford, 1910. Vol. iv of the Oxford translation of Aristotle. The tenth Book is " spurious beyond question."
9. Paul Gohlke. German translation of Books I-X. (" Tierkunde " ; sect. VIII. 1 of *Die Lehrschriften*), Paderborn, 1949, 2nd ed. 1957.
10. J. Tricot. French translation of Books I-X, with introduction and notes, in two volumes. Paris, 1957. Mainly from Dittmeyer's text. Follows D'Arcy Thompson closely. (See also pp. lv, lvii.)

STUDENTS OF ARISTOTLE'S ZOOLOGY

BESIDE those who have edited and translated the *H.A.*, numerous scholars, some of them outstanding

ARISTOTLE

personalities of their times, have contributed to its elucidation, and in preference to a bare recital of their names or mention of them in an abbreviated form in the *apparatus*, I append here a few notes on their careers, and on those of the earlier editors.

Nicolaus Leonicus Thomaeus (Niccolò Leonico Tomeo), 1456–1531. Born at Venice. Professor of philosophy at Padua (of which he became a citizen) 1485–1495 and with brief intervals for the remainder of his life. He was the first person to lecture there on the Greek text of Aristotle (1497); he was a strong supporter of the new philosophy. Among his friends were Copernicus, Erasmus, Reginald Pole, Cuthbert Tonstall; among his pupils, Pietro Bembo. He was a versatile writer. His ten *Dialogi* deal with philosophical, moral, and other subjects; one is on dice-playing. He published a Latin translation, with commentary, of Aristotle's *Parva naturalia* (Venice, 1522); and his *Opuscula* (Venice, 1525) contains paraphrases of Aristotle's *De incessu* and *De motu animalium*, a text of the *Mechanica* in his own Latin translation, with commentary, *Quaestiones amatoriae* (these were later translated into French), *Quaestiones naturales*, and his translation of Proclus' commentary on Plato's *Timaeus*. The Juntine edition of Theophrastus, like that of Aristotle's *De animalibus*, was founded on his work (1527). His *De varia historia libri iii.*, written in his youth, he did not publish until 1531. His tame crane, which he had kept for 40 years, shortly predeceased him.

Augustinus Niphus (Agostino Nifo), dates uncertain, latter part of 15th century and first part of 16th. Professor of philosophy at Padua (1492–1495), at Naples, Rome, Bologna, Pisa (1519). A prolific writer. Was deputed by Leo X (Giovanni de' Medici) to defend the catholic doctrine of immortality against Pomponazzi; in consequence he was authorized to call himself by the name of the Medici. Edited Aristotle's *Physics* with Averroes' commentary, Venice, 1495. Produced numerous commentaries on Aristotle (including *Expositiones in omnes Aristotelis libros de historia animalium . . . de partibus animalium . . . ac de generatione animalium,*

HISTORIA ANIMALIUM

Venice, 1546), which were frequently reprinted; the 1654 Paris edition is in 14 volumes.

Julius Caesar Scaliger (Della Scala), 1484–1558. Philologist, philosopher, naturalist, and poet. Born at Riva on the Lago di Garda; a kinsman of the emperor Maximilian. Studied under A. Dürer; fought at the battle of Ravenna 1512. Studied medicine and natural history at Bologna and Padua 1514–1519. Spent 42 years in Italy, then settled at Agen on the Garonne; became a French citizen 1528. Denounced Erasmus in an oration published 1531. Author of much Latin verse (five volumes between 1533 and 1547), and of *De causis linguae Latinae* (1544), the first Latin grammar on scientific principles. His commentary on Theophrastus' *De causis plantarum* and his edition of Aristotle's *H.A.* appeared posthumously; so too did his *Poëtice* (in seven books, 1561) on poetic theory. His best-known philosophical work is his *Exotericarum exercitationum libri xv. de subtilitate* (1557), dedicated to Cardanus.

Simon Grynaeus, 1493–1541. Son of a Swabian peasant. Scholar and theologian. Studied at Vienna. Professor of Greek at Heidelberg 1524, and of Latin also 1526. Discovered the MS. of the first five books of the fifth decade of Livy, 1527. Left Heidelberg owing to his religious views 1529, and settled in Basel, where he became professor of Greek. Reorganized the university of Tübingen; returned to Basel before 1536. Represented the Swiss divines at the Worms Conference between Catholics and Protestants, 1540. He had considerable influence. His chief works are his Latin versions of Plutarch, Aristotle, and Chrysostom.

Petrus Victorius (Piero Vettori), 1499–1585. Student at Pisa 1514; professor of Latin at Florence (later also of Greek and of moral philosophy) 1538–1585. The greatest scholar of his day in Italy. Edited with commentaries Sophocles, Aeschylus, parts of Cicero, Aristotle's *Ethics*, *Rhetoric*, *Poetics*, *Politics*, *De partibus animalium*, Xenophon's *Memorabilia*, Terence, Sallust, etc., and produced 38 volumes of *Variae lectiones* (25 published in 1533, 13 in 1569). Composer (1564) of the epitaph for Michelangelo's catafalque.

Adrianus Turnebus (Adrian Tournebu), 1512–1565. Professor of Greek at Paris 1547; director of the Royal

Press at Paris 1552–1556, where he printed texts of Greek and Latin authors, including Aeschylus, Sophocles, and Aristotle. Author of commentaries on parts of Cicero's works, on Varro *De lingua Latina*, and on Book I of Horace's *Odes*; translator of Oppian *De venatione*, and of some of the minor works of Aristotle, Theophrastus and Plutarch. The first 24 books of his *Adversaria*, containing explanations and emendations of passages in classical authors, were published in 1564–1565, the remaining six books in 1573 (Paris). His *Poemata* were published in 1580. He was a man of considerable influence on French literature, and according to Montaigne " he knew more, and knew it better, than any man of his century, or for ages past."

J. B. Camotius (Giovanni Battista Camozzi), 1515–1581 (or 1591). Taught philosophy at the University of Bologna (1550) and in Macerata (1555); called to Rome by Pius IV and taught Greek there, 1558; a friend of Cosimo I, Grand Duke of Tuscany. Author of three books of commentary on Theophrastus' *Metaph.* i, Venice, 1551; edited Aristotle, 6 voll., Venice, 1551–1558; translated Psellus' commentary on Aristotle's *Physics* 1554 and Alexander Aphrodisiensis' commentary on Aristotle's *Meteorology* 1556, etc.

Conradus Gesnerus (Konrad von Gesner), 1516–1565. " Totius historiae naturalis parens ac veluti promptuarium " (Tournefort), described by Cuvier as " the German Pliny." A prolific writer; best known to his contemporaries as a botanist. His *Historia animalium*, which marks the starting-point of modern zoology, was published at Zürich in four volumes (viviparous quadrupeds, 1551; oviparous quadrupeds, 1554; birds, 1555; fishes and aquatic animals, 1558; a fifth volume, on serpents, was published later). " The historie of foure-footed beastes " (1607) and " The historie of serpents " (1608) by Edward Topsell, of Christ's College, are largely translations of the corresponding parts of Gesner's *Historia*. Gesner was also a lover of mountains : his *Descriptio Montis Fracti sive Montis Pilati* was published in 1555. The periodical *Gesnerus*, for the history of medicine and natural science, was founded by Jean Strohl of Zürich (1886–1942); first issue 1943.

Hieronymus Mercurialis (Hieronimo Mercuriali), 1530–1606.

HISTORIA ANIMALIUM

Professor at Padua 1569, at Bologna 1587, at Pisa 1592. Editor of Hippocrates, Venice, 1588. Author of many medical treatises, of commentaries on Hippocrates, and of *Variarum lectionum libri iv.*, *in quibus complurium maximeque medicinae scriptorum infinita paene loca . . . declarantur*, Venice, 1571. A fifth book was added in the 1576 Basel edition, a sixth in the 1585 Paris edition, and six further chapters in the 1588 Venice edition.

Friedrich Sylburg, 1536–1596. Studied at Marburg, Jena, Geneva, and Paris. Was at Frankfurt 1583–1591, at Heidelberg 1591–1596. Completed Xylander's edition of Pausanias 1584, edited Dionysius of Halicarnassus, Apollonius *De syntaxi*, three volumes of *Scriptores historiae Romanae*, etc., as well as the works of Aristotle.

Julius Pacius (Giulio Pace or Pacio), 1550–1635. Jurist, philologist, and philosopher. Educated at Padua; accepted the reformed religion; taught at Geneva; professor at Heidelberg (where he taught law) 1585; later taught at Montpellier and Valenza; was professor at Padua 1618. An active collector of manuscripts.

Isaac Casaubon, 1559–1614. Born in Geneva, of Huguenot parents. Studied and lectured at Geneva 1578–1596; lectured at Montpellier 1596–1598; visited Paris 1598, and settled there 1600; in 1604 was appointed a librarian in the Royal Library. A close friend of Joseph Scaliger. Was invited to England 1610 by Abp. Bancroft, and became a prebendary at Canterbury; visited Oxford and Cambridge, and became a friend of James I and Lancelot Andrewes, Bishop of Ely. He became naturalized, and spent $3\frac{1}{2}$ years in England (" the island of the blest "). Buried in Westminster Abbey. Edited Strabo (1587), the New Testament (1587), Polyaenus (*editio princeps*, 1589), Aristotle (1590), commentary on Theophrastus *Characters* (1592), annotations on Suetonius (1595, reprinted down to 1736); edited Athenaeus (1597), *Historiae Augustae scriptores* (1603), Polybius (1609); 60 voll. of *Adversaria* were deposited by his son in the Bodleian. *Life* by Mark Pattison.

Felix Accorambonus (Felice Accoramboni), of Gubbio; son of Hieronimo, a medical writer, who taught medicine at Padua. Felix taught philosophy at Rome, wrote commentaries on Galen *De temperamentis* and Theo-

ARISTOTLE

phrastus *De plantis*, and was the author of *Interpretatio obscuriorum locorum et sententiarum omnium operum Aristotelis*, Rome, 1590. A new edition, entitled *Vera mens Aristotelis, id est lucidissima et eruditissima in omnia Aristotelis opera explanatio*, Rome, 1603.

Guillaume Du Val, 1572–1646. Besides his edition of the works of Aristotle (Paris, 1619), he was the author of a number of religious works; his *Phytologia, sive philosophia plantarum*, was published after his death (Paris, 1647).

Petavius (Denys Pétau), S.J., 1583–1652. Professor at Reims, and of dogmatic theology at Paris 1621–1643. A prolific writer and a poet. Author of several works on chronology, and of *Uranologion, sive Systema variorum authorum qui de sphaera ac sideribus eorumque motibus graece commentati sunt*, Paris, 1630, the appendix to which, *Accesserunt variarum dissertationum libri octo ad authores illos intellegendos . . . utiles*, contains some matter relevant to *H.A.* Editor of Epiphanius, Julian the Apostate, Synesius, Themistius, etc.

Claudius Salmasius (Claude de Saumaise), 1588–1653. Born at Semur-en-Auxois in Burgundy, discoverer (at the age of 19) of the Heidelberg MS. of the *Anthologia Palatina*. Another prolific writer. Editor and annotator of classical texts. His most remarkable work is the *Plinianae exercitationes in Caii Julii Solini Polyhistora*, Paris, 1629. " Non homini sed scientiae deest quod nescivit Salmasius " (Balzac). In 1649 he wrote the *Defensio pro Carolo I*, which drew a reply from Milton.

Johannes Beckmann, 1739–1811. " Begründer der technologischen Wissenschaft." Taught at St Petersburg 1763–1765; professor of philosophy 1766 and of oeconomics 1770 at Göttingen. Wrote a history of inventions and discoveries in 5 voll. (1780–1805), which was translated into English (four editions, 1797, 1814, 1817, 1846) and a botanical lexicon (1801). Author also of *De historia naturali veterum libellus primus*, Göttingen, 1766, and editor and annotator of [Aristotle] *Liber de mirabilibus auscultationibus*, Göttingen, 1786.

Armand-Gaston Camus, 1740–1804. A Jansenist, a strong republican and revolutionist. Elected to the States General 1789, and attracted attention by his speeches against social inequalities. He was elected to the

HISTORIA ANIMALIUM

National Convention, and in the course of discharging a commission was handed over as a prisoner to the Austrians 1793; two years later he was exchanged for the daughter of Louis XVI. He refused to take part in the Napoleonic régime. In 1796 he was restored to his office as Archivist, to which he had been appointed in 1789 by the Constituent Assembly. His chief work is the *Code justiciaire, ou Recueil des décrets de l'Assemblée nationale* . . ., 1792.

Johann Gottlob Schneider, 1750-1822. Professor (1776-1811) at Frankfurt on Oder and Breslau, and university librarian at Breslau. Editor (1797-1798) of the first independent Greek (Greek-German) lexicon since Stephanus (1572); Liddell-Scott-Jones is one of its direct descendants. Author of several zoological works, and editor of the zoological works of Aelian and Aristotle (*H.A.*), also of Aristotle's *Politics*, of Theophrastus, Nicander, Oppian, Vitruvius, Xenophon, Pindar, etc.

August Immanuel Bekker, 1785-1871. Philologist and textual critic. Professor at Berlin 1810-1871; travelled in Europe (including England) collating MSS. 1810-1821. Published a large number of critical editions of classical texts, including Plato (1816-1823), Oratores Attici (1822), Aristotle (1831-1836), Aristophanes (1828), Livy (1829-1830), Tacitus (1831), Homer (1843 and 1858), and 25 volumes of the *Corpus scriptorum historiae byzantinae*. Some of his editions were first published in England. " He could be silent in seven languages."

Arend Friderich Augustus Wiegmann, 1802-1841. Author of botanical works, and (jointly) of a *Handbuch der Zoologie*, which went through seven editions between 1832 and 1871, of *Herpetologia Mexicana* (1834), and of *Observationes zoologicae criticae in Aristotelis historiam animalium*, Leipzig, 1826, which deals with several passages in the first two chapters of Book II of *H.A.*

Constantin W. Lambert Gloger, author of many books on animals, especially birds, among them *Dissertatio inauguralis, sistens disquisitionum de avibus ab Aristotele commemoratis specimen I*, Breslau, 1830, and *Vollständiges Handbuch der Naturgeschichte der Vögel Europa's*, Breslau, 1834.

Carl Jacob Sundevall, 1801-1875, curator of the zoological museum at Stockholm, author of several zoological

ARISTOTLE

works, including *Die Thierarten des Aristoteles*, Stockholm, 1863 (originally published in Swedish, Stockholm, 1862), a classified glossary of the animals mentioned by Aristotle; he speaks highly of Gloger's *Dissertatio* (see above).

In addition to the editors and commentators mentioned above or elsewhere in this volume, the following have offered suggestions on textual points:

Valentin Rose, *De Aristotelis librorum ordine et auctoritate commentatio*, Berlin, 1854. [p. 231]

Henry Jackson, *Journal of Philology*, XXXII (1913), p. 302.

Herbert Paul Richards, *Journal of Philology*, XXXIV (1918), p. 251.

L. A. W. C. Venmans, *Mnemosyne*, LV (1927), pp. 184-186.

E. Janssens, *Revue belge de philologie et d'histoire*, XII (1933), pp. 613-615.

Ludwig Radermacher, *Wiener Studien*, LXIII (1949), pp. 84 f.

D'Arcy W. Thompson, *Classical Quarterly*, XXIX (1945), pp. 54 ff., beside the suggestions in his translation of *H.A.*

Gunnar Rudberg, *Eranos*, 49 (1951), pp. 31-34.

Among recent writers the following should be mentioned:

Otto Körner (1858-1935), author of

> *Die homerische Thierwelt*, Berlin, 1880, rev. edn. Munich, 1930.
>
> *Die homerische Tiersystem u. seine Bedeutung für die zoologische Systematik des Aristoteles*, Wiesbaden, 1917.
>
> *Über . . . die Verwertung homerischer Erkenntnisse . . . in der Tiergeschichte des Aristoteles*, in *Sudhoffs Archiv für Geschichte der Medizin*, XXIV (1931), pp. 185-201. K. holds that Aristotle accepted Homer's evidence as on a par with actual observation.

D'Arcy Wentworth Thompson (1860-1948), son of the professor of Greek at Queen's College, Galway; himself

professor of Natural History at the University of St
Andrews, 1884–1948, President of the Classical Association 1929, Linnaean Medallist, etc. Translator and
annotator of *H.A.*, 1910 ; author of *Growth and Form*,
Cambridge, 1917, 2nd edn. 1942, an epoch-making work
in the study of biological forms ; *A Glossary of Greek
Birds*, 1895, 2nd edn. 1936 ; *A Glossary of Greek Fishes*,
1947.

Carl Rudolf Burckhardt (1866–1908), a member of an old
and distinguished Basel family. He held strong views
about the writing of the history of biology and of zoology,
and both wrote on this subject and lectured on it at
Basel university, where he was professor from 1894 ; in
1899–1900 he was lecturing on the *H.A.* itself. In 1907
he was scientific director of the zoological station of the
Berlin Aquarium at Rovigno. An appreciation by
Gottlob Imhof of his significance for comparative
anatomy and the history of biology is given in *Zoologische Annalen*, III (1910), pp. 156-176, where the list of
his publications from 1889 to 1908 contains 63 items, the
last being *Aristoteles und Cuvier* (*Zool. Ann.* III, pp. 69-
77). The first article in the first number of *Zoologische
Annalen*, Würzburg, I (1905), pp. 1-28, is his *Das erste
Buch der aristotelischen Tiergeschichte*, in which he gives
a detailed analysis of the first six chapters of *H.A.* Book
I, with a large folding table showing their contents. B.
argues that the true beginning of the treatise is at 487 a
11, and that the preceding paragraphs, of which the
subject is the parts of animals, should be placed at the
end of chapter 1 (488 b 8), where they would naturally
lead on to chapter 2. Among other papers by him are :

> *Das koische Tiersystem, eine Vorstufe der zoologischen
> Systematik des Aristoteles*, in *Verhandlungen der
> naturf. Gesellsch. in Basel*, XV (1904), pp. 377-413, in
> which he claims that there is a close correspondence
> between the way in which animals are listed in
> [Hippocrates] περὶ διαίτης, ii and the arrangement
> adopted by Aristotle (the μέγιστα γένη ; see Notes,
> §§ 9 ff.), and argues for the existence of an earlier
> " Coan Tiersystem " underlying both : Aristotle's
> advance was the development of a logical principle
> of systematic arrangement.

ARISTOTLE

Zur Geschichte der biologischen Systematik, ibid. XVI (1903 [*sic*]), pp. 388-440.
Über antike Biologie, in *34. Jahresheft d. Vereins schweiz. Gymnasiallehrer* (1904).

August Steier, *Aristoteles und Plinius : Studien zur Geschichte der Zoologie,* Würzburg, 1913, a separate edition of three articles in *Zoologische Annalen,* IV and V (1912 and 1913), one of which is *Zoologische Probleme bei Aristoteles und Plinius* (*Zool. Ann.* V (1913), pp. 267-305).

Wilhelm Kroll (1869–1939), editor of Pauly-Wissowa, author of numerous works on classical writers, including *Zur Geschichte der aristotelischen Zoologie,* in *Sitzungsberichte der Akademie der Wissenschaften in Wien,* Philol.-hist. Klasse, CCXVIII (1940), pp. 3-30. K. deals chiefly with the use made by Pliny of *H.A.* and of later zoological writings.

Hans Heinrich Balss, b. 1886. Head Keeper of the Bavarian Zoological Collection at Munich 1927 until his retirement in 1951, editor of a selection from Aristotle's biological writings, with Greek text and German translation,[a] Munich, 1943, and of a similar collection of Greek and Latin texts on ancient astronomy, Munich, 1949 ; author of *Die Zeugungslehre und Embryologie in der Antike* (a conspectus of ancient embryology), in *Quellen und Studien zur Geschichte der Naturw. u. der Medizin,* V (1936), and of *Praeformation u. Epigenese in der griechischen Philosophie,* in *Archeion,* IV (1923), pp. 319-325 ; *Studien über Aristoteles als vergleichenden Anatom,* in *Archeion,* V (1924), pp. 5-11 ; *Albertus Magnus als Zoologe,* Munich, 1928 ; *Alb. M. als Biologe,* Stuttgart, 1947.

Willem Karel Kraak, *Vogeltrek in de oudheid, en het bijzonder bij Aristoteles,* Amsterdam, 1940.

Bernhard Peyer, *Über die zoologische Schriften des Aristoteles,* in *Gesnerus,* III (1946), pp. 58-71, gives an interesting and useful sketch of Aristotle's biological work,

---

[a] *Aristoteles: Biologische Schriften* (Tusculum-Bücher). Of 124 pp. of Greek text, about 50 consist of passages from the *H.A.* (accompanied by A.-W.'s translation, modified). There is an Introduction of 28 pp. (printed as a " Nachwort ") and 14 pp. of Notes on the subject-matter. The selection includes many of Aristotle's distinctive theories, and his most remarkable observations, and the whole provides an excellent introduction to his biological writings.

HISTORIA ANIMALIUM

drawing attention to his remarkable observations of the *Mustelus laevis*. He reproduces the diagram of the embryo's placentoid structure from Nicolaus Steno (*Ova viviparorum spectantes observationes*, 1675), and Johannes Müller's drawing of the same, from his famous paper *Über den glatten Hai des Aristoteles*, Berlin, 1840 and 1842; see *H.A.* 565 b 1 ff., and *cf. G.A.* (Loeb ed.), 754 b 33, n.

The following will also be found useful:

Francis Joseph Cole, *A History of Comparative Anatomy*, London, 1944 (pp. 28 ff.).

Authenticity

A work such as the *H.A.*, which is professedly an assemblage of factual information, is of its very nature susceptible to interpolation, and interpolations are naturally more difficult to detect in such a work than in one which sets forth a continuous chain of argument. It should also be recognized that not all passages which are clearly out of place, or in some way appear to be inappropriate to their context, are automatically to be regarded as spurious. It is possible that some may have been added marginally by Aristotle himself as his own memoranda or as a convenient way of recording statements drawn from elsewhere. An example is the passage about the martichoras at 501 a 23 ff. recorded on the authority of Ktesias, which is regarded as an interpolation by A.-W. and Dittmeyer. A longer passage is the description of the chamaeleon at 503 a 15 ff., which Regenbogen claims (see n. *ad loc.*) comes from Theophrastus; he cites it as an example of the way in which additions could have been made to the *H.A.* by succeeding generations of Peripatetics; another example, he suggests, is the remark at 522 b 24

Short passages.

ARISTOTLE

Book VIII. (where see n.) referring to King Pyrrhus. Upon such passages opinions may differ, and whatever decision one comes to must to a large extent be subjective, since it is impossible to say whether they are or are not integral parts of the treatise. There are, however, some passages which quite clearly do not form an integral part of the treatise, as for instance chapters 21 to 27 of Book VIII,[a] as will readily be seen by reference to the Summary (pp. xcii f.). But it cannot be claimed that Books VII and IX, which are decisively rejected by A.-W. and Dittmeyer, fail to qualify on this score, for the Summary shows that Book VII and some portions of Book IX do in fact fulfil part of the whole scheme as outlined in the introductory chapters of Book I. If Books VII and IX are to be rejected, the rejection must be for other reasons.

Book VII. Aubert and Wimmer were the first editors to question the authenticity of Book VII, and, while admitting that in general the style is not dissimilar to that of Aristotle, so that its having passed as genuine for

[a] According to Thompson (note on 603 a 30), chapters 21-30 of Book VIII " bear evidence of an alien hand." Sir W. D. Ross (*Aristotle*, p. 12) writes " Book X and probably also Books VII, VIII. 21-30, and IX are spurious, and date in all probability from the third century B.C." Jaeger categorically asserts (*Aristotle*, Eng. tr. p. 329) that the *H.A.* " shows the clearest traces of different authors ; the last books are by younger members of the School, who appear as continuing, completing, and even correcting and criticising, the work of the master." The kind of evidence he has in mind is shown by his later remark (p. 441, n.), "As an example of [post-Aristotelian origin indicated by numerous technical terms foreign to Aristotle] one can point to the un-Aristotelian, spurious books of the *H.A.*, whose origin in particular cases can still be illuminated more precisely by such study of words."

so long is not surprising, base their objection to it on a number of passages which they consider to be un-Aristotelian, and obscure and confused in expression. Dittmeyer endorses this opinion, and suggests that the Book may have been the work of some medical writer or an " iatrosophist "; he cites Kühlewein's demonstration [a] that a good deal of the matter contained in it is taken from the Hippocratic writings. The main objection, however, appears to be based on its style rather than on its subject-matter. It is considered to be of doubtful authenticity by D'Arcy Thompson, who remarks (note on 581 a 9) that nearly one-half of its contents can be closely paralleled by passages in *G.A.* III and IV.

In the ninth Book A.-W. find inconsistencies, irrelevancies and repetitions, and some un-Aristotelian obscurities of style; it may, they think, have been put together from notes left by Aristotle, but it is a disorderly composition and some of it is " careless bungling " (*zum Theil gedankenloses Machwerk*). Dittmeyer follows them in rejecting it, and endorses Joachim's view [b] that it was put together by some Peripatetic at the beginning of the third century, incorporating matter from Theophrastus.

Book IX.

The most recent French translator, J. Tricot (Paris, 1957), is much more conservative. He admits that Book VII is carelessly put together and shows in some places poverty of thought and of expression, but he points out that such faults are found elsewhere in Aristotle, and the borrowings from other writers might well have been made by Aristotle himself. The

[a] H. Kühlewein, in *Philologus*, 42 (1884), p. 127; *cf.* Poschenrieder, *loc. cit.*, p. lvii, n. *a.*

[b] H. Joachim, *De Theophrasti libris περὶ ζώων*, Bonn, 1892.

ARISTOTLE

Book X.   arguments against Book IX he considers to a large extent arbitrary.

With Book X the case is different. Here, as with Books VII and IX, it is possible to take into account the factors of subject-matter, theory, and style, and in consequence it has come to be very generally agreed that Book X is not a genuine work of Aristotle. Whether this is so or not, its subject, the causes of sterility, quite clearly shows that it is not an integral part of the treatise. In addition, it is absent from all but one of the oldest Greek manuscripts of *H.A.*, the first few words only of it appearing, subjoined to the end of the 7th (their 9th) Book. Nevertheless, it must have been present in the Greek MS. from which the Arabic translation was made in the ninth century, since Michael Scot includes it in his Latin version made from the Arabic, and a book on this subject is listed among the works of Aristotle by Diogenes Laertius. If it is established [a] that Diogenes' list rests ultimately upon the authority of Aristo, fourth head of the Lyceum after Aristotle, this would imply that a work on this subject was attributed to Aristotle as early as the last part of the third century B.C., though it would not prove the identity of that work with *H.A.* Book X. The Book was omitted by Gaza from his translation of *H.A.*, its authenticity was questioned by Camus, and it was condemned by Schneider, A.-W., and Dittmeyer.[b]

[a] See P. Moraux, *Les Listes anciennes des ouvrages d'Aristote*, 1951.
[b] Leonardus Spengel, *De Aristotelis libro decimo historiae animalium*, Heidelberg, 1842, believed the book to be genuine, but supposed the existing Greek text to have been translated from William of Moerbeke's Latin version in the 14th or 15th century.

HISTORIA ANIMALIUM

Dittmeyer, accepting that it dates from the third century, thinks it was written by some "iatrosophist," who drew from the *G.A.* and the lost *Problems*, and also from the Hippocratic treatises.[a] Rudberg, who discusses the Book at considerable length, and prints Michael's version of it,[b] takes a similar view, and believes it was compiled in the time of Strato. The text was, he thinks, not transmitted with that of *H.A.* (it is, as we saw above, absent from three of the four best MSS.), but was appended to it by some Syriac or Arabic scholar owing to the similarity of its subject-matter to that of Book VII, which at that time was the last (ninth) Book of *H.A.* Rudberg further holds that the translation of this Book attributed to William of Moerbeke (which he also prints) was not made by him, but was supplied by a later hand when the Greek text had been discovered after the existence of a tenth book had become known through Michael's version. Robin[c] says of it, "le dernier est douteux." As before, Tricot is more cautious, and considers that its discrepancy with *G.A.* on the theory of human reproduction (a strong argument with those who reject the Book) may be accounted for by a change of opinion on Aristotle's part; he also rejects the suggestion that it dates from the time of Strato, since if the list mentioned above rests on the authority of Aristo, he could

---

[a] He refers to Fr. Poschenrieder, *Die naturwissenschaftlichen Schriften des Aristoteles in ihrem Verhältnis zu den Büchern der hippokratischen Sammlung* (Bamberg, 1887), p. 33.

[b] Gunnar Rudberg, *Zum sogenannten zehnten Buch der aristotelischen Tiergeschichte*, Uppsala, 1911.

[c] Léon Robin, *La Pensée grecque*², Paris, 1948, p. 362. *Cf.* Ross, quoted above, p. liv, n. *a*.

hardly have made the mistake of attributing to Aristotle a work written by, or composed in the time of, Strato, one of his predecessors in the Lyceum, who lived less than a century before his own time. The text of the tenth Book is printed by A.-W. but without textual notes, translation, or commentary, and it is omitted as " spurious beyond question " by D'Arcy Thompson from his translation. It is omitted also from the present edition.

Date of the Treatise

It might seem that, taking account of the nature of the treatise itself, and of the questions which have been raised about the authenticity of some of its parts, the dating of *H.A.* would present considerable difficulties. W. W. Jaeger, in his book on Aristotle's development (the original German edition was published in 1923), took the view that " all indications point to a late date for the origin of the philosopher's zoological works " (Eng. trans., p. 330). This view, however, was not founded upon an examination of the zoological works themselves; Jaeger dated them late because he had constructed his scheme of Aristotle's development without taking them into consideration, and because he believed on other grounds that this kind of activity was characteristic of Aristotle's later years. D'Arcy Thompson, in the prefatory note to his translation of *H.A.*, published in 1910, had remarked that the frequent references in that work to places in or near Lesbos suggested that " Aristotle's natural history studies were carried on, or mainly carried on, in his middle age, between his

two periods of residence in Athens."[a] A detailed study of the place-names in *H.A.*, undertaken by Sir Desmond Lee,[b] combined with other converging evidence,[c] shows convincingly that much of Aristotle's zoological research was done during his middle years, especially during the two years which he spent in Lesbos (345/4 to 343/2), when he was aged about forty. It seems likely, however, that a treatise such as *H.A.* would be continually receiving additions,[d] and perhaps the references to elephants are the result of information received through Alexander's expedition.[e]

Much work, of course, still remains to be done on the *H.A.*, especially on its relation to the other zoological works as well as to other works in the Aristotelian corpus. Even within the supposedly genuine parts of the *H.A.* itself there are numerous discrepancies, to some of which I have drawn attention in the notes, but a thorough investigation of this subject is beyond the scope of the present edition.

This Edition

No new collations of the manuscripts have been made for the present edition, and it seems doubtful whether

[a] *Cf.* Thompson, *Aristotle as a Biologist*, 1913, p. 12.

[b] H. D. P. Lee, " Place-names and the Date of Aristotle's Biological Works," *C.Q.* 42 (1948), pp. 61 ff.

[c] See the references conveniently given by G. E. R. Lloyd, *op. cit.* (*supra*, p. xxiii, n. *a*), pp. 59 f.

[d] P. Louis (*op. cit.* pp. 302 ff.) thinks that *H.A.* contains developments which belong to different stages in Aristotle's career, and that a thorough study of his classification of animals would go some way towards clarifying the chronology of the scientific treatises, as well as of that of the various parts of these treatises. See also G. E. R. Lloyd, *op. cit.*

[e] A supposition approved by Jaeger, *op. cit.* p. 330.

ARISTOTLE

much further improvement of the text can be derived from the Greek MSS., of which Dittmeyer made a thorough investigation, including a full collation of C^a, A^a and D^a. The text has, however, been thoroughly worked through, and account has been taken of the proposals of various scholars for its improvement. Where I have accepted the reading of Bekker's edition I have not normally given the MSS. variants, which will be found in his and Dittmeyer's *apparatus*; but (except for minutiae) I have endeavoured to indicate all departures from Bekker's text, and to record all readings and proposals of importance. It is not feasible to retain the same length of printed line as in Bekker's edition, but the marginal figures have been so placed as to produce the smallest possible discrepancy with Bekker's numeration.[a] The text has been reparagraphed throughout for convenience of reading and reference.

With regard to Michael Scot's translation, I have followed the same practice as I did for my editions of *P.A.* and *G.A.* (see *G.A.* p. xxx). I have transcribed passages containing places which some previous editor or I myself had already felt for some reason to be doubtful; and the pertinent parts of these, where they have anything to contribute, I have given in the *apparatus*.

I have adopted the same principles of translation as in the two previous volumes (*P.A.* and *G.A.*), and notes have been provided on important, interesting,

[a] M. Paul Moraux's complaint in his review of the Loeb *Posterior Analytics* and *Topics* (*L'Antiquité Classique*, XXXI (1962), p. 338) is understandable, but to speak of an " inconvénient majeur " is an exaggeration. To adhere exactly to Bekker's lay-out in every line of the text would result in a typographical monstrosity.

or difficult passages, but I have not thought it necessary to point out where Aristotle has gone wrong. The three Tables on pp. xci ff. are intended to supplement the index, which is to be included in vol. iii ; pending its appearance, it is hoped that the Tables may to some extent serve also as an index. Elucidation of some of the more important terms used by Aristotle is provided in the Notes on Terminology, pp. lxii ff. ; these may be supplemented if necessary by reference to the Introduction and Appendices in the Loeb edition of *G.A.*

It is a great pleasure to acknowledge the help and encouragement which I have received during the preparation of the translation ; the opportunity for undertaking it I owe to my former teacher and colleague the late Dr. W. H. D. Rouse. The whole of the original draft of the translation in this volume was read by Dr. Joseph Needham, F.R.S., of Gonville and Caius College, and by Dr. Sydney Smith, of St. Catharine's College ; I am indebted to Dr. Needham for many valuable references in the notes. For the identification of many of the animals mentioned by Aristotle I have relied upon D'Arcy Thompson's *Glossary of Greek Birds*[2] (1936) and *Glossary of Greek Fishes* (1947), to which the reader is referred for further information ; there still remain, however, many uncertainties. My indebtedness to others will be apparent in the course of the work itself.

A. L. P.

INSTITUTE FOR ADVANCED STUDY
    PRINCETON
    3rd February 1964

NOTES ON TERMINOLOGY

Parts, " uniform " and " non-uniform "

Two sorts of " parts."

(1) The term " part," which occurs in the title of the treatise *De partibus animalium* (or, as Aristotle himself calls it, at *G.A.* 782 a 21, the treatise on the *Causes of the Parts of Animals*), includes considerably more than is normally included by the English phrase " part of the body." For instance, we should not normally call blood a " part," but Aristotle applies the term to all the constituent substances of the body as well as to the limbs and organs. For Aristotle, all the parts are either " uniform " or " non-uniform." At the very beginning of *H.A.* he states this distinction, giving definitions of the two kinds of " parts," and he makes use of it in the course of his subsequent treatment, as will be seen from the Summary (pp. xcii f.). Aristotle lists a number of the uniform parts at the beginning of Book III, ch. 2. The term " organ," as applied to the non-uniform parts, corresponds closely to Aristotle's own description of them at *P.A.* 647 b 23 as τὰ ὀργανικὰ μέρη, " the instrumental parts." His remarks at the beginning of *H.A.* may be supplemented by what he tells us elsewhere (*P.A.* 647 b 22 ff.):

(a) Some of the uniform parts are the material out of which the non-uniform are constructed (*i.e.*, each non-uniform or instrumental part is made out of bones, sinews, flesh, etc.);

(b) some uniform parts, *viz.*, the " fluid " ones,[a] serve as nourishment for those in class (a), since all growth is derived from fluid matter;

(c) some uniform parts are " residues "[b] from those in class (b), *e.g.*, faeces and urine.

The heart is the only " part " which comes under both headings (*P.A.* 647 a 25 ff.): it is made out of one uniform part only, but at the same time it has essentially a definite configuration, and thus it is also a non-uniform part.

Identity and difference.

(2) So far as the individuals of any one type of animal are concerned, their parts are of course identical; as Aristotle says, any one man's eye or nose is the same as any other man's, any one horse's leg is the same as any other

---

[a] See § 13.   [b] See §§ 22 ff.

HISTORIA ANIMALIUM

horse's leg, and so on. But when we pass on to consider larger groups, *e.g.*, Birds, of which there are a great number of kinds, although all kinds of birds have a beak, the beak of one kind of bird differs from that of another ; and when we proceed still further, and compare the various larger groups themselves, such as Birds compared with Fishes, although we still find a correspondence, *e.g.*, between the feathers of birds and the scales of fishes, the correspondence is not so close. This applies to uniform parts as well as to non-uniform.

(3) In a sense, then, the parts of the various animals are "the same," in another sense they are not the same, but different ; and the two modes of difference are described by Aristotle in the following way. (*a*) The minor mode of difference he calls difference " by the more and less," or difference " by excess and deficiency " (see *H.A.* 486 b 7)—what we should call difference of degree. Examples of such difference are : some birds have a short beak or short feathers, others have long ones ; cartilage is of the same nature as bone, but differs from it " by the more and less " (516 b 32) ; the hedgehog's spines are a harder and stiffer version of hair (517 b 22), and so on. (*b*) When the divergence becomes wider, as for instance between different groups of animals such as Birds and Fishes, the difference can no longer be described as being " by the more and less," but the parts now are the same only τῷ ἀνάλογον : they correspond only " by analogy," they are merely the " counterparts " of each other. Examples are (486 b 18 ff.) : bone and fish-spine, nail and hoof, hand and claw, a bird's feather and a fish's scale. *The "more and less."*

"Counterparts."

(4) When he describes these two modes of difference at *H.A.* 486 a ff., Aristotle links them closely with the contrast between γένος and εἶδος ; but he is not consistent in maintaining the distinction between " the more and less " and " analogy " ; and we find a parallel inconsistency in his use of γένος and εἶδος (see next §). Thus, at *H.A.* 588 a 22 ff. some animals are said to differ in certain respects from Man by " the more and less," while others differ from Man " by analogy." At *H.A.* 486 b 18 (as we have seen, § 3), bone is said to differ from fish-spine *not* by " the more and less " but " by analogy," whereas at 516 b 13 ff. we have each of the *Inconsistent usage.*

lxv

ARISTOTLE

two modes of difference applied *within* a single group: the bones of birds are said to differ but little among themselves (presumably by " the more and less "), and those of fishes differ among themselves " by analogy." Similarly (*ibid.*), the bones of some oviparous quadrupeds are " more bony," those of others are " more like fish-spine," which (according to 486 b 14 ff.) would imply difference " by analogy " *within* the group. The contrast between the two modes of difference is therefore in practice not confined to γένη versus εἴδη, and therefore we are not surprised to find a parallel inconsistency in Aristotle's use of the terms γένος and εἶδος themselves.

Γένος *and* εἶδος [a]

Usage. (5) These terms, which have given rise to the well-known terms *genus* and *species*, are used by Aristotle in his zoological works in what appears to the modern reader a somewhat confusing way; and it is therefore important from the outset to be on one's guard against expecting to find that they invariably, or indeed often, bear the technical meanings which have come to be attached to their modern descendants. Indeed, in the *H.A.* Aristotle's meaning is most frequently conveyed in translation by avoiding the terms *genus* and *species*, though there are a few passages in the *H.A.* and elsewhere in the zoological works where a contrast of this sort, or something like it, seems to be intended by Aristotle. In the *H.A.*, these passages all occur in introductory portions of the treatise [b]; and although their outlook agrees with what we find in the logical works, in the zoology they are isolated, and in the main body of it Aristotle does not observe the distinction between γένος and εἶδος. The question how these passages come to be in the zoology is an interesting one, and raises the further question of the relation of the logical to the zoological treatises; but these questions cannot be discussed here.

Indiscriminate use of (6) It is, however, important to observe that when Aristotle

[a] Much of this and the following Note is based on Mr. D. M. Balme's article " γένος and εἶδος in Aristotle's biology," *C.Q.* LVI (n.s. XII), 1962, pp. 81 ff., where the subject is very fully treated.
[b] They are: 486 a 16 ff., b 17 ff., 488 b 30 ff., 490 b 16 ff., 491 a 18, 497 b 9 ff., 505 b 26 ff., 539 a 27.

uses γένος and εἶδος in this definitely contrasted way, he means by γένος what in two passages of the *H.A.* he calls μέγιστον γένος (see §§ 9 ff. below). This is clear from his statement at 486 a 23, " By γένος I mean, *e.g.*, Bird and Fish "—two groups which in the later passages he calls μέγιστα γένη. These two groups we should not now normally describe as *genera*, and hence it is somewhat misleading even in such contexts to translate by *genus* and *species*, though it is not easy to find satisfactory alternatives. Otherwise, he uses γένος not only of any of the kinds of animals which fall under the μέγιστα γένη, *e.g.*, dog (658 a 29), but also of the various kinds of dogs (574 a 16); and although it is true that when speaking of a group which contains sub-groups or varieties he tends to use the term γένος, there are so many instances of his applying the term γένος to a group or type which he elsewhere calls εἶδος, that he evidently does not intend to reserve γένος for larger groups, or even for groups which contain varieties. In fact, any group or type which he calls εἶδος he can also call γένος. Probably the word γένος retains a good deal of its fundamental connotation of kinship (" kind "), and εἶδος its connotation of visible shape (" form "); and this may be responsible for Aristotle's more frequent use of γένος even where εἶδος would have done equally well. In fact, one might consider γένος a more appropriate word for use in zoology, unless for some special reason emphasis needs to be laid upon the visible form; and of course εἶδος is a word by no means confined to zoology, any more than " form " is so confined in English: both words occur in various contexts with various shades of meaning. To sum up, Aristotle so often applies the two terms γένος and εἶδος indiscriminately to one and the same type of animal that obviously his choice of one term or the other in these cases has been a matter of indifference.

(7) From the statements about γένος and εἶδος and the modes of difference in the passages referred to above (§ 4), we should get the following scheme:

Animals which are the same εἴδει: their parts are the same.

Animals which are the same γένει, but different εἴδει: their parts differ by " the more and less."

the two terms.

The theoretical scheme.

ARISTOTLE

Animals which differ γένει: their parts are the same only τῷ ἀνάλογον.

No taxonomic structure. (8) This scheme, however, as I have pointed out, is not adhered to in the main part of the treatise, and, as we have seen, Aristotle uses all four terms much more widely; in consequence they possess no precise and settled connotation. In particular, no attempt is made (*cf.* §§ 10 ff. below) to work out a detailed or continuous series of gradations of γένη, and there is no indication of any taxonomic hierarchy with successive subdivisions. From the μέγιστα γένη down to any actual type of animal, the term γένος is equally applicable.

The μέγιστα γένη, *the " main groups "*

References. (9) Aristotle speaks of certain μέγιστα γένη, or " main groups " of animals, in the following passages:
(*a*) At 490 b 7 ff. he says that the main groups of animals are Birds, Fishes, and Cetacea (all of which are blooded), and Testacea, Crustacea, Cephalopods, and Insects (all of which are bloodless). Beyond these, he says, the γένη of the remaining animals are not μεγάλα: we do not find one εἶδος containing many εἴδη. But then, at line 31, he says there are many εἴδη of the γένος Viviparous Quadrupeds; yet Viviparous Quadrupeds were not included in the list given a few lines earlier.
(*b*) At 505 b 26 he says that the μέγιστα γένη differ from the remaining γένη of the other animals in that they are blooded whereas the latter are bloodless. They are: the Viviparous Quadrupeds, the Oviparous Quadrupeds, Birds, Fishes, Cetacea. (Man cannot be reckoned as a μέγιστον γένος, because he is a single type on his own; but he is, of course, to be included among the Blooded animals.) This passage seems to deny the four Bloodless γένη mentioned in the previous passage the title of μέγιστον γένος.
(*c*) We should add to these the passage already referred to in § 6, 486 a 23, where Aristotle says " By γένος I mean for example Bird and Fish." Here Bird and Fish are described as γένη merely, but in this passage Aristotle is concerned with the contrast of γένος and εἶδος.

Purpose of (10) The fact that he is inconsistent in his use of the term

lxviii

μέγιστον γένος, and in listing the groups of animals to which it applies, suggests that the purpose of the classification is practical and not theoretical. (Nor is it exhaustive: some animals are not catered for by it at all: Man, for example, comes under no μέγιστον γένος; nor do Serpents.) A glance at the Summary on pp. xcii f. will show how Aristotle makes practical use of the μέγιστα γένη for arranging what he has to record about the various animals. He uses the wide divisions Blooded and Bloodless, and within these he treats of the μέγιστα γένη and the animals which fall under no μέγιστον γένος.

the classification.

(11) We may also recall here what I pointed out earlier (p. vii), that some of the μέγιστα γένη are " nameless ": Birds, Fishes, and Cetacea have names in Greek, but Aristotle has to use descriptive adjectives for the Bloodless γένη (*e.g.*, τὰ μαλακόστρακα means " the soft-shelled animals "). Between the μέγιστα γένη and the ultimate species (such as mullet, hedgehog) there are, with one exception, no group names: the exception is τὰ λόφουρα, the bushy-tailed animals, a name intermediate between Viviparous Quadrupeds and horse, mule, etc. (The σελάχη, the cartilaginous fishes, may perhaps be reckoned as providing another such intermediate name.) Aristotle points out the lack, but he does not attempt to make it good. He is clearly at this stage not intending to produce any taxonomic scheme. He is taking the obvious large, main groups as a convenient basis for marshalling his material. Taxonomy, if it is to be attempted at all, must come at a later stage, when all the differences have been collected and recorded. Neither Blooded and Bloodless, nor the μέγιστα γένη, can pretend to be more than preliminary classifications.[a]

Lack of intermediate stages.

Ἀρχή

(12) This term, meaning literally " source " or " beginning," is difficult to translate effectively. It is often represented by " principle " or " first principle." There is, however, really but little difficulty about this term, for the context will usually indicate what its connotation is.

---

[a] Comments on the μέγιστα γένη will also be found in the footnotes to 490 b 22, 491 a 6, and 505 b 32. See also Introd. pp. vii ff.

ARISTOTLE

A few examples of its use may be given. (1) Often, as at *G.A.* 715 a 6, it is a principle or source of " movement " (ἀρχὴ τῆς κινήσεως). Hence obviously (2) the Motive (Efficient) Cause may be described as an ἀρχή, and so may the other Causes, including Matter; and for the same reasons the sexes also are ἀρχαί (*cf. H.A.* 590 a 3 f., and § 58); so is semen. (3) An ἀρχή is something which though small in itself is of great importance and influence as being the source or starting-point upon which other things depend and which causes great changes (κινήσεις) in them (see *G.A.* 716 b 3, 763 b 23 ff., 766 a 14 ff.). The ultimate ἀρχή of an animal is its heart (*e.g.* ; *G.A.* 766 a 35 ff.; *cf.* p. xxx); and at *H.A.* 511 b 10 we read, " the nature of the blood and of the blood-vessels looks as if it is an ἀρχή "). At *H.A.* 547 b 12 Aristotle speaks of the ἀρχή of spontaneously generated Testacea taking its rise (see § 27) in muddy places (*cf.* 539 a 18); and at 561 a 10 of the ἀρχή of an egg. There are also ἀρχαί which are external to the animal, *e.g.* the sun and moon (*G.A.* 777 b 24).

Τὸ ὑγρὸν καὶ τὸ ξηρόν, " *fluid substance and solid substance* "

Meaning. (13) These are two of the four fundamental substances in Aristotle's physical theory, the other two being τὸ θερμόν and τὸ ψυχρόν, " hot substance " and " cold substance." Following Ogle in his translation of the *P.A.*, I use the renderings " fluid " and " solid," as being more in conformity with the definitions given by Aristotle himself than " moist " and " dry," which are sometimes used. Actually, neither pair of English words fully expresses the Greek meanings. Aristotle's definitions of them (at *G. & C.* 329 b 30) are : " ὑγρόν is that which is not bounded by any boundary of its own but can readily be bounded ; ξηρόν is that which is readily bounded by a boundary of its own but can with difficulty be bounded " : at the end of each definition there should of course be understood " by a boundary imposed from without." The application of these terms as adjectives, *e.g.*, " fluid " to flesh, blood, fat, etc., " solid " to skin, sinew, bone, etc. (see *H.A.* 487 a 2 ff.), indicates that

these things exhibit the qualities concerned and are largely composed of the corresponding substances.

(14) The two substances, the "fluid" and the "solid," as Aristotle tells us in the passage referred to above (*G. & C.*), are "passive" (παθητικά), while the other two, the "hot" and the "cold," are "active," "creative" (ποιητικά). The "hot" is that which brings together, causes to combine, things which are of the same kind; the "cold" is that which brings together both cognate things and things of different kinds. The two passive substances thus tend to serve as "matter," and the two active ones as instruments of "form" and "movement."

The "hot" and the "cold."

(15) Each of the four so-called "elements" in Aristotle's physics, viz., Fire, Air, Water, and Earth, consists of a pair of these four fundamental substances (*G. & C.* 330 b 3).

The four "elements."

Fire is hot and ξηρόν
Air is hot and ὑγρόν
Water is cold and ὑγρόν
Earth is cold and ξηρόν [a]

We have here an excellent illustration of the difficulty of finding satisfactory translations for ξηρόν and ὑγρόν.

Κύημα, "*fetation*"

(16) Aristotle's definition of κύημα (at *G.A.* 728 b 34) is "the first (or primary) mixture of male and female"; and although the term is often so used, it is also used by Aristotle to include more than this. Actually, it covers all stages of the living creature's development from the time when the "matter" is first "informed" to the time when the creature is born or hatched. Hence we find κύημα applied to the embryo or fetus of Vivipara, to the "perfect" eggs of birds (*cf. H.A.* 489 b 7) and the "imperfect" eggs of Cephalopods, etc. (the last-named are still so called after deposition, *G.A.* 733 a 24), to the roe of fishes (741 a 37) and to larvae (758 b 13); indeed, the larva is compared to the earliest stage of the κύημα in viviparous animals (758 a 33).

Meaning.

[a] Hence the terms "earthy" and "solid" are to some extent interchangeable (*cf.* p. xv).

ARISTOTLE

Transla- (17) There is no English word which conveys the wide range
tion. of the term κύημα, and therefore I introduced in my
edition of the *G.A.* the term " fetation," by which I invariably translate it.

Nourishment, residues, etc.

(18) Aristotle's reference at *H.A.* 489 b 8 to the two kinds of
ingredient in an egg recalls his doctrine of the two grades
of " nourishment " in the living animal. A short account
of his theory of nourishment is therefore appropriate.

Concoc- (19) After mastication, the food passes into the stomach,
tion. where it is " concocted " [a] by means of the " natural (or
vital) heat " resident there. Any living thing (anything
" with Soul in it ") possesses " natural heat," and the
chief seat of the Soul and the source of the vital heat
is the heart (or its counterpart). But also, every part
of the body as well has its own natural heat, derived
from the heart through the blood; thus, the stomach
concocts the nourishment before passing it on to the
heart, and other parts may concoct it still further when
the heart has sent it on to them. Beside the stomach,
the liver and the spleen assist in the concoction of the
nourishment (*P.A.* 670 a 20 ff.).

Blood. (20) Having received its first stage of concoction in the
stomach, the nourishment passes on to the heart, where
it undergoes its most important stage of concoction, and
is thereby turned into blood, the " ultimate nourish-
"Pneuma- ment " for the whole body (*P.A.* 647 b 5). It is probable
tization" of that, in Aristotle's view, an important part of this process
blood. was the " pneumatization " of the blood, *i.e.*, the charg-
ing of it with Σύμφυτον Πνεῦμα (see *G.A.* Loeb edition,
App. B) and with the special " movements " requisite
to enable it (*a*) to maintain the " being " of the animal
as such and (*b*) to supply its growth.

Two grades (21) Corresponding to these two functions of nourishment,
of nourish- Aristotle distinguishes two grades of nourishment (*G.A.*
ment. 744 b 33 ff.). The first-grade nourishment (*a*), which is

[a] The Greek word for concoction is the same as that employed to
denote the process of ripening or maturing of fruit, corn, and the like by
means of heat—also that of baking and cooking; indeed, the processes
are regarded by Aristotle as being fundamentally identical. It is also
applied by him to the " maturing " of the embryo.

described as "nutritive" and "seminal," provides the whole animal and its parts with "being"; the second-grade (b) is described as "growth-promoting," and causes increase of bulk. At *H.A.* 489 b 8 we find a parallel distinction with regard to the nourishment in the egg: the embryo develops from part of the egg (*i.e.*, the part which is concerned with the animal's "being"); the remainder it uses as food to promote its growth.

(22) This distinction between the two grades of nourishment enables us also to distinguish the different classes of "residue." In the development of the embryo, it is the leavings of the first-grade nourishment, or "nutritive residue," left over after the "supreme parts"—flesh and the other sense-organs—have been provided for, which are used to form the bones and sinews; the second-grade, inferior, nourishment (which is taken in from the mother or from outside) is used to form nails, hair, horns, etc. The latter is more "earthy" than the former; indeed, with such residue in mind Aristotle can say (*G.A.* 745 b 19) that "residue is unconcocted substance, and the most unconcocted substance in the body is earthy substance." Residues:

(23) Generally, more blood is produced than is required for the purposes mentioned in § 21, and the surplus may then undergo a further stage of concoction, and Nature is often able to turn it to some useful purpose. These are the *useful* residues: examples are semen, menstrual fluid, milk. Marrow is produced when "the surplus of bloodlike nourishment is shut up in the bones" and concocted by their heat (*P.A.* 652 a 5, a 20). Sometimes, when the nourishment is particularly abundant, the surplus blood is concocted into fat, such as lard and suet. Also, some of the blood, reaching the extremities of the vessels in which it is carried, makes its way out in the form of nails, claws, and hair. (a) useful;

(24) Residues may appear at various stages: they may appear before, as well as after, the nourishment has been turned into blood; and then they are residues of "nourishment at its first stage"; thus (*P.A.* 653 a 2; *cf. De somno* 458 a 1 ff.), after a meal the nourishment rises as vapour through the vessels to the brain, where it is cooled, and then condenses into *phlegma* and *ichor* (serum). But both of these, it seems, may also be *useless* (b) ambiguous;

lxxiii

ARISTOTLE

residues, for at *H.A.* 511 b 10 f. *phlegma* is mentioned in company with dung, and at 487 a 6 in company with the excretions from the belly and the bladder, though perhaps it is most often a residue of the *useful* nourishment (*G.A.* 725 a 14). *Ichor*, too, the "watery part of the blood," is sometimes unconcocted blood, sometimes corrupted blood (*P.A.* 653 a 2). For *ichor* see also note on 489 a 24.

(*c*) useless; (25) Residues then are "the surplus of nourishment" (*G.A.* 724 b 26); but there are useless as well as useful residues, for residues may come either from the useful or the useless nourishment (725 a 4). Among *useless* residues are the excrements; these are natural useless residues; but there are also some unnatural ones, as has already been hinted. Among them perhaps should be included bile (mentioned in company with *phlegma* at *H.A.* 511 b 10), which serves no useful purpose. It is a residue produced by the liver; it is a "colliquescence" (σύντηγμα), resulting from decomposition proceeding contrary to nature.

(*d*) the (26) Some of the most important residues are the generative
generative   residues, semen and menstrual fluid—natural and useful
residues.   residues, for which Nature has set apart special places in the body. The difference between them is one of degree of concoction: semen is a residue of the final stage of useful nourishment (*G.A.* 726 a 26); so is menstrual fluid, but the female has not sufficient natural heat to carry the concoction far enough to produce semen. Like the blood, of which it is a more fully concocted form, semen derives its character primarily from the heart, where the blood is "pneumatized" and charged with the requisite "movements" (see § 20); like blood, therefore, it is a vehicle of Soul.

Συνίστασθαι

Meaning (27) The meaning of this verb is perfectly clear, but there is
and usage.   no one convenient word in English by which to translate it. Hence there is the danger that Aristotle's thought may become obscured through the use of varying words to represent it in different passages. It does not occur as frequently in the *H.A.* as in the *G.A.*, where it is

commonly used to describe the initial stage in the "constituting" of the embryo; and there too the active voice, συνιστάναι, is used to describe the action of the semen in " constituting " the embryo. (Aristotle compares the action of rennet in " setting " milk, and uses the same verb to describe it.) The verb is thus associated with the imposition of form upon matter in the production of living creatures; and we find it also used of the spontaneous formation of Testacea (at *H.A.* 547 b 12; the noun σύστασις is used two lines later) and of certain Insects (at 556 b 26 f.). At 547 b 12 it is the ἀρχή (see § 12) of the creature which is said to " take its rise " or to " take shape " (*cf.* also 539 a 18). To " take shape " is indeed sometimes a possible way of translating the word, but it is not feasible to adopt any one consistent translation, and I have therefore often thought it advisable to draw attention to the use of the word by means of a footnote. See further, *G.A.* (Loeb ed.), Introd. §§ 54 ff.

Ἐπαμφοτερίζειν, " *dualize* "

(28) As this word expresses something distinctive in Aristotle's thought, and is applied to phenomena which render difficult any tidy classification of animals, it seems best to use consistently an equally distinctive word to translate it. I have chosen " dualize " for this purpose. Aristotle uses it to express the situation when an animal (or any other object) " plays a double game," " runs with the hare and hunts with the hounds " (*L.-S.-J.*), manages to be in some respects on both sides of whatever fence is under consideration. *Meaning.*

(29) Perhaps the most obvious application of ἐπαμφοτερίζειν is to pairs of anatomical alternatives, as when Aristotle is faced with deciding whether the sea-anemone is an animal or a plant (it dualizes, *P.A.* 681 b 1), whether the pig as a class is solid-hoofed or cloven-hoofed (it dualizes, *H.A.* 499 b 12, b 21), whether the hermit-crab belongs to the Crustacea or the Testacea (it dualizes, 529 b 24). Apes, monkeys and baboons dualize between man and quadrupeds (502 a 16), or between biped and quadruped (*P.A.* 689 b 32); the ostrich dualizes between bird and quadruped (*P.A.* 697 b 14). The seal *Usage:* (*a*) *animals, etc.;*

ARISTOTLE

dualizes between land-animals and water-animals, but although it breathes and does not take in water, and sleeps and breeds on land, it spends most of its time in the water and feeds there, so it must be reckoned among aquatic creatures (566 b 27 ff.; *cf.* § 40 below). Furthermore, it shares with fishes (the verb here is ἐπαλλάττειν, 501 a 22) the characteristic which most of them possess of being saw-toothed. What is the criterion to be? Is it to be anatomical, or dietary, or ecological? (The problem gives rise to an important discussion at 589 a 10 ff. on what is meant by ἔνυδρον and πεζόν; see below, §§ 33-60.) The seal is indeed an inveterate dualizer. It is mentioned again at *P.A.* 697 b 1 ff. as dualizing anatomically between ἔνυδρα and πεζά, just as the bat dualizes anatomically between πτηνά and πεζά (or τετράποδα).

(*b*) parts of animals;
(30) Dualizing is also to be observed in the parts of animals. Thus, the intermediate legs of many-footed creatures dualize between the extreme legs at front and back with regard to their manner of bending (*H.A.* 498 a 18); liver and spleen appear to dualize between being μονοφυῆ and διφυῆ, though actually they are διφυῆ (*P.A.* 669 b 15); in ovoviparous animals such as the Selachia the position of the uterus dualizes as between the position in viviparous and the position in oviparous animals (511 a 25; *cf. G.A.* 719 a 12).

(*c*) other uses.
(31) The notion can, however, be applied also to pairs of characteristics which are not primarily or noticeably anatomical: *e.g.*, Man dualizes between being μονοτόκον and πολυτόκον (*H.A.* 584 b 28), or between these and ὀλιγοτόκον as well: his dualizing ranges " over all the γένη " (*G.A.* 772 b 1 f.). Some animals dualize in being both solitary and gregarious (488 a 2; *cf.* a 7). Certain fishes are found both in deep and in shallow water; they dualize (598 a 15).

(32) The notion can even be applied to inanimate objects: water and air dualize τῷ ἄνω καὶ κάτω (*Physics* 205 a 29); and there are some things which are neither always in motion nor always at rest, but dualize (*Physics* 259 a 25).

Implications.
(33) It is interesting to note that the result of the discussion at Book VIII. 2 (589 a 10 ff.) already referred to (§ 29) on the meaning of ἔνυδρον and πεζόν is in effect to minimize the importance of the place of actual habitat and

to emphasize the importance of physical differences in the composition of the animals themselves, *e.g.*, whether they need to take in air for the purpose of cooling their heat; and these are determined by the κρᾶσις, the " blend " of their bodies, *i.e.*, by the most fundamental physical facts about them in Aristotle's view (*cf.* above, Introd. pp. xv ff.). This indeed is what determines an animal's character as πεζόν or/and ἔνυδρον, as well as certain other characteristics which it exhibits.

Κρᾶσις, Συμμετρία

(34) The notion of κρᾶσις (blend) was not invented by Aristotle, nor was he the last to make use of it. It is found in earlier philosophic writing, and in the Hippocratic corpus, and the term has given rise to such modern terms as temper, temperature and temperament. In the Hippocratic treatise π. διαίτης the theory is expounded that the human organism, body and " soul " alike, is compounded of Fire and Water (which means, ultimately, out of the hot, the cold, the solid and the fluid substances), and in ch. 35 of the first Book we find a list of the different varieties of blend (κρῆσις, σύγκρησις) of Fire and Water which may be found in the " soul " of different individuals; upon this blend its health and sensitivity depend. A similar belief is found in Aristotle. At *P.A.* 650 b 28 we read that in an animal whose heart has a watery blend the way is already prepared for a timorous disposition. Man is the most intelligent animal: this proves his εὐκρασία—the excellence of his blend: the heat in the heart is purest in man (*G.A.* 744 a 29). Best of all are those animals whose blood is hot and thin and clear: thickness and heat make for strength, whereas thinness and coldness make for sensitivity and intelligence. The same applies to the counterpart of blood in bloodless animals (*P.A.* 648 a 3 ff.); that is why bees and ants are intelligent (650 b 19 ff.).

(35) On the purely physical level, health depends upon κρᾶσις. Melancholics (*i.e.*, those who have too much black bile) are always in need of medical attention, because their body is in a state of irritation owing to its blend (*Eth. Nic.* 1154 b 13). Health and well-being, says

ARISTOTLE

Aristotle, we consider to lie in the κρᾶσις and συμμετρία of hot things and cold, either with regard to each other or with regard to the surrounding environment (*Phys.* 246 b 4 ff.). The definition of health as the συμμετρία of hot things and cold is twice mentioned in the *Topics* (139 b 21, 145 b 8). Indeed, the nature (φύσις) of many things can ultimately be traced back to these two ἀρχαί (see § 12), the hot and the cold (*P.A.* 648 a 24). In generation, says Aristotle, male and female need συμμετρία towards each other, because all that is produced by art or by nature exists in virtue of some due proportion (λόγῳ τινί ἐστιν) : in this case, the heat must be σύμμετρος (*G.A.* 767 a 17 ff.), and at 777 b 28 we read that heatings and coolings μέχρι συμμετρίας τινὸς ποιοῦσι τὰς γενέσεις, after that they produce dissolutions. The purpose of the brain, which is the coldest of all the parts in the body, is to counterbalance the heat of the heart, *i.e.*, to achieve measure and the mean (τὸ μέτριον καὶ τὸ μέσον) ; it makes the heat in the heart well-blent (εὔκρατον). The brain, however, needs a moderate amount of heat, and this is supplied by small blood-vessels which run up to the membrane round it ; this blood must be thin and clear. Fluxes occur when the parts round the brain are too cold for a σύμμετρος κρᾶσις (*P.A.* 652 a, b). Further, Nature has placed some of the senses in the head because the blend of the blood there is σύμμετρος, and suitable for keeping the brain warm and for providing quiet and accuracy for the senses (*P.A.* 686 a 9). The liver also contributes greatly towards maintaining a good blend (εὐκρασία) of the body, and health, for next to the heart it contains more blood than any other internal organ. Some animals (*e.g.*, the toad and tortoise) have a poor liver, on a par with the poor blend of their bodies (*P.A.* 673 b 25). At *G.A.* 777 b 7 we read that the reason why an animal is long-lived is that its blend is about the same as that of the air surrounding it (*cf.* passage from *Physics* above), and at 767 a 31 that the condition of the body depends upon the blend of the surrounding air and of the foods taken in, especially upon the nourishment supplied by the water (and all foods contain water). The Hippocratic treatise π. ἀρχαίης ἰητρικῆς is based upon the theory that the human body and its foods consist of a large number

of distinctive substances (δυνάμεις), and maintains that for health there must be, both in the body and in its foods, a proper blend of these ingredients.

(36) These quotations are sufficient to show clearly the meaning of κρᾶσις. κρᾶσις, or rather good κρᾶσις (εὐκρασία), is, in fact, a special case of συμμετρία, commensurateness, right proportion, proper balance, a condition when every part, ingredient, or other factor concerned is at the right strength, or of the right size, or present in the right amount, in relation to all the others. (In modern English " symmetry " has come to be applied to a much narrower field.) Hence we find the adjective σύμμετρος and the noun συμμετρία often associated with κρᾶσις, as we have seen.

(37) From the examples it will also be seen that there can be κρᾶσις not only in living bodies (and " souls "), but also in their foods, and in their environment (*e.g.*, the air); indeed, the word κρᾶσις can also be used in the sense of climate, as at *H.A.* 606 b 3, where again heat and cold are involved (the regions concerned are described as " having bad winters "); and from this it is an easy step to the meaning of " temperature."

(38) In the important discussion in *H.A.* Book VIII, chapter 2, it is the κρᾶσις of animals' bodies of which Aristotle is speaking (see 590 a 14).

Πεζόν and ἔνυδρον, and their implications

(39) As stated above (§§ 29, 33), the discussion in *H.A.* VIII. 2 about what is meant by πεζόν and ἔνυδρον has some very important implications. The term ἔνυδρον, as Aristotle himself there shows, is used in several senses ; but the term πεζόν too is ambiguous. What is its primary connotation ? It might seem to be that of living, breeding, feeding, and walking on land, and hence possessing feet. But since all (at least, all two- and four-footed) animals which satisfy these conditions also breathe (*i.e.*, take in air for the purpose of cooling their heat), this function comes to be inseparably and essentially and even predominantly implied by πεζόν. Thus, at *P.A.* 669 a 9 ff. (quoted in § 50 c, d below) we read that while some dualizers which are πεζά and take in air spend

Meaning of πεζόν.

ARISTOTLE

most of their time in the water, some of those in the water *partake to such an extent of the* πεζὴ φύσις *that the* τέλος of their being (? or staying) alive lies *in their breath*. Indeed, just before this (at 668 b 33) we read that any γένος of animals has a lung because it is πεζόν : blooded animals are so hot that they need cooling from outside. Furthermore, at *Topics* 144 b 33 ff. it is pointed out that πεζόν does not primarily denote *locality*. Thus, to describe an animal as πεζόν means first and foremost that it takes in air for the purpose of cooling : this is necessary because the animal has so much blood, and this is a sure sign that the animal's κρᾶσις is *hot*.

Meaning of ἔνυδρον. (40) Πεζόν therefore would on a first examination appear not to be on a par with ἔνυδρον, in that it has not merely, or has not primarily, a connotation of locality. The very form of the word ἔνυδρον, on the other hand, seems unavoidably to involve locality. But in fact we find that this is not so, or at least, not without an important qualification. At *Resp.* 477 a 8 we read that animals which are ἔνυδρα in their φύσις are obliged to *get their food out of the water* ; and this again is in agreement with the passage of the *Topics* referred to above, where it is pointed out that ἔνυδρον does not denote locality (ἔν τινι or ποῦ) any more than πεζόν does, but quality (ποιόν τι). The seal is a case in point : it dualizes in more than one respect (§ 29), but what is decisive in favour of its being ἔνυδρον is that it spends most of its time in the sea *and feeds there* (566 b 30 f.). Why, then, is an animal " obliged to get its food from the water ? " Aristotle gives his answer when dealing with certain animals that dualize, in this instance animals which take in air for cooling but live and feed in the water : the reason is that in the σύστασις of their γένεσις they partake to some extent (τι) of that stuff (ὕλη) *out of which they get their food*, for that which is κατὰ φύσιν is agreeable to each and every animal (*H.A.* 590 a 9 ff.) : it is in fact due to " the κρᾶσις of their body " (589 b 23). In other words, there is a considerable amount of ὑγρόν in their κρᾶσις in addition to the predominant θερμόν.[a]

---

[a] The use of the term κρᾶσις of course implies in itself more than one ingredient, even when the animal is " hot " ; and in all these discussions it is advisable to bear in mind Aristotle's statement at *Meteor.* 359 b 32 ff. : " τὸ ὑγρόν does not exist without τὸ ξηρόν (ἔστι δ᾽ οὔτε τὸ

HISTORIA ANIMALIUM

further relevant point is that no animal which takes in water and has gills (except the cordylus) feeds on land (589 a 23, 589 b 24).

(41) There is a close parallel between those dualizers which live in the water and yet *partake* to such an extent of the πεζὴ φύσις that breathing is essential to them (*P.A.* 669 a 9 ff.; see § 39) on the one hand, and those πεζά which breathe but live and feed in the water because in the course of their formation they *partake* to some extent (τι) of τὸ ὑγρόν (*H.A.* 590 a 9 ff.; see § 40). Aristotle's remarks in the context of the latter passage seem to suggest that the two ingredients θερμόν and ὑγρόν are fairly evenly matched.

(42) At *H.A.* 589 a 19 ff. Aristotle says that animals which take in air and breed on land, yet get their food from watery places and spend most of their time in the water "appear to be the only animals that dualize—one could consider them as being both πεζά and ἔνυδρα." In view of the number of occasions on which he describes animals as dualizing (see §§ 29, 30), this statement may seem surprising. The reason why he awards pride of place to this particular type of dualizing may perhaps be that its source can be easily and directly traced back to a fundamental factor, viz., the κρᾶσις of the animals' bodies. See also § 58. *Dualizers par excellence.*

(43) Aristotle's first reference in the *H.A.* to the πεζόν/ἔνυδρον difference is at 487 a 16 ff. in the first Book. There are, he says, two senses in which animals can be ἔνυδρα (in fact he mentions three): (1) because they have their βίος and τροφή in the water, and take in and emit water and cannot live if deprived of it (*e.g.*, most fishes); (2) because, although they have their τροφή and διατριβή in the water, yet do not take in water but air, and breed away from the water; examples are otter, beaver, crocodile (footed animals), shearwater and "plunger" (winged), water-snake (footless). (3) Some have their τροφή in the water, and cannot live out of it, yet take in neither air nor water; examples are the sea-anemone and shellfish. *Πεζόν and ἔνυδρον in H.A. I.*

(44) Of land-animals (χερσαῖα: in this passage Aristotle is

ὑγρὸν ἄνευ τοῦ ξηροῦ . . .), nor τὸ ξηρόν without τὸ ὑγρόν: all these things are spoken of in respect of the excess" [*sc.*, of one or the other].

lxxxi

using the word πεζόν to mean " footed " as contrasted with winged and footless), we can distinguish (1) those which take in and emit air (*e.g.*, man and all χερσαῖα which have a lung); (2) those which do not take in air, but live and get their food on land (*e.g.*, wasps, bees, and other insects). Further, (3), many of the χερσαῖα " as has been stated " [*i.e.*, he implies these are dualizers], get their food from the water; no ἔνυδρον, however, which takes in sea-water gets its food from the land.

Lastly, some animals begin by living in the water, and then change their form and live away from the water.

Πεζόν and ἔνυδρον in *H.A.* VIII.
(45) In the discussion in Book VIII, Aristotle begins by saying that one way in which animals are divided up is κατὰ τοὺς τόπους : some are πεζά and some are ἔνυδρα. This difference can, however, be variously interpreted, as follows :

(1) Those which take in air are called πεζά, those which take in water, ἔνυδρα.
(2) Some animals do not take in either, but their nature is such that they are sufficiently provided for with regard to the blend of the cooling which comes from air or water, and are called πεζά and ἔνυδρα respectively.
(3) Those which feed and spend their time in the air are called πεζά, those which feed and spend their time in water, ἔνυδρα.

Aristotle then goes on to point out that these divisions do not seen entirely satisfactory, since they lead to describing certain animals as both πεζά and ἔνυδρα. Division (1), therefore, so far as ἔνυδρα are concerned, must be distinguished into (a) animals which take in water for the same reason that others take in air, *viz.*, for cooling ; and (b) those which take in water because they cannot help doing so while feeding. There is also a further case, (c), animals which take in *air*, but owing to the blend of their body and their βίος are ἔνυδρα.

Factors variously combined.
(46) It is clear, then, that the situation is not a simple one, and that several factors are concerned, which are not always found in the same combinations (*e.g.*, not all animals that take in air spend their whole time on land), and it therefore needs looking into. This is exactly the sort of inquiry in which, as we saw (Introd. pp. vi, x), Aristotle is especially interested : to discover *the various ways in which differences are found combined*, with a

view to ascertaining the "causes." In the present field, the following considerations are involved:
(1) the methods by which animals are cooled (κατάψυξις);
(2) where they feed (τροφή);
(3) why those which take in water do so;
(4) where they spend their time (διατριβή);
(5) where they breed (τόκος);
of which there are the following varieties (symbols in square brackets as in § 47 below):

κατάψυξις:
    A cooled by air-intake [A/C]
    B cooled by water-intake [W/C]
    C cooled by the surrounding air [E/C]
    D cooled by the surrounding water [E/C]
    E self-cooled by the σύμφυτον πνεῦμα [S/C]

τροφή:
    F feed in the air (*i.e.*, on land)
    G feed in the water

water-intake:
    H for cooling (= *B* above)
    K during feeding (not for cooling)

διατριβή:
    L spend no time in the water
    M spend some time in the water
    N spend the whole time in the water

τόκος:
    O breed in the air (*i.e.*, on land)
    P breed in the water.

Clearly, several different combinations of these characteristics are possible, and some of them invariably carry others with them, thus

All *C* are *F*, *L*, and *O*;
All *D* are *G*, *N*, and *P*;

but not all combinations are reciprocal, thus

All *K* are *N*, but not all *N* are *K*, and so on.

Some combinations result in perfectly straightforward cases, others in complicated ones, as the following analysis will show. In all cases the ultimate explanation (the "cause") is to be found in the animal's original κρᾶσις—its "blend" or physical composition.

(47) I have endeavoured to make this analysis as complete as possible by taking into account statements from other parts of the Aristotelian corpus in addition to *H.A.* I have not, however, always included place of feeding and breeding where these are obvious or when mention of them is omitted by Aristotle.

ARISTOTLE

In the following paragraphs these abbreviations are used for ease of identification and reference:

A/C = air-cooled (by taking in air)
W/C = water-cooled (by taking in water)
E/C = environment-cooled (by air or water, but not by taking either in)
S/C = self-cooled

[R] indicates that the description satisfies the definition of πεζόν or ἔνυδρον given in § 39 or § 40.

Analysis of cases.
(48) *Straightforward cases* (*H.A.* 589 a 12 f., b 14 f.)
  I. (a) [A/C] Animals which take in air for cooling and feed in it (*i.e.*, on land) are called πεζά. [R]
     (b) [W/C] Animals which take in water for cooling and live and feed in it are called ἔνυδρα. [R]
  II. [E/C] Animals which take in neither but whose nature is such that they derive sufficient cooling—the required κρᾶσις—from one or the other are called πεζά or ἔνυδρα according as they live and feed in air or water (*H.A.* 589 a 13 ff.). [ἔνυδρα, R ; πεζά inapplicable]
  III. [S/C] Animals which are self-cooled by the σύμφυτον πνεῦμα (*P.A.* 669 a 1).

(49) *Complicated cases* (*H.A.* 589 a 18 ff.)
Some animals take in air and breed on land, but feed and live in the water. (The case of animals which take in water and feed and live on land can be ignored; only one such is known, the cordylus (newt): *H.A.* 589 b 24 ff.; *cf. Resp.* 476 a 5, *P.A.* 695 b 25, and even this, together with the frog, is described at *H.A.* 487 a 27 simply as a marsh-dwelling ἔνυδρον.) Although these animals take in air, they must have access to the water if they are to continue living. Some spend their whole time in the water. Some actually take in water as well as air, and so on. Hence the various senses of ἔνυδρον must be more precisely distinguished.

Various senses of ἔνυδρον.
(50) The different senses in which animals can be said to be ἔνυδρα are to be distinguished as follows (*H.A.* 589 b 13 ff.):
  IV. (a) [W/C] = I. (b) above. The animal takes in water for cooling, and has gills for the purpose. [R, because no animal with gills feeds on land; see § 40.]
      (b) The animal takes in water incidentally and per-

force while feeding—*not* for the sake of cooling—and discharges it either (i) [A/C] through a blowhole or (ii) [E/C] by other means. [R]

(c) [A/C] These animals are ἔνυδρα " on account of the κρᾶσις of their bodies and their βίος " (589 b 22 f.): they take in air and breed on land, but live in the water because " in the σύστασις of their γένεσις they receive a share of the stuff [ὕλη] whence they derive their food " (590 a 8 ff.). [R]

A closely parallel passage is at *P.A.* 669 a 9 ff., where however these animals are considered in their role as πεζά: " Many animals dualize in their nature: of those which are πεζά and [? = *i.e.*] take in air some spend most of their time in the water, and "—the next phrase suggests that we should here make a fourth subheading, since the animals which Aristotle goes on to mention obviously spend their whole time in the water—

(d) [A/C] " some of those in the water partake to such an extent of the πεζὴ φύσις that the τέλος of their being alive [? or " staying alive ": τοῦ ζῆν] lies in their breath." [R]

(51) It must not, of course, be supposed that these headings and sub-headings are all mutually exclusive. Thus Crustacea and Cephalopods are described both by II—they derive sufficient cooling from contact with the water—and by IV. (b) (ii)—they take in water during the course of feeding, not for the sake of cooling. Similarly, Cetacea are described both by IV. (b) (i) and IV. (d). *Examples of cases.*

Examples of these various cases are given in *H.A.* VIII. 2 and elsewhere:

I. (a) These are the " most perfect " animals, viviparous, hotter and more fluid and not earthy in their nature. They have a soft lung (not spongy or stiff) well supplied with blood: such a lung is the proof of natural heat. They are air-cooled, and owing to their great heat they need a great deal of respiration to cool them. They take in air and feed in it (*G.A.* 732 b 30, *P.A.* 668 b 32, *Resp.* 470 b 24 ff., 477 a 16 ff.).

(b) Blooded animals, which have no lung but gills

ARISTOTLE

instead. These are cooled by water-intake, and they feed in the water. These are the Selachia, and Fishes generally. (*Resp.* 476 a 1 ff.)

(52) II. This heading covers the following, all of which are bloodless:

(i) Crustacea and Cephalopods, which have no lung (being bloodless) and little heat, therefore need but little cooling: the surrounding water is sufficient to cool them (*Resp.* 475 b 5 ff., 476 b 30 ff.). Further examples are the sea-anemone and shellfish; see § 43 (3).

(ii) Some small short-lived bloodless animals (probably certain insects are meant) can obtain sufficient cooling from the surrounding air or water (*Resp.* 474 b 25 ff., 475 b 15).

(iii) Some of the longer-lived insects are hotter and need more cooling. This is provided by their having a split under the diazoma, which enables them to be cooled through the membrane there, which is quite thin. Examples are: bees, wasps, etc., cockchafers, cicadas (*Resp.* 474 b 31 ff.; *cf.* § 44 (2)).

(Self-cooled animals.) (53) III. In the passage just referred to (*Resp.*), Aristotle appears to imply that these creatures are cooled by the surrounding air, and that the thinness of the membrane enables the air to exert a greater cooling effect; but he at once goes on (475 a 7 ff.) to mention the friction produced against this membrane by the rising and subsiding of what he here calls the ἔμφυτον πνεῦμα (this friction is responsible for the humming noise produced by bees etc.; *cf.* 456 a 19, and see n. on *H.A.* 535 b 12); [the swelling and subsiding of the σύμφυτον πνεῦμα in bloodless animals and insects (bees, wasps, flies, etc.) is mentioned at 456 a 11 ff.]; and at 475 a 10 ff. Aristotle goes on to compare the rising and subsiding movement caused by the ἔμφυτον πνεῦμα with the movement caused by the lung in animals that take in air, and by the gills in fishes; and says that in insects this movement produces sufficient cooling for them. This is in agreement with his statement at *P.A.* 669 a 1 f., that bloodless animals can cool themselves by the σύμφυτον πνεῦμα.

None of the animals which come under this heading takes in air or water, and they are all bloodless.

(54) In elucidation of the preceding paragraph it should be noted that certain animals which breathe (*viz.*, oviparous quadrupeds and birds) can remain under water for a long time because their θερμόν is scanty and can therefore be sufficiently cooled over a long period by *the mere movement of the lung* (*P.A.* 669 a 36 f.). A similar statement is found at *Resp.* 470 b 20 ff. : in the oviparous quadrupeds, once the lung has become inflated it can *of itself by its movement* produce a cooling effect, enabling the animal to remain submerged (the same is repeated at 475 a 21 ff.). It seems probable that Aristotle regarded the action of the σύμφυτον πνεῦμα in insects as parallel to these cases.

(55) IV. (a) = I. (b) above. *Cf.* § 43 (1). (Ἔνυδρα.)

  (b) This heading includes : (i) Cetacea (whales, dolphins, etc.), which are blooded, footless animals, cooled by the intake of air (*Resp.* 476 b 13 ; for their discharge of the water they take in see *Resp.* 476 b 25) ; (ii) Crustacea and Cephalopods, bloodless animals without lungs ; see above, II. (i) (*Resp.* 475 b 5 ff., 476 b 30 ff. ; for their discharge of the water see also *H.A.* 527 b 18).

  (c) Certain Ovipara : quadrupeds and scaly animals such as both kinds of tortoise, lizards, serpents etc.; frogs, and < water- > birds (*cf.* § 43 (2)). These have a spongy lung with little blood in it or none ; hence they need less respiration ; hence they can remain a long time under water because their θερμόν is scanty and can be sufficiently cooled over a long period by the mere movement of the lung ; see above, § 54 (*Resp.* 470 b 12, 475 a 20, *P.A.* 669 a 27). Other animals which spend much or most of their time in the water are crocodiles and watersnakes (*Resp.* 475 b 27 ff.).

  In addition, there are certain Vivipara, such as the otter and beaver (see § 43 (2)), and the seal, a "deformed" quadruped (*Resp., loc. cit.* ; see also *P.A.* 657 a 23 f., and § 29 above); and the

elephant, a fantastic quadruped, which is a special and very interesting case (*P.A.* 658 b 33—659 a 33). In his nature this animal is both ἐλῶδες (swamp-haunting) and πεζόν: he gets his food from the water and spends time there; but being a land-animal and blooded he has to breathe; and since owing to his enormous size he cannot, like other blooded viviparous animals, get quickly out of the water on to land, he must be equally at home in both places: that is why he has his long nose.

All those under (c) are air-cooled.

(d) This applies to the Cetacea, which spend all their time in the water, and feed and breed there.

Empedocles' view incorrect.
(56) It is clear from this schedule that several different combinations of the factors listed in § 46 are found, and that even although taking in air often carries with it breeding on land, it does not invariably do so. Nor does the taking in of air necessarily decide an animal's habitat—or, to put it another way, it is not, as Empedocles believed, on account of the heat of their φύσις that some animals are ἔνυδρα and some πεζά (*Resp.* 477 b 9, 478 a 7 ff.). The context of this passage shows that Aristotle means to deny Empedocles' contention that certain animals took to the water simply because their φύσις was so hot and they thought this was a good way of counteracting it. In fact, there are hot, blooded, viviparous, lunged, air-cooled animals both on land and in water; and (477 b 10 f.) some ἔνυδρα are cold, being completely bloodless (Crustacea, etc.), or have but little blood.

(57) Similarly, we find that animals which have different methods of cooling may all feed in the same place. Thus we find feeding in the water

(a) whales, etc., which are hot, blooded, lunged, and A/C;
(b) fishes, which though blooded are less hot, and W/C;
(c) Cephalopods and Crustacea, which are cold, bloodless, lungless, and E/C.

All these find their food in the water.

Comparison of this difference
(58) Thus there are some animals which must unavoidably after all be labelled " dualizers," because they exhibit the essential characteristic of both πεζά and ἔνυδρα: they

HISTORIA ANIMALIUM

take in air for cooling and they feed in the water. They with male are for that reason, as we have seen (§ 42), the dualizers and female *par excellence*; and Aristotle cannot help saying that something has gone wrong with them (*H.A.* 589 b 28): their φύσις has, as it were, got warped (ὡσπερανεὶ διεστράφθαι); and he compares with πεζόν and ἔνυδρον the case of male and female: some males become feminine in appearance, and some females masculine. Also, he says, it is clear that some small item in an animal's original σύστασις, if this item is ἀρχή-like (see § 12), will determine its being male or female. This is a particularly interesting comparison for the following reasons: the difference between male and female, according to Aristotle, is one of greater and less heat (*G.A.* 765 b 15 ff., 775 a 5 ff.; *cf.* 728 a 18, 738 a 36); hot and cold are important ἀρχαί (see § 35); male and female are ἀρχαί (see *G.A.* 716 b 10, 763 b 23). Thus these dualizers, in addition to what we may consider the main ingredient in their σύστασις while they are being formed (*i.e.*, in their κρᾶσις), receive a portion of another ingredient which is identical with that whence they get their food (590 a 9).

(59) The idea that an animal's σύστασις and its place of habi- Correspontat should closely correspond is clearly expressed in the dence of *De Respiratione*, 477 a 27 ff., where we read that some κρᾶσις and living things' σύστασις contains "more earth" (*e.g.*, (1) *De res-* plants), that of others "more water" (*e.g.*, the ἔνυδρα), *piratione;* that of fliers " more air," and that of πεζά " more fire," and that each of these have their appointed station in the places proper (or congenial, οἰκείοις) to them. Here the question of conservation comes in, and Aristotle draws a distinction between the ὕλη of which an animal consists and the ἕξις or state of that ὕλη (477 b 18 ff.); here, however, we are not concerned with ἕξεις. If a thing were constructed out of wax or ice, Nature would not place it in heat in order to conserve it, for τὸ θερμόν melts things which have been constituted by its opposite. Similarly, τὸ ὑγρόν destroys things which have been constituted by θερμόν and ξηρόν (*e.g.*, water destroys salt). Assuming then that the ὕλη for all σώματα is τὸ ὑγρόν and τὸ ξηρόν [these are the two " passive " δυνάμεις; see § 14 and Introd. p. xxii] then it is reasonable that things which have been constituted out of [*sic*; not " by "]

ARISTOTLE

ὑγρόν (and perhaps cold too) are ἐν ὑγροῖς, and things constituted out of ξηρόν are ἐν ξηρῷ (*e.g.*, trees, which are ξηρά, grow in earth, which is ξηρόν, and not in water). The φύσεις, therefore, of ὕλη are of the same character as the places they are in, fluid ones in water, solid ones in earth, hot ones in air. (In this way Aristotle manages to by-pass the difficulty of saying what animals have their abode in fire; however, both Vivipara and birds are " hot " ; see above, §§ 51, 54, and Introd. pp. xv f.). So φύσις is conserved best of all ἐν τοῖς οἰκείοις τόποις (477 b 16 ; *cf. H.A.* 590 a 10 f., " that which is κατὰ φύσιν is agreeable to each and every animal " : this refers to the ingredients of which it is composed : see § 40) ; and ὕλη itself cannot be opposed to the places it is in. (The reverse may be true of ἕξεις, but this is here irrelevant.) At *G.A.* 761 a, b Aristotle seems at first sight to have been less successful in by-passing the question of fire-dwellers. He establishes (761 a 21 ff.) a correlation between earth (= ξηρόν) and plants, water (= ὑγρόν) and Testacea (later, 761 b 13, ἔνυδρα generally), air and πεζά, and is then at a loss to find a class of animals corresponding to fire : this γένος, he says, must be looked for on the moon. He does, however, substantially modify the rigidity of the scheme by pointing out that as seawater contains " a share of all the parts, ὑγροῦ καὶ πνεύματος καὶ γῆς," it can contain a quota of all the creatures which come into being in each of those realms ; so we need not be surprised to find plants and πεζά in the water, and the fact that it contains earth will allow even for the presence in it of Testacea, which are cold and earthy (761 b 6, 762 a 28). What is more, Aristotle goes so far as to say (762 a 19) that in earth water is present, and in water *pneuma* is present, and in all *pneuma* soul-heat is present. This would indeed seem to offer a solution to all possible difficulties. It is however, not clear how far such a drastic universalization is envisaged by Aristotle in the *H.A.* ; indeed in the *G.A.* its immediate purpose is to facilitate the explanation of the spontaneous generation of Testacea ; and it is perhaps safer and fairer, so far as the *H.A.* is concerned, to assume that what Aristotle felt should be the normal situation was a general correspondence between an animal's κρᾶσις and its place of abode, though

2) *De Gen. An.*

(60) To sum up, we may observe that the phenomena connected with an animal's character as πεζόν or/and ἔνυδρον, its place of abode, of feeding, etc., which have been examined in the preceding paragraphs, and the phenomena connected with its manner of reproduction (see Introd. pp. xv ff.), are parallel manifestations of one and the same thing, *viz.*, the κρᾶσις of its original σύστασις. They, and certain other manifestations, such as the natural posture of the body, the number of organs of locomotion, sex, colour of hair (*G.A.* 784 a 23 ff.), and so forth, are all controlled by, and are dependent upon, the interrelation and συμμετρία of the four substances which are fundamental in Aristotle's physical scheme. We can see from these things something of the way in which Aristotle conceives the " material cause " to operate in zoology.

Two parallel manifestations of κρᾶσις.

THE TABLES

I HAVE provided below three separate tables of contents, which give increasingly detailed information, to enable the reader to see the structure of the treatise and to help him to find his way about in it. These and the Index (to be included in the third volume) are intended to be mutually supplementary. The three tables give :
1. A scheme of the treatise.
2. A summary of the contents of the main part of the work.
3. Details of the subjects dealt with in the various sections.

The " HISTORIA " itself, *i.e.* the collection of facts about animals, begins about three-quarters of the way through chapter 6 of Book I (at 491 a 14). It is preceded by some INTRODUCTORY matter (chs. 1-5 and the first half of ch. 6), and by a short but most important statement about the METHOD and PURPOSE of the undertaking (middle of ch. 6, 491 a 7 ff.). The introductory matter and the statement I have summarized in the first table ; the second and third tables deal only with the " Historia " proper.

Both in the text and in these tables I have for convenience adhered to the generally accepted order of the Books, as in the Berlin edition and in most printed texts, although in the MSS. the present Book VII follows the present Book IX.

HISTORIA ANIMALIUM

1. Scheme of the *Historia animalium*

BK.	CH.	
I	1	INTRODUCTORY

        The two sorts of " parts " : (*a*) uniform ; (*b*) non-uniform (=instrumental).

        The degrees of difference in (i) the instrumental parts ;
                                        (ii) the uniform parts.

        Other kinds of differences :
           Specimens of differences in manner of life,
                                  in activities, and
                                  in dispositions.

    2-3 Some instrumental parts are common to all (or most) animals, though exhibiting differences.

      4 Fluid is an essential uniform part for all animals : it and the parts in which it occurs exhibit differences.

      5 Specimens of differences in methods of reproduction, in means of locomotion.

      6 The main classes of animals (but precise and exhaustive classification is not possible).

      6 The METHOD and PURPOSE of the undertaking : To ascertain the differences that animals exhibit—to record the *facts* about all of them, as a necessary preliminary to discovering the *causes* thereof.

Then follows the HISTORIA itself, *i.e.*, the records of the differences arranged under three main headings (a fourth, B in the list below, is very brief, and contains a few differences not otherwise classified) :

A. Differences in the PARTS (Bk. I. 6 to Bk. IV. 7).
B. Differences not otherwise classified (Bk. IV. 8-11).
C. Differences relating to REPRODUCTION (Bk. V. 1 to Bk. VII. 12).
D. Differences relating to ACTIVITIES and MANNER OF LIFE, which vary according to the animals' DISPOSITIONS and DIET (Bk. VIII. 1 to Bk. IX. 50 (49B)).

ARISTOTLE

2. Summary

The following table summarizes the contents of headings A to D in the foregoing table.

(M) indicates that a section deals exclusively or chiefly with Man,

(P) indicates that the treatment is seriatim by " parts,"

(G) indicates that the treatment is seriatim by groups or kinds of animals, though some such series are less complete than others.

A. Differences of PARTS

BOOK

I. 6–III. 1	A. 1. Instrumental parts of Blooded animals :
	   (i) Vivipara : (*a*) external (M, P); (*b*) internal (M, P);
	             (*c*) external (P).
	   (ii) Ovipara : (*a*) external (G).
	   (iii) All : (*b*) internal (P).

III. 2-22	A. 2. Uniform parts of Blooded animals (P).
IV. 1-7	A. 3. Parts of Bloodless animals (G) : (*a*) external ; (*b*) internal.

B. Differences not otherwise classified

IV. 8	B. 1. Number of sense-faculties (G) : (i) Blooded animals; (ii) Bloodless animals.
9	B. 2. Voice (G): (i) Bloodless animals; (ii) Blooded animals.
10	B. 3. Sleep and dreaming (G).
11	B. 4. Distinction (or absence of distinction) of sex (G).
	B. 5. Some secondary sex-differences.

C. Differences of REPRODUCTION

V. 1	The various methods of reproduction (summary).
V. 2-7	C. 1. Coition (G) : (i) Blooded animals ; (ii) Bloodless animals.
8-13	C. 2. Breeding seasons (G).
14	C. 3. Ages and signs of maturity (G).
	C. 4. Methods of reproduction (G) :
15-32	(i) Bloodless animals (G);
V. 33–VI. 37	(ii) Blooded animals (G), and ultimately
VII. 1-12	Man.

D. Differences of DIET and DISPOSITION

VIII	Introductory :
1	Degrees of difference in dispositions. The *scala naturae* : from plants upwards the stages

		are continuous. Differences are related to the manner in which animals produce and rear their young, and to the nature of their food.
	2	Differences of habitat: differences within " aquatic " and " terrestrial " are occasioned by the purpose for which animals take in air or water, by their bodily " blend," and by the nature of their food.
VIII. 2-11		D. 1. *Differences of diet* (and habitat) (G).
		(Note.—In chs. 12-20 and 28-30 most of the differences recorded are not inherent natural differences *between* various kinds of animals, but differences brought about *in* this or that kind of animal by such external factors as changes of temperature or of locality.
		Chs. 21-26 appear not to be an integral part of the scheme.)
	12-17	Differences in manner of life due to changes of temperature, season, etc.; Migration; Hibernation (and estivation) (G).
	17	Differences in sloughing habits.
	18-20	Differences in healthiness due to seasons and weather (G).
	21-26	Diseases of certain quadrupeds.
	27	Enemies of bees.
	28-29	Differences in the occurrence, physique, and dispositions of animals due to locality and climate.
	30	Differences in healthiness of marine animals due to breeding seasons. Seasonal changes of colour in Fishes.
IX. 1		D. 2. *Differences of disposition according to sex.*
		D. 3. Enmities between animals, due to *diet and manner of life.*
	2	D. 4. Enmity and sociability of Fishes.
	3-48	D. 5. *Differences, chiefly of intelligence and instinctive habits* (i.e., *of dispositions and activities*) (G). (Particular attention is given to Birds (chs. 7-36) and to Bees (ch. 40).)
	49	As conditions (*pathē*) can modify activities, so activities can modify dispositions, can even cause physical changes, *e.g.*, in fowls.
	50	Physical changes due to castration. Rumination.
	49B	Birds' seasonal colour-changes, and dust- and water-baths.

ARISTOTLE

3. Detail of Contents

The following table gives details of the contents under headings A, C, and D. It is unnecessary to give further details for B.

The numbers in brackets indicate chapters.

A. PARTS

Bk. I *A. I. Instrumental parts of Blooded animals:* (i) *Vivipara*
(a) *external parts* [*of Man*]
(7) head ; (8) face ; (9) eyebrows and eyes ; (11) ears, nose, jaws, lips, mouth, tongue, etc.; (12) neck ; trunk, front parts, and (15) rear parts. Correspondence between upper and lower, etc. The arms and their parts, the legs and their parts. Man's unique posture. Flexion of the limbs. Position of the sense-organs.

A. 1. (i) (b) *internal parts* [*of Man*]
(16) head ; brain, etc. ; oesophagus, windpipe, lung, stomach ; (17) heart, diaphragm, liver, spleen, gall-bladder ; kidneys and bladder.

Bk. II (1) Recapitulation of the kinds of difference.

A. 1. (i) (c) *external parts* (resumed) [chiefly viviparous quadrupeds and Man]
(1) head, limbs ; peculiarities of the elephant ; chest, flexion of limbs, stationary posture, movement of limbs in walking, tail, hair ; peculiarities of the camel ; legs, feet, horns and hucklebones, breast, generative organs, relative sizes of the parts according to age ; teeth ; peculiarities of the martichoras ; (7) mouth ; (8) apes, etc., which have some features in common with Man, some with quadrupeds.

A. 1. (ii) *Ovipara* (a) *external parts*
(10) Quadrupeds ; (11) peculiarities of the chamaeleon ; (12) Birds ; (13) Fishes (but some are viviparous) ; (14) serpents, etc.

A. 1. (iii) *Blooded animals generally* (b) *internal parts* [Man, Viviparous quadrupeds, Cetacea, Oviparous quadrupeds, serpents, Birds, Fishes].
(15) oesophagus and windpipe, diaphragm, lung, spleen, gall-bladder ; (16) kidneys and bladder ; (17) heart, liver, spleen, stomach, and intestines.

Bk. III (1) generative organs, male and female.
A. 2. Uniform parts of Blooded animals
(2) Blood, flesh, etc., and their counterparts.
Heart and blood-vessels in Man ; (5) sinews ; (6) fibres ; (7) bones ; (8) cartilage ; (9) horn, nail, claw ; (10) hair ; (11) skin ; (12) change of colour ; (13) membranes ; (14) Omentum ; (15) bladder ; (16) flesh ; (17) fat and suet ; (19) blood ; (20) marrow, milk ; (21) rennet ; (22) semen.

HISTORIA ANIMALIUM

Bk. IV *A. 3. Parts of Bloodless animals*
(1) The four classes of Bloodless animals: Cephalopods, Crustacea, Testacea, Insects.
Cephalopods: external parts, internal parts.
(2) Crustacea: external parts; (3) internal parts.
(4) Testacea and intermediate animals: external parts, internal parts.
(7) Insects: external parts, internal parts.

B. Miscellaneous

Bk. IV (8–11) For details see the Summary.

C. Reproduction

Bk. V (1) Programme of the order in which the groups are to be discussed (not begun until ch. 15, and then not strictly adhered to): Testacea, Crustacea, Cephalopods, Insects; Fishes viviparous and oviparous, Birds, footed Ovipara and Vivipara, Man.
The various methods of reproduction.
C. 1. Methods of coition: (i) *Blooded animals:* (2) birds and viviparous quadrupeds; (3) oviparous quadrupeds; (4) serpents, etc.; (5) fishes, including Selachia; Cetacea;
(ii) *Bloodless animals:* (6) Cephalopods; (7) Crustacea; (8) Insects.
C. 2. Breeding seasons:
(8-11) general; birds (special reference to halcyon), insects, fishes including Selachia; (12) Cephalopods, Testacea; (13) wild and domesticated birds (pigeons).

C. 3. Ages and signs of maturity
(14) Voice, etc., and times of sexual maturity, in Man and viviparous quadrupeds.

C. 4. Reproduction: (i) *Bloodless animals*
(15) Testacea (without copulation; spontaneous generation); starfish and hermit-crab; (16) sea-anemone, sponges; (17) Crustacea; (18) Cephalopods; (19) Insects; various methods of reproduction: some larviparous (some grubs derived from copulation, some without); animals found in snow and fire; (20) ichneumon-wasp; (21) bees, honey and honeycomb; (23) wasps; (24) humble-bee; (25) ants; (26) land-scorpion; (27) spiders; (28) grasshoppers; (29) locust; (30) cicada; (31) insects not generated from grubs; fleas, lice, etc.; (32) clothes-moth, cheese-mite, book-scorpion, basket-worm, fig-wasp.

C. 4. Reproduction: (ii) *Blooded animals*
(*a*) *Quadruped Ovipara:* (33) tortoise, etc., lizard, crocodile.
(*b*) *Footless Ovipara:* (34) serpents and (ovoviviparous) viper.

Bk. VI (*c*) *Footed Ovipara (Birds):* (1) laying seasons and habits; (2) eggs and wind-eggs; display of pigeons; (3) development of the embryo; double-yolked eggs; (4) pigeons;

xcvii

ARISTOTLE

(5) vulture ; (6) eagles and other large birds ; (7) cuckoo (8) incubation habits of certain birds ; (9) peafowl.
(d) *Marine animals:* (10) the Selachia ; their embryology ; (12) Cetacea : dolphin, whale, porpoise, seal ; (13) Oviparous fishes : general observations ; (14) carp, *chalkis*, *glanis*, etc. ; (15) Spontaneously-generated Fishes ; (16) abnormality of the eel ; (17) spawning seasons.
(e) *Viviparous quadrupeds:* (18) General. Effects of pairing on disposition. Menstruation. Particular animals : swine ; (19) sheep, goat ; (20) dog ; (21) cattle ; (22) horse ; (23) ass ; (24) mule ; (25) camel ; (27) elephant ; (28) wild swine ; (29) deer ; (30) bear ; (31) lion ; (32) hyena ; (33) hare ; (34) fox ; (35) wolf, cat, etc. ; (36) Syrian " mule " ; (37) mouse.

Bk. VII (f) *Man:* (1) signs of puberty ; (2) catamenia ; (3) signs of conception ; effluxion ; abortion ; (4) pregnancy, twins —quintuplets, superfetation ; (5) lactation, periods of fecundity, individual differences in women, transmitted characteristics ; (7) impregnation ; (8) the embryo ; (9) parturition ; (10) obstetrics ; (11) milk ; (12) convulsions in infants.

D. Diet and Disposition

Bk. VIII Introductory (1, 2). See Summary.

D. 1. Diet (and habitat)
(2) Testacea, sea-anemones, limpet, sea-turtles, Crustacea, Cephalopods, Fishes, eels ; (3) Birds ; (4) Oviparous quadrupeds, serpents ; (5) Viviparous quadrupeds : wolf, hyena, bear, lion ; lake- and river-animals : beaver, otter, etc. ; (6) pig ; (7) cattle ; (8) horse and mule ; (9) elephant, camel ; (10) sheep, goats ; (11) Insects.

Differences due to changes of season, etc.
(12) (i) Migration : Birds, Fishes ; (14) (ii) Hibernation and estivation : Insects, serpent, lizard, etc., Fishes ; (16) Birds ; (17) Viviparous quadrupeds : porcupine, bear, dormouse ; (iii) Sloughing : Oviparous quadrupeds, serpents, vipers, Insects, Crustacea.

Favourable and healthy seasons
(18) Birds ; (19) Fishes ; (20) Testacea.

Diseases of viviparous quadrupeds and some remedies.
(21) pig ; (22) dog and others ; (23) cattle ; (24) horses ; (25) asses ; (26) elephants.

Favourable and healthy seasons (resumed)
(27) Insects. Enemies of the bee.

Effect of locality and climate on occurrence, physique and disposition
(28, 29). Also, Interbreeding of animals.

Effect of pregnancy on health :
(30) Testacea, Crustacea, Cephalopods, Fishes.

HISTORIA ANIMALIUM

Bk. IX *Dispositions*
(1) The varieties of dispositions.
D.2. Differences due to sex : Man and viviparous quadrupeds; dogs, bears, leopards, Crustacea.

D. 3. Enmities between animals, due to diet and manner of life.
(2) *D. 4. Enmity and sociability of Fishes.*

D. 5. Intelligence and habits (dispositions and activities) :
(3) sheep, goats; (4) cattle, horses; (5) deer; (6) bear and other wild animals, ichneumon, crocodile, tortoise, etc., hedgehog, marten.
Birds : (7) swallow, pigeons; (8) quail, partridge; (9) woodpecker; (10) crane, pelican; (11) stone-curlew, hawk, vulture, etc.; (12) wagtail, web-footed birds, *chalkis*; (13) jay, etc.; (14) halcyon; (15) hoopoe; (16) reedwarbler; (17) *krex*, etc.; (18) herons; (19) blackbirds; (20) thrushes; (21) bluebird; (22) oriole, shrike; (23) *pardalos, kollyrion*; (24) daws; (25) larks; (26) woodcock; (27) ibis; (28) horned owls; (29) cuckoo; (30) *kypselos* (= *apous*) goatsucker; (31) ravens; (32) eagles; (33) Scythian bird; (34) owls, phene; (35) petrel; (36) hawks.
Marine animals : (37) Fishes, Cephalopods (esp. cuttlefish, octopus, nautilus).
(38) Industrious Insects; (39) spiders, etc.; (40) combbuilding Insects : bees; (41) wasps; (42) anthrenas; (43) humble-bees.
Disposition of (44) lion, etc.; (45) bison; (46) elephant; (47) camel (the King of Scythia's mare); (48) dolphin.

(49, 50, 49B). See the Summary.

ABBREVIATIONS

The following abbreviations have been used for the titles of the zoological works :

> *H.A. Historia animalium*
> *P.A. De partibus animalium*
> *G.A. De generatione animalium*
> *Resp. De respiratione*

While this volume was in the press, the Budé edition of *H.A.* i-iv was published: *Aristote, Histoire des animaux*, tome I, livres I-IV. Texte établi et traduit par Pierre Louis. Paris, 1964.

SIGLA

Manuscripts cited throughout

A (=A^a) Marcianus graecus Z 208.
C (=C^a) Laurentianus LXXXVII. 4.
P   Vaticanus graecus 1339.
D (=D^a) Vaticanus graecus 262.

Manuscripts occasionally cited

Rhen.   Rhenani, nunc Parisinus supp. graecus 212.
m   Parisinus graecus 1921.
Ambr.   Ambrosianus 46 (I. 56 sup.)
E (=E^a) Vaticanus graecus 506.

Bekker attached the small ^{a} to his symbols for four of these manuscripts (as shown above) to distinguish them from other manuscripts for which he used the corresponding plain symbols. Since these other manuscripts are not relevant to *H.A.*, in order to simplify the appearance of the *apparatus criticus* I have omitted the small ^{a}.

Where required, A^1=first hand of A, A^2=later hand(s) of A.

Readings and emendations

Σ   Michael Scot's Latin translation (either its actual words, or the original Greek reading clearly implied) from my own transcription.
Gul.   William of Moerbeke's Latin translation.
vulg.   The reading of Bekker's Berlin edition.
A.-W.   Wimmer, in Aubert and Wimmer's edition.
Buss.   Bussemaker.
Camot.   Camotius.
Cs.   Camus.
Dt.   Dittmeyer.
Pi.   Piccolos.
Scal.   Scaliger.
Sn.   Schneider.
Sylb.   Sylburg.
Th.   D'Arcy Thompson.

Other names are unabbreviated.

[ ]   Denote words wrongly placed or incorporated into the text.
⟨ ⟩   Denote (*a*) in the Greek text, words or parts of words supplied conjecturally;
            (*b*) in the English, either the translation of words supplied in the Greek, or words required to complete the sense.

THE TRADITIONAL ORDER of the works of Aristotle as they appear since the edition of Immanuel Bekker (Berlin, 1831), and their division into volumes in this edition.

	PAGES
I. The Categories (Κατηγορίαι)	1-15
On Interpretation (Περὶ ἑρμηνείας)	16-24
Prior Analytics, Books I-II ('Αναλυτικὰ πρότερα)	24-70
II. Posterior Analytics, Books I-II ('Αναλυτικὰ ὕστερα)	71-100
Topica, Books I-VIII (Τοπικά)	100-164
III. On Sophistical Refutations (Περὶ σοφιστικῶν ἐλέγχων)	164-184

(The foregoing corpus of six logical treatises is known also as the *Organon*).

(For pages 184-313 see volumes IV-VI.)

On Coming-to-be and Passing-away (Περὶ γενέσεως καὶ φθορᾶς)	314-338
On the Cosmos (Περὶ κόσμου)	391-401
IV. Physics, Books I-IV (Φυσική)	184-224
V. Physics, Books V-VIII (Φυσική)	224-267
VI. On the Heavens, Books I-IV (Περὶ οὐρανοῦ)	268-313

(For pages 314-338 see volume III.)

VII. Meteorologica, Books I-IV (Μετεωρολογικά)	338-390

(For pages 391-401 see volume III.)

THE TRADITIONAL ORDER

	PAGES
VIII. On the Soul, Books I-III (Περὶ ψυχῆς)	402-435
Parva naturalia:	
On Sense and Sensible Objects (Περὶ αἰσθήσεως)	436-449
On Memory and Recollection (Περὶ μνήμης καὶ ἀναμνήσεως)	449-453
On Sleep and Waking (Περὶ ὕπνου καὶ ἐγρηγόρσεως)	453-458
On Dreams (Περὶ ἐνυπνίων)	458-462
On Prophecy in Sleep (Περὶ τῆς καθ' ὕπνον μαντικῆς)	462-464
On Length and Shortness of Life (Περὶ μακροβιότητος καὶ βραχυβιότητος)	464-467
On Youth and Old Age. On Life and Death (Περὶ νεότητος καὶ γήρως. Περὶ ζωῆς καὶ θανάτου)	467-470
On Respiration (Περὶ ἀναπνοῆς)	470-480
On Breath (Περὶ πνεύματος)	481-486
IX. Historia Animalium, Books I-III (Περὶ τὰ ζῷα ἱστορίαι)	486-523
X. Historia Animalium, Books IV-VI (Περὶ τὰ ζῷα ἱστορίαι)	523-581
XI. Historia Animalium, Books VII-X (Περὶ τὰ ζῷα ἱστορίαι)	581-639
XII. Parts of Animals, Books I-IV (Περὶ ζῴων μορίων)	639-697
On Movement of Animals (Περὶ ζῴων κινήσεως)	698-704
Progression of Animals (Περὶ πορείας ζῴων)	704-714
XIII. Generation of Animals, Books I-V (Περὶ ζῴων γενέσεως)	715-789
XIV. Minor Works:	
On Colours (Περὶ χρωμάτων)	791-799
On Things Heard (Περὶ ἀκουστῶν)	800-804
Physiognomics (Φυσιογνωμονικά)	805-814
On Plants, Books I-II (Περὶ φυτῶν)	815-830

THE TRADITIONAL ORDER

		PAGES
	On Marvellous Things Heard (Περὶ θαυμασίων ἀκουσμάτων)	830-847
	Mechanical Problems (Μηχανικά)	847-858
	(For pages 859-930 see volume XV.)	
	(For pages 930-967 see volume XVI.)	
	On Invisible Lines (Περὶ ἀτόμων γραμμῶν)	968-972
	The Situations and Names of Winds (Ἀνέμων θέσεις καὶ προσηγορίαι)	973
	On Melissus, Xenophanes and Gorgias (Περὶ Μελίσσου, Περὶ Ξενοφάνους, Περὶ Γοργίου)	974-980
XV.	Problems, Books I-XXI (Προβλήματα)	859-930
XVI.	Problems, Books XXII-XXXVIII (Προβλήματα)	930-967
	(For pages 968-980 see volume XIV.)	
	Rhetoric to Alexander (Ῥητορικὴ πρὸς Ἀλέξανδρον)	1420-1447
XVII.	Metaphysics, Books I-IX (Τὰ μετὰ τὰ φυσικά)	980-1052
XVIII.	Metaphysics, Books X-XIV (Τὰ μετὰ τὰ φυσικά)	1052-1093
	Oeconomica, Books I-III (Οἰκονομικά)	1343-1353
	Magna Moralia, Books I-II (Ἠθικὰ μεγάλα)	1181-1213
XIX.	Nicomachean Ethics, Books I-X (Ἠθικὰ Νικομάχεια)	1094-1181
	(For pages 1181-1213 see volume XVIII.)	
XX.	Athenian Constitution (Ἀθηναίων πολιτεία)	—
	Eudemian Ethics, Books I-VIII (Ἠθικὰ Εὐδήμεια)	1214-1249
	On Virtues and Vices (Περὶ ἀρετῶν καὶ κακιῶν)	1249-1251
XXI.	Politics, Books I-VIII (Πολιτικά)	1252-1342

THE TRADITIONAL ORDER

PAGES

(For pages 1343-1353 see volume XVIII.)

XXII. " Art " of Rhetoric (Τέχνη ῥητορική) . 1354-1420

(For pages 1420-1447 see volume XVI.)

XXIII. Poetics (Περὶ ποιητικῆς) . . . 1447-1462
[Longinus], On the Sublime
[Demetrius], On Style

HISTORIA ANIMALIUM

ΤΩΝ ΠΕΡΙ
ΤΑ ΖΩΙΑ ΙΣΤΟΡΙΩΝ

Α

I Τῶν ἐν τοῖς ζῴοις μορίων τὰ μέν ἐστιν ἀσύνθετα, ὅσα διαιρεῖται εἰς ὁμοιομερῆ, οἷον σάρκες εἰς σάρκας, τὰ δὲ σύνθετα, ὅσα εἰς ἀνομοιομερῆ, οἷον ἡ χεὶρ οὐκ εἰς χεῖρας διαιρεῖται οὐδὲ τὸ πρόσωπον εἰς πρόσωπα.

Τῶν δὲ τοιούτων ἔνια οὐ μόνον μέρη ἀλλὰ καὶ μέλη καλεῖται. τοιαῦτα δ' ἐστὶν ὅσα τῶν μερῶν ὅλα ὄντα ἕτερα μέρη ἔχει ἐν αὑτοῖς, οἷον κεφαλὴ καὶ σκέλος καὶ χεὶρ καὶ ὅλος ὁ βραχίων καὶ ὁ θώραξ· ταῦτα γὰρ αὐτά τ' ἐστὶ μέρη ὅλα, καὶ ἔστιν αὐτῶν ἕτερα μόρια.

Πάντα δὲ τὰ ἀνομοιομερῆ σύγκειται ἐκ τῶν ὁμοιομερῶν, οἷον χεὶρ ἐκ σαρκὸς καὶ νεύρων καὶ ὀστῶν.

Ἔχει δὲ τῶν ζῴων ἔνια μὲν πάντα τὰ μόρια ταὐτὰ ἀλλήλοις, ἔνια δ' ἕτερα. ταὐτὰ δὲ τὰ μὲν εἴδει τῶν μορίων ἐστίν, οἷον ἀνθρώπου ῥὶς καὶ ὀφθαλμὸς ἀνθρώπου ῥινὶ καὶ ὀφθαλμῷ, καὶ σαρκὶ σὰρξ

---

[a] " Parts "; see Notes on Terminology, § 1.
[b] See Notes, § 1.

ARISTOTLE

HISTORIA ANIMALIUM

BOOK I

THE parts [a] which are found in animals are of two kinds : (*a*) those which are incomposite, *viz.*, those which divide up into uniform portions,[b] for example, flesh divides up into flesh ; (*b*) those which are composite, *viz.*, those which divide up into non-uniform portions, for example, the hand does not divide up into hands, nor the face into faces.

<small>I. INTRODUCTION. Two sorts of "parts": (a) uniform, (b) non-uniform (=instrumental).</small>

Some of these are called not only parts but limbs. These are the parts which, while being complete and entire, contain other (different) parts within themselves, for example, head, leg, hand, the arm as a whole, and the chest : all these are complete and entire parts, which have other parts belonging to them.

Now all the non-uniform parts are composed out of the uniform ones, for example, the hand is composed of flesh, sinews and bones.

With regard to animals, some have all their parts mutually identical, some have parts of a different character. Some parts are specifically identical, for example, one man's nose and eye are identical with another's nose and eye, one's flesh with another's

<small>Degrees of difference in the instrumental parts;
(i) the in-</small>

3

ARISTOTLE

486 a

καὶ ὀστῷ ὀστοῦν· τὸν αὐτὸν δὲ τρόπον καὶ ἵππου καὶ τῶν ἄλλων ζῴων, ὅσα τῷ εἴδει ταὐτὰ λέγομεν ἑαυτοῖς· ὁμοίως γὰρ ὥσπερ τὸ ὅλον ἔχει πρὸς τὸ ὅλον, καὶ τῶν μορίων ἔχει ἕκαστον πρὸς ἕκαστον. τὰ δὲ ταὐτὰ μέν ἐστιν, διαφέρει δὲ καθ' ὑπεροχὴν καὶ ἔλλειψιν, ὅσων τὸ γένος ἐστὶ ταὐτόν. λέγω δὲ γένος οἷον ὄρνιθα καὶ ἰχθύν· τούτων γὰρ ἑκάτερον ἔχει διαφορὰν κατὰ τὸ γένος, καὶ ἔστιν εἴδη πλείω ἰχθύων καὶ ὀρνίθων.

Διαφέρει δὲ σχεδὸν τὰ πλεῖστα τῶν μορίων ἐν

486 b

αὐτοῖς[1] παρὰ τὰς τῶν παθημάτων ἐναντιώσεις, οἷον χρώματος καὶ σχήματος, τῷ τὰ μὲν μᾶλλον ταὐτὰ[2] πεπονθέναι τὰ δ' ἧττον, ἔτι δὲ πλήθει καὶ ὀλιγότητι καὶ μεγέθει καὶ σμικρότητι καὶ ὅλως ὑπεροχῇ καὶ ἐλλείψει. τὰ μὲν γάρ ἐστι μαλακόσαρκα αὐτῶν τὰ δὲ σκληρόσαρκα, καὶ τὰ μὲν μακρὸν ἔχει τὸ ῥύγχος τὰ δὲ βραχύ, καὶ τὰ μὲν πολύπτερα τὰ δ' ὀλιγόπτερά ἐστιν. οὐ μὴν ἀλλ' ἔνιά γε καὶ ἐν τούτοις ἕτερα ἑτέροις μόρια ὑπάρχει, οἷον τὰ μὲν ἔχει πλῆκτρα τὰ δ' οὔ, καὶ τὰ μὲν λόφον ἔχει τὰ δ' οὐκ ἔχει. ἀλλ' ὡς εἰπεῖν τὰ πλεῖστα καὶ ἐξ ὧν μερῶν ὁ πᾶς ὄγκος συνέστηκεν, ἢ ταὐτά ἐστιν ἢ διαφέρει τοῖς τ'[3] ἐναντίοις καὶ καθ' ὑπεροχὴν καὶ ἔλλειψιν· τὸ γὰρ μᾶλλον καὶ ἧττον ὑπεροχὴν ἄν τις καὶ ἔλλειψιν θείη.

---

[1] αὑτοῖς AD, A.-W., Dt.: *in eis* Σ: αὐτοῖς vulg.
[2] ταὐτὰ A.-W., τὰ αὐτὰ P: ταῦτα D, *istud* Σ: αὐτὰ AC, vulg.
[3] τ' C: om. vulg.

---

[a] I have translated γένος here by "genus," but it will be seen from the definition which immediately follows that "genus" has not its modern technical meaning. It would in fact be misleading to translate γένος and εἶδος normally by "genus" and "species"; and I have used terms such as

flesh, one's bone with another's bone; and the same applies to the parts of a horse, and of such other animals as we consider to be specifically identical; for as the whole is to the whole, so every part is to every part. In other cases, they are, it is true, identical, but they differ in respect of excess and defect: this applies to those whose genus[a] is the same; and by genus I mean, for example, bird and fish: each of these exhibits difference with respect to genus, and of course there are numerous species both of fishes and of birds.

Now, generally speaking, the differences exhibited in animals by most of the parts lie in the contrasting oppositions of their secondary characteristics,[b] *e.g.*, of colour or shape: some exhibit the same characteristic, but to a greater or less degree; some differ in respect of possessing more or fewer of a particular feature; some in respect of its greater or smaller size—*i.e.*, generally, they differ by way of excess and defect.[c] An illustration: the flesh of some animals is soft in consistency, of others, hard; some have a long bill, others a short one; some have many feathers, some few. Furthermore, even among the animals we are discussing, some have parts which are absent from others: *e.g.*, some have spurs, some have not; some have crests, others have not. But, in general, most of the parts, *i.e.*, those out of which the main bulk of the body is composed, are either identical or differ by way of opposition, *i.e.*, by excess and defect—for we may consider " the more and less " as being the same as " excess and defect."

" group," " class," " kind," and the like. See further, Notes, §§ 4 ff.

[b] These παθήματα, or secondary sex-characteristics, are dealt with at length in *G.A.* Book V.   [c] See Notes, § 3.

"Ενια δὲ τῶν ζῴων οὔτε εἴδει τὰ μόρια ταὐτὰ ἔχει οὔτε καθ' ὑπεροχὴν καὶ ἔλλειψιν, ἀλλὰ κατ' ἀναλογίαν, οἷον πέπονθεν ὀστοῦν πρὸς ἄκανθαν καὶ ὄνυξ πρὸς ὁπλὴν καὶ χεὶρ πρὸς χηλὴν καὶ πρὸς πτερὸν λεπίς· ὃ γὰρ ἐν ὄρνιθι πτερόν, τοῦτο ἐν τῷ[1] ἰχθύι ἐστὶ λεπίς.

Κατὰ μὲν οὖν τὰ[2] μόρια ἃ ἔχουσιν ἕκαστα τῶν ζῴων, τοῦτόν τε τὸν τρόπον ἕτερά ἐστι καὶ ταὐτά, καὶ ἔτι τῇ θέσει τῶν μερῶν· πολλὰ γὰρ τῶν ζῴων ἔχει μὲν ταὐτὰ μέρη, ἀλλὰ κείμενα οὐχ ὡσαύτως, οἷον μαστοὺς τὰ μὲν ἐν τῷ στήθει τὰ δὲ πρὸς τοῖς μηροῖς.

Ἔστι δὲ τῶν ὁμοιομερῶν τὰ μὲν μαλακὰ καὶ ὑγρά, τὰ δὲ ξηρὰ καὶ στερεά, ὑγρὰ μέν, ἢ ὅλως ἢ ἕως ἂν ᾖ ἐν τῇ φύσει, οἷον αἷμα, ἰχώρ, πιμελή, στέαρ, μυελός, γονή, χολή, γάλα ἐν τοῖς ἔχουσι, σάρξ τε καὶ τὰ τούτοις ἀνάλογον, ἔτι ἄλλον τρόπον τὰ περιττώματα, οἷον φλέγμα, καὶ τὰ ὑποστήματα τῆς κοιλίας καὶ κύστεως· ξηρὰ δὲ καὶ στερεὰ οἷον νεῦρον, δέρμα, φλέψ, θρίξ, ὀστοῦν, χόνδρος, ὄνυξ, κέρας [ὁμώνυμον γὰρ τὸ μέρος,[3] ὅτι τῷ σχήματι καὶ τὸ ὅλον λέγεται κέρας],[4] ἔτι ὅσα ἀνάλογον τούτοις.

Αἱ δὲ διαφοραὶ τῶν ζῴων εἰσὶ κατά τε τοὺς βίους καὶ τὰς πράξεις καὶ τὰ ἤθη καὶ τὰ μόρια,

[1] τῷ codd. : om. vulg.
[2] τὰ A¹C : om. vulg.
[3] τὸ μέρος A : πρὸς τὸ μέρος PC : πρὸς τὸ γένος D, vulg.
[4] secludenda. ὅτι ... λέγεται Th., ὅταν ... λέγηται vulg. : totum om. Σ.

[a] See Notes, § 13.
[b] See note on 489 a 23.   [c] See Notes, §§ 22 ff.

HISTORIA ANIMALIUM, I. i

Some animals, however, have parts which are not specifically identical, nor differing merely by excess and defect: these parts correspond only " by analogy ": of which an example is the correspondence between bone and fishspine, nail and hoof, hand and claw, feather and scale: in a fish, the scale is the corresponding thing to a feather in a bird.

Very well, then: with regard to the parts they possess the various animals are identical or different in the manner just described; and the same holds good with regard to the position of their parts: many animals have identical parts, but differently placed; some, for instance, have their teats on the breast, some near the thighs.

We can also distinguish differences among the uniform parts: some are soft and fluid,[a] others are solid and firm. The fluid ones may be such without qualification, or else fluid only so long as they remain in the natural organism: examples of the latter are blood, serum,[b] lard, suet, marrow, semen, gall, milk (in those that have it), flesh, and analogous parts; so too, in another way, the residues,[c] *e.g.*, phlegm, and the excretions of the belly and the bladder. Examples of solid and firm parts are: sinew, skin, bloodvessel, hair, bone, gristle, nail, horn [this can be included, because here the part has the same name as the whole, which is also called " horn " on account of its shape]; to these must be added parts which are analogous to them. *(ii) the uniform parts.*

Further differences exhibited by animals are those which relate to their manner of life, their activities, and their dispositions, as well as their parts.[d] Let us *Other kinds of differences:*

---

[d] Aristotle is therefore not concerned exclusively with morphology.

ARISTOTLE

περὶ ὧν τύπῳ μὲν εἴπωμεν πρῶτον, ὕστερον δὲ περὶ ἕκαστον γένος ἐπιστήσαντες ἐροῦμεν.

Εἰσὶ δὲ διαφοραὶ κατὰ μὲν τοὺς βίους καὶ τὰ ἤθη καὶ τὰς πράξεις αἱ τοιαίδε, ᾗ τὰ μὲν ἔνυδρα αὐτῶν ἐστι τὰ δὲ χερσαῖα, ἔνυδρα δὲ διχῶς, τὰ μὲν ὅτι τὸν βίον καὶ τὴν τροφὴν ποιεῖται ἐν τῷ ὑγρῷ καὶ δέχεται τὸ ὑγρὸν καὶ ἀφίησι, τούτου δὲ στερισκόμενα οὐ δύναται ζῆν, οἷον πολλοῖς συμβαίνει τῶν ἰχθύων· τὰ δὲ τὴν μὲν τροφὴν ποιεῖται καὶ τὴν διατριβὴν ἐν τῷ ὑγρῷ, οὐ μέντοι δέχεται τὸ ὕδωρ ἀλλὰ τὸν ἀέρα, καὶ γεννᾷ ἔξω. πολλὰ δ' ἐστὶ τοιαῦτα καὶ πεζά, ὥσπερ ἐνυδρὶς καὶ λάταξ καὶ κροκόδειλος, καὶ πτηνά, οἷον αἴθυια καὶ κολυμβίς, καὶ ἄποδα, οἷον ὕδρος. ἔνια δὲ τὴν μὲν τροφὴν ἐν τῷ ὑγρῷ ποιεῖται καὶ οὐ δύναται ζῆν ἐκτός, οὐ μέντοι δέχεται οὔτε τὸν ἀέρα οὔτε τὸ ὑγρόν, οἷον ἀκαλήφη καὶ τὰ ὄστρεα. τῶν δ' ἐνύδρων τὰ μέν ἐστι θαλάττια, τὰ δὲ ποτάμια, τὰ δὲ λιμναῖα, τὰ δὲ τελματιαῖα, οἷον βάτραχος καὶ κορδύλος.

Τῶν δὲ χερσαίων τὰ μὲν δέχεται τὸν ἀέρα καὶ ἀφίησιν, ὃ καλεῖται ἀναπνεῖν καὶ ἐκπνεῖν, οἷον ἄνθρωπος καὶ πάντα ὅσα πνεύμονα ἔχει τῶν χερσαίων· τὰ δὲ τὸν ἀέρα μὲν οὐ δέχεται, ζῇ δὲ καὶ τὴν τροφὴν ἔχει ἐν τῇ γῇ, οἷον σφὴξ καὶ μέλιττα καὶ τὰ

---

[a] For the purpose of cooling. (On the meaning of ἔνυδρον see Notes, §§ 39 ff.) In fact, all the non-selachian fishes take in sea-water continually, and excrete salt through the gills, in order to control the osmotic pressure inside the body. (The selachian fishes have an osmotic pressure which is the same as that of the sea-water.)

[b] The meaning (and translation) of ὄστρεον has to some extent to be determined from the context. This is not always easy (and perhaps not always important), as in the present

describe these first in general outline, and then we will go on to speak of the various kinds, giving special attention to each.

Here are examples of differences in respect of manner of life, dispositions, and activities. Some are water-animals, others land-animals. There are two ways of being water-animals. Some both live and feed in the water, take in water and emit it,[a] and are unable to live if deprived of it: this is the case with many of the fishes. Others feed and live in the water; but what they take in is air, not water; and they breed away from the water. Many of these animals are footed: *e.g.*, the otter, the beaver, the crocodile; many are winged, *e.g.*, the shearwater and the plunger; many are footless, *e.g.*, the water-snake. Some animals, although they get their food in the water and cannot live away from it, still take in neither water nor air: examples are the sea-anemones and shellfish.[b] Again, some water-animals live in the sea, some in rivers, some in lakes, some in marshes (*e.g.*, the frog and the newt).[c]

Specimen examples of differences in manner of life, activities,

Similarly, there are various kinds of land-animals: some take in air and emit it (processes known as inhalation and exhalation); examples are man and all land-animals that have lungs. Others do not take in any air, although they live and get their food on land,

passage, where the more particularized rendering " oysters " might do equally well. It is clear from 490 b 11 (where see note) that ὄστρεον was used as a popular term for what Aristotle more precisely designates ὀστρακόδερμα (" Testacea "), *i.e.*, animals with potsherd-shells; and we may justifiably translate by the popular English term " shellfish " (*cf.*, *e.g.*, 487 b 9), if we remember that in such cases " shellfish " will not include, *e.g.*, crabs (Aristotle's " soft-shelled " animals, Grustacea).

[c] But see Notes, §§ 40, 49.

9

ARISTOTLE

ἄλλα ἔντομα. (καλῶ δ' ἔντομα ὅσα ἔχει κατὰ τὸ σῶμα ἐντομάς, ἢ ἐν τοῖς ὑπτίοις ἢ ἐν τούτοις τε καὶ τοῖς πρανέσιν.)

Καὶ τῶν μὲν χερσαίων πολλά, ὥσπερ εἴρηται, ἐκ τοῦ ὑγροῦ τὴν τροφὴν πορίζεται, τῶν δ' ἐνύδρων καὶ δεχομένων τὴν θάλατταν οὐδὲν ἐκ τῆς γῆς. ἔνια δὲ τῶν ζῴων τὸ μὲν πρῶτον ζῇ ἐν τῷ ὑγρῷ, ἔπειτα μεταβάλλει εἰς ἄλλην μορφὴν καὶ ζῇ ἔξω, 5 οἷον ἐπὶ τῶν ἀσκαρίδων· ⟨καὶ γὰρ αὗται τὸ μὲν πρῶτον⟩ ἐν τοῖς ποταμοῖς ⟨ζῶσιν, ἔπειτα μεταβάλλει αὐταῖς ἡ μορφὴ καὶ⟩ γίγνεται ἐξ αὐτῶν ⟨ἡ ἐμπὶς καὶ ζῇ ἔξω⟩.[1]

Ἔτι τὰ μέν ἐστι μόνιμα τῶν ζῴων, τὰ δὲ μεταβλητικά. ἔστι δὲ τὰ μόνιμα ἐν τῷ ὑγρῷ· τῶν δὲ χερσαίων οὐδὲν μόνιμον. ἐν δὲ τῷ ὑγρῷ πολλὰ τῷ προσπεφυκέναι ζῇ, οἷον γένη ὀστρέων πολλά. 10 δοκεῖ δὲ καὶ ὁ σπόγγος ἔχειν τινὰ αἴσθησιν· σημεῖον δ' ὅτι χαλεπώτερον ἀποσπᾶται, ἂν μὴ γένηται λαθραίως ἡ κίνησις, ὥς φασιν. τὰ δὲ καὶ προσφύεται καὶ ἀπολύεται, οἷόν ἐστι γένος τι τῆς καλουμένης ἀκαλήφης· τούτων γάρ τινες νύκτωρ ἀπολυόμεναι

---

[1] correxi, Dt., Σ, alios secutus. ἐπὶ τῶν ἐν τοῖς ποταμοῖς ἐμπίδων (ἀσπίδων A¹C, ἀσκαρίδων Karsch, Dt.)· γίνεται γὰρ ἐξ αὐτῶν ὁ οἶστρος vulg.: γίνεται . . . οἶστρος secl. A.-W., Dt.: sicut quod nominatur grece ambidas, nam ipsum manet in fluminibus prius, deinde mutatur forma eius et fiet ex eo animal quod dicitur astaniz et vivit extra Σ. cf. et 551 b 27 seqq.

---

[a] In our Greek mss. this passage has been corrupted by an annotation which has ousted the original text. The received text as in Bekker's edition reads: "The gnats [so PDA²; *aspides* A¹C] are an instance of this, for out of them comes into being the gadfly." Karsch conjectured *askarides*

e.g., wasps, bees, and the other insects. (I call insects those creatures which have insections on their bodies, whether on their underside only, or both on their underside and on their backs.)

Now, as I have said, many land-animals get their food from the water; but no water-animal which takes in sea-water gets its food from the land. Yet again, some animals spend the first part of their lives in the water, and then change to another shape and live away from the water. The bloodworms are an instance of this: at first they live in the rivers, then their shape changes and the gnat develops out of them and lives away from the water.[a]

Furthermore, some animals are stationary, others move about. The stationary ones are found in the water; no land-animal is stationary. In the water we find many animals which live by sticking fast to some object: many kinds of shellfish [b] do this. We must include the sponge in this connexion, for so far as we can see it has sensation of a sort: this is indicated by the fact that it is more difficult to dislodge, unless the effort to do so is made surreptitiously—or so it is alleged. Some animals moor themselves, and then break loose; examples are a certain kind of sea-anemone as it is called: for some of these break loose

for *aspides*, and this alteration is adopted by Dittmeyer, who brackets the second part of the sentence, which is obviously wrong. Michael Scot's Latin version clearly shows what has happened, and I have ventured to restore the Greek text from it, with reference to *H.A.* 551 b 26 ff., where a fuller account of the matter is given.—The observation of the metamorphosis of larvae living in the water is remarkable. Dr J. Needham informs me that the metamorphosis of mosquitoes from larvae in the water is also clearly recognized in the *Huai Nan Tzu* book, ch. 17, of 2nd-cent. B.C. China.

[b] See notes on 487 a 26 and 490 b 11.

νέμονται. πολλὰ δ' ἀπολελυμένα μέν ἐστιν ἀκίνητα δέ, οἷον ὄστρεα καὶ τὰ καλούμενα ὁλοθούρια. τὰ δὲ νευστικά, οἷον ἰχθύες καὶ τὰ μαλάκια καὶ τὰ μαλακόστρακα, οἷον κάραβοι. τὰ δὲ πορευτικά, οἷον τὸ τῶν καρκίνων γένος· τοῦτο γὰρ ἔνυδρον ὂν τὴν φύσιν πορευτικόν ἐστιν. τῶν δὲ χερσαίων ἐστὶ τὰ μὲν πτηνά, ὥσπερ ὄρνιθες καὶ μέλιτται, καὶ ταῦτ' ἄλλον τρόπον ἀλλήλων, τὰ δὲ πεζά. καὶ τῶν πεζῶν τὰ μὲν πορευτικά, τὰ δ' ἑρπυστικά, τὰ δ' ἰλυσπαστικά. πτηνὸν δὲ μόνον οὐδέν ἐστιν, ὥσπερ νευστικὸν μόνον ἰχθύς· καὶ γὰρ τὰ δερμόπτερα πεζεύει, καὶ νυκτερίδι πόδες εἰσί, καὶ τῇ φώκῃ κεκολοβωμένοι πόδες.

Καὶ τῶν ὀρνίθων εἰσί τινες κακόποδες, οἳ διὰ τοῦτο καλοῦνται ἄποδες· ἔστι δ' εὔπτερον τοῦτο τὸ ὀρνίθιον. σχεδὸν δὲ καὶ τὰ ὅμοια αὐτῷ εὔπτερα μὲν κακόποδα δ' ἐστίν, οἷον χελιδὼν καὶ δρεπανίς· ὁμοιότροπά τε γὰρ καὶ ὁμοιόπτερα πάντα ταῦτα, καὶ τὰς ὄψεις ἐγγὺς ἀλλήλων. φαίνεται δ' ὁ μὲν ἄπους πᾶσαν ὥραν, ἡ δὲ δρεπανὶς ὅταν ὕσῃ τοῦ θέρους· τότε γὰρ ὁρᾶται καὶ ἁλίσκεται, ὅλως δὲ καὶ σπάνιόν ἐστι τοῦτο τὸ ὄρνεον. πορευτικὰ δὲ καὶ νευστικὰ πολλὰ τῶν ζῴων ἐστίν.

Εἰσὶ δὲ καὶ αἱ τοιαίδε διαφοραὶ κατὰ τοὺς βίους καὶ τὰς πράξεις. τὰ μὲν γὰρ αὐτῶν ἐστιν ἀγελαῖα

---

[a] An important observation, although it did not suggest to Aristotle that locomotion on land preceded locomotion in the air.

[b] It will be seen from 490 a 6 ff. that Aristotle recognizes three different kinds of winged animals: those whose wings are (1) feathered (πτερωτόν), as birds; (2) membranous (πτιλωτόν), as bees; (3) dermatous (δερμόπτερα), as bats. I have adopted the translation " dermatous " for the third in

at night to look for food. Many animals, though unattached, are stationary, *e.g.*, oysters, and the holothuria as they are called. Some are swimmers, *e.g.*, fishes, and Cephalopods, and Crustacea (*e.g.*, crayfish). Some walk about, *e.g.*, the whole class of crabs, which although water-animals have the natural faculty of being able to walk. Some land-animals are winged, *e.g.*, birds and bees (winged in different ways, these); some are footed. Of footed animals, some walk, some creep, some wriggle. No animal is merely able to fly (winged),[a] as the fish is merely able to swim. Even the dermatous-winged [b] animals can walk about, and bats have feet, and the seal has stunted feet.

Some birds have feet which are not much use, and this accounts for their being called *apodes* [c] (footless). The small bird concerned is a very good flier; and practically all those which resemble it are good fliers and have poor feet: examples are the swallow [d] and the *drepanis*.[e] All these are of similar habits and similar plumage, and resemble one another closely in appearance. The *apous* can be seen throughout the season, but the *drepanis* only when it has been raining, in summer: then it is seen and caught, but on the whole it is a rare bird. Many animals combine ability to walk and ability to swim.

Here are some further differences with respect to animals' manner of life and activities. Some are

default of a more satisfactory term. The exact modern term would be " patagial," but this fails to convey what Aristotle's term does, *viz.*, the notion that the wings are made of *skin*.

[c] Probably the swift.

[d] Probably the rock- or cliff-swallow, *Hirundo rupestris.*

[e] Lit., the sickle-bird; perhaps the Alpine swift, *Cypselus melba.*

ARISTOTLE

488 a τὰ δὲ μοναδικά, καὶ πεζὰ καὶ πτηνὰ καὶ πλωτά, τὰ δ' ἐπαμφοτερίζει. καὶ τῶν ἀγελαίων [καὶ τῶν μοναδικῶν]¹ τὰ μὲν πολιτικὰ τὰ δὲ σποραδικά ἐστιν· ἀγελαῖα μὲν οὖν οἷον ἐν τοῖς πτηνοῖς τὸ τῶν περιστερῶν γένος καὶ γέρανος καὶ κύκνος (γαμψώνυ-
5 χον δ' οὐδὲν ἀγελαῖον), καὶ τῶν πλωτῶν πολλὰ γένη τῶν ἰχθύων, οἷον οὓς καλοῦσι δρομάδας, θύννοι, πηλαμύδες, ἀμίαι²· ὁ δ' ἄνθρωπος³ ἐπαμφοτερίζει.

Πολιτικὰ δ' ἐστὶν ὧν ἕν τι καὶ κοινὸν γίγνεται πάντων τὸ ἔργον, ὅπερ οὐ πάντα ποιεῖ τὰ ἀγελαῖα.
10 ἔστι δὲ τοιοῦτον ἄνθρωπος,⁴ μέλιττα, σφήξ, μύρμηξ γέρανος. καὶ τούτων τὰ μὲν ὑφ' ἡγεμόνα ἐστὶ τὰ δ' ἄναρχα, οἷον γέρανος μὲν καὶ τὸ τῶν μελιττῶν γένος ὑφ' ἡγεμόνα, μύρμηκες δὲ καὶ μυρία ἄλλα ἄναρχα. καὶ τὰ μὲν ἐπιδημητικὰ καὶ τῶν ἀγελαίων καὶ τῶν μοναδικῶν, τὰ δὲ ἐκτοπιστικά. καὶ τὰ
15 μὲν σαρκοφάγα, τὰ δὲ καρποφάγα, τὰ δὲ παμφάγα, τὰ δὲ ἰδιότροφα, οἷον τὸ τῶν μελιττῶν γένος καὶ τὸ τῶν ἀραχνῶν· τὰ μὲν γὰρ μέλιτι καί τισιν ἄλλοις ὀλίγοις τῶν γλυκέων χρῆται τροφῇ, οἱ δ' ἀράχναι ἀπὸ τῆς τῶν μυιῶν θήρας ζῶσιν, τὰ δὲ ἰχθύσι χρῶνται τροφῇ. καὶ τὰ μὲν θηρευτικά, τὰ δὲ θη-
20 σαυριστικὰ τῆς τροφῆς ἐστι, τὰ δ' οὔ. καὶ τὰ μὲν οἰκητικὰ τὰ δ' ἄοικα, οἰκητικὰ μὲν οἷον ἀσπάλαξ, μῦς, μύρμηξ, μέλιττα, ἄοικα δὲ πολλὰ τῶν ἐντόμων καὶ τῶν τετραπόδων. ἔτι τοῖς τόποις τὰ μὲν τρωγλοδυτικά, οἷον σαύρα, ὄφις, τὰ δ' ὑπέργεια, οἷον

---

¹ secl. Sn.: ἐπαμφοτερίζει καὶ τῶν ἀγ. καὶ τῶν μοναδικῶν. ⟨τῶν δ' ἀγ.⟩ τὰ μὲν πολ. Gohlke.
² ἀμίαι AC: ἅμιαι PD, vulg.
³ ὄνος (sc. piscis) tent. Th.
⁴ ἀνθρήνη coni. Th.

HISTORIA ANIMALIUM, I. i

gregarious, some solitary: this applies to footed animals, winged ones, and swimmers alike; others are dualizers.[a] Some of the gregarious animals are social,[b] whereas others are more dispersed. Examples of gregarious animals are: birds—the pigeon class, the crane, the swan (*N.B.*: no crook-taloned bird is gregarious); swimmers—many groups of fishes, *e.g.*, those called migrants, the tunnies, the *pelamys*, and the bonito.[c] And man dualizes.

The social animals are those which have some one common activity; and this is not true of all the gregarious animals. Examples of social animals are man, bees, wasps, ants, cranes. Some of them live under a ruler, some have no ruler; examples: cranes and bees live under a ruler, ants and innumerable others do not. Some of the gregarious and also some of the solitary animals remain in one situation, others roam about. Further, some are carnivorous, some frugivorous, others omnivorous; some have a diet peculiar to themselves, as for instance bees and spiders: bees feed on honey and a few other sweet materials, spiders live by catching flies; some animals live on fish. Again, some catch their food, others store it up; others do not. Some provide themselves with habitations, some have none; examples of the former are the mole, the mouse, the ant, and the bee; of the latter, most insects and quadrupeds. Further, they differ in place of habitation: some live in holes underground, *e.g.*, lizards and snakes; some live above

---

[a] *i.e.*, show both characteristics. I use the term "dualize" to avoid disguising an important verb of Aristotle's, ἐπαμφοτερίζειν. See Notes, §§ 28 ff.

[b] *i.e.*, live in close communities.

[c] *Pelamys* and bonito are varieties of tunny. *Cf.* note on 506 b 13.

ARISTOTLE

488 a

25 ἵππος, κύων. [καὶ τὰ μὲν τρηματώδη τὰ δ᾽ ἄτρητα].[1]

Καὶ τὰ μὲν νυκτερόβια, οἷον γλαύξ, νυκτερίς, τὰ δ᾽ ἐν τῷ φωτὶ ζῇ. ἔτι δ᾽ ἥμερα καὶ ἄγρια, καὶ τὰ μὲν ἀεὶ [οἷον ἄνθρωπος καὶ ὀρεὺς ἀεὶ ἥμερα, τὰ δ᾽][2] ἄγρια, ὥσπερ πάρδαλις καὶ λύκος· τὰ δὲ καὶ ἡμεροῦσθαι δύναται ταχύ, οἷον ἐλέφας· [ἔτι ἄλλον τρό-
30 πον·][3] πάντα γὰρ ὅσα ἥμερά ἐστι γένη, καὶ ἄγριά ἐστιν, οἷον ἵπποι, βόες, ὕες, ἄνθρωποι,[4] πρόβατα, αἶγες, κύνες. καὶ τὰ μὲν ψοφητικά, τὰ δὲ ἄφωνα, τὰ δὲ φωνήεντα, καὶ τούτων τὰ μὲν διάλεκτον ἔχει τὰ δὲ ἀγράμματα, καὶ τὰ μὲν κωτίλα τὰ δὲ σιγηλά,

488 b

τὰ δ᾽ ᾠδικὰ τὰ δ᾽ ἄνῳδα· πάντων δὲ κοινὸν τὸ περὶ τὰς ὀχείας μάλιστα ᾄδειν καὶ λαλεῖν. καὶ τὰ μὲν ἄγροικα, ὥσπερ φάττα, τὰ δ᾽ ὄρεια, ὥσπερ ἔποψ, τὰ δὲ συνανθρωπίζει, οἷον περιστερά.

Καὶ τὰ μὲν ἀφροδισιαστικά, οἷον τὸ τῶν περδί-
5 κων καὶ ἀλεκτρυόνων γένος, τὰ δὲ ἀγνευτικά, οἷον τὸ τῶν κορακοειδῶν ὀρνίθων γένος· ταῦτα γὰρ σπανίως ποιεῖται τὴν ὀχείαν. καὶ τῶν θαλαττίων τὰ μὲν πελάγια, τὰ δ᾽ αἰγιαλώδη, τὰ δὲ πετραῖα. ἔτι τὰ μὲν ἀμυντικὰ τὰ δὲ φυλακτικά· ἔστι δ᾽ ἀμυντικὰ μὲν ὅσα ἢ ἐπιτίθεται ἢ ἀδικούμενα ἀμύνεται,
10 φυλακτικὰ δ᾽ ὅσα πρὸς τὸ μὴ παθεῖν τι ἔχει ἐν αὑτοῖς ἀλεωρήν.

Διαφέρουσι δὲ καὶ ταῖς τοιαῖσδε διαφοραῖς κατὰ

---

[1] secl. Dt. : habet Σ.
[2] seclusi : γίννος loco ἄνθρωπος Dt. : ὄνος καὶ ἵππος Th.
[3] seclusi.
[4] ἄνθρωποι codd. : om. vulg. : ὄνοι coni. Pi.

---

[a] This phrase is suspected or excised by some editors,

ground, *e.g.*, horses and dogs; [and some burrow holes for themselves, while others do not.] *ᵃ*

Some animals are nocturnal, as the owl and the bat; some live by daylight. Further, we find some are tame, others wild; some are always [tame, *e.g.*, man and the mule; some always] *ᵇ* wild, as the leopard and the wolf; some can also be quickly tamed, *e.g.*, the elephant; for any kind of animal which is tame exists also in a wild state,*ᶜ e.g.*, horses, oxen, swine, men, sheep, goats, dogs. Again, some emit a noise, some are mute; some have a voice; and of the latter some are articulate and others inarticulate; some are always chattering, some tend to be quiet; some are tuneful, some are not. But it is common to all of them to sing or chatter most of all about the time of mating. Again, some animals live in the open countryside, *e.g.*, the ringdove; some live on the mountains, *e.g.*, the hoopoe; some frequent human habitation, *e.g.*, the pigeon.

Again, some are specially prone to copulation, *e.g.*, the partridge and the domestic fowl; others are less prone to it, *e.g.*, the tribe of crows, which infrequently copulate. Of marine animals, some live well out at sea, others by the shore, others on the rocks. Again, some are pugnacious when attacked, others merely put up a defence: the former are such as set upon other animals or fight back when ill treated; the latter possess some means for defending themselves against ill treatment.

Now here are the sorts of ways in which animals and dispositions.

partly because of the strained meaning which has to be given to the two adjectives in the Greek.

ᵇ These words are probably an interpolation, and I have therefore marked them for omission.

ᶜ Cf. P.A. 643 b 5.

τὸ ἦθος. τὰ μὲν γάρ ἐστι πρᾶα καὶ δύσθυμα καὶ οὐκ ἐνστατικά, οἷον βοῦς, τὰ δὲ θυμώδη καὶ ἐνστατικὰ καὶ ἀμαθῆ, οἷον ὗς ἄγριος, τὰ δὲ φρόνιμα καὶ δειλά, οἷον ἔλαφος, δασύπους,[1] τὰ δ' ἀνελεύθερα καὶ ἐπίβουλα, οἷον οἱ ὄφεις, τὰ δ' ἐλευθέρια[2] καὶ ἀνδρεῖα καὶ εὐγενῆ, οἷον λέων, τὰ δὲ γενναῖα καὶ ἄγρια καὶ ἐπίβουλα, οἷον λύκος· εὐγενὲς μὲν γάρ ἐστι τὸ ἐξ ἀγαθοῦ γένους, γενναῖον δὲ τὸ μὴ ἐξιστάμενον ἐκ τῆς αὑτοῦ φύσεως.

Καὶ τὰ μὲν πανοῦργα καὶ κακοῦργα, οἷον ἀλώπηξ, τὰ δὲ θυμικὰ καὶ φιλητικὰ καὶ θωπευτικά, οἷον κύων, τὰ δὲ πρᾶα καὶ τιθασσευτικά, οἷον ἐλέφας, τὰ δ' αἰσχυντηλὰ καὶ φυλακτικά, οἷον χήν, τὰ δὲ φθονερὰ καὶ φιλόκαλα, οἷον ταώς. βουλευτικὸν δὲ μόνον ἄνθρωπός ἐστι τῶν ζῴων. καὶ μνήμης μὲν καὶ διδαχῆς πολλὰ κοινωνεῖ, ἀναμιμνήσκεσθαι δ' οὐδὲν ἄλλο δύναται πλὴν ἄνθρωπος.

Περὶ ἕκαστον δὲ τῶν γενῶν τά τε περὶ τὰ ἤθη καὶ τοὺς βίους ὕστερον λεχθήσεται δι' ἀκριβείας μᾶλλον.

II Πάντων δ' ἐστὶ τῶν ζῴων κοινὰ μόρια, ᾧ δέχεται τὴν τροφὴν καὶ εἰς ὃ δέχεται· ταῦτα δ' ἐστὶ ταὐτὰ καὶ ἕτερα κατὰ τοὺς εἰρημένους τρόπους, ἢ κατ' εἶδος ἢ καθ' ὑπεροχὴν ἢ κατ' ἀναλογίαν ἢ τῇ θέσει διαφέροντα. μετὰ δὲ ταῦτα ἄλλα κοινὰ μόρια ἔχει τὰ πλεῖστα τῶν ζῴων πρὸς τούτοις, ᾗ ἀφίησι τὸ περίττωμα τῆς τροφῆς [καὶ ᾗ λαμβάνει][3]· οὐ γὰρ

---

[1] δασύπους secl. Dt.
[2] Sn., vulg.: ἐλεύθερα codd.
[3] secl. edd.

[a] The expression in the present sentence is condensed, since " identical " and " diverse " do not both apply to all the modes specified; I have therefore translated the καί by

differ from each other in regard to disposition. Some are gentle, and sluggish, and not inclined to be aggressive, *e.g.*, the ox ; others are ferocious, aggressive and stubborn, *e.g.*, the wild boar ; some are intelligent and timid, *e.g.*, the deer and the hare ; others are mean and scheming, *e.g.*, serpents ; others are noble and brave and high-bred, *e.g.*, the lion ; others are thorough-bred, wild and scheming, *e.g.*, the wolf. (By high-bred is meant one which comes of a good stock, by thorough-bred one which remains true to its own type.)

Again, some are mischievous and wicked, *e.g.*, the fox ; others are spirited and affectionate and fawning, *e.g.*, the dog ; some are gentle and easily tamed, *e.g.*, the elephant ; others are bashful and cautious, *e.g.*, the goose ; some are jealous and ostentatious, like the peacock. The only animal which is deliberative is man. Many animals have the power of memory and can be trained ; but the only one which can recall past events at will is man.

With regard to the dispositions and manner of life of each several kind of animals we will speak in more detail later on.

All animals have in common the part by which they take in food and the part into which they take it. These parts respectively are either identical, or diverse, in the ways already described [a] : *viz.*, specifically, or in respect of excess, or by analogy, or differing in position. In addition to these, the majority of animals have other parts in common as well—first, the parts by which they discharge the residue [b] that comes from their food : I say the majority, for not

II Some instrumental parts common to all or most animals, though exhibiting differences.

---

[a] " or " : " and/or " would be more precise. See above, 486 b 15 ff., and 486 a 22, n.   [b] See Notes, §§ 22 ff.

ARISTOTLE

πᾶσιν ὑπάρχει τοῦτο. καλεῖται δ' ᾗ μὲν λαμβάνει, στόμα, εἰς ὃ δὲ δέχεται, κοιλία· τὸ δὲ λοιπὸν πολυώνυμόν ἐστιν. τοῦ δὲ περιττώματος ὄντος διττοῦ, ὅσα μὲν ἔχει δεκτικὰ μόρια τοῦ ὑγροῦ περιττώματος, ἔχει καὶ τῆς ξηρᾶς τροφῆς, ὅσα δὲ ταύτης, ἐκείνης οὐ πάντα. διὸ ὅσα μὲν κύστιν ἔχει, καὶ κοιλίαν ἔχει, ὅσα δὲ κοιλίαν ἔχει, οὐ πάντα κύστιν ἔχει. ὀνομάζεται γὰρ τὸ μὲν τῆς ὑγρᾶς περιττώσεως δεκτικὸν μόριον κύστις, κοιλία δὲ τὸ τῆς ξηρᾶς.

III Τῶν δὲ λοιπῶν πολλοῖς ὑπάρχει ταῦτά τε τὰ μόρια καὶ ἔτι ᾗ τὸ σπέρμα ἀφιᾶσιν· καὶ τούτων ἐν οἷς μὲν ὑπάρχει γένεσις ζῴων τὸ μὲν εἰς αὑτὸ ἀφιέν, τὸ δ' εἰς ἕτερον. καλεῖται δὲ τὸ μὲν εἰς αὑτὸ ἀφιὲν θῆλυ, τὸ δ' εἰς τοῦτο ἄρρεν. ἐν ἐνίοις δ' οὐκ ἔστι τὸ ἄρρεν καὶ θῆλυ· ᾗ καὶ τῶν μορίων τῶν πρὸς τὴν δημιουργίαν ταύτην διαφέρει τὸ εἶδος· τὰ μὲν γὰρ ἔχει ὑστέραν τὰ δὲ τὸ ἀνάλογον.

Ὅσα μὲν οὖν ἀναγκαιότατα μόρια τοῖς ζῴοις τὰ μὲν πᾶσιν ἔχειν συμβέβηκε τὰ δὲ τοῖς πλείστοις, ταῦτ' ἐστίν.

Πᾶσι δὲ τοῖς ζῴοις αἴσθησις μία ὑπάρχει κοινὴ μόνη ἡ ἁφή, ὥστε καὶ ἐν ᾧ αὕτη μορίῳ γίγνεσθαι πέφυκεν, ἀνώνυμόν ἐστιν· τοῖς μὲν γὰρ ταὐτὸ τοῖς δὲ τὸ ἀνάλογόν ἐστιν.

IV Ἔχει δὲ καὶ ὑγρότητα πᾶν ζῷον, ἧς στερισκόμενον ἢ φύσει ἢ βίᾳ φθείρεται. ἔτι ἐν ᾧ γίγνεται,

---

[a] See *G.A.* 716 a 17 ff.
[b] A " part " ; see Notes, § 13.

all animals have such parts. The part by which they take in their food is known as the mouth; that into which they take it, the belly; the remaining parts have many different names. Now as the residue is twofold, those animals which have parts to receive the fluid residue have also a part for the ⟨residue from the⟩ solid nutriment; but those which have the latter do not all have the former. Hence, all animals which have a bladder have a bowel as well; but not all which have a bowel have a bladder. (Bladder, of course, is the name for the part which receives the fluid residue, bowel the name for the part which receives the solid residue.)

Of the remaining animals many have, in addition to the parts mentioned, a part by which they emit semen; and of those animals which generate, one will emit semen into itself, one into another individual. The former is known as "female," the latter as "male"[a]—though in some animals male and female are not found. Hence the parts which serve this function differ in form, and some animals have a uterus, others an analogous part.

So then, I have now mentioned the parts which are essential to animals, both those which all animals must possess, and those which most must possess.

One of the senses, and only one, is common to all animals, *viz.*, touch; hence there is no particular name for the part in which this sense has its natural place: it is an identical part in some animals, an analogous part in others.

Now every animal contains fluid,[b] and if it is deprived of this either in the natural course or forcibly, it perishes. And further, there must be some receptacle in which this fluid exists: that means another

III

IV Fluid an essential "part" for all animals: differences

τοῦτο ἄλλο. ἔστι δὲ ταῦτα¹ τοῖς μὲν αἷμα καὶ φλέψ, τοῖς δὲ τὰ² ἀνάλογον τούτων· ἔστι δ' ἀτελῆ ταῦτα, οἷον τὸ μὲν ἲς τὸ δ' ἰχώρ.

Ἡ μὲν οὖν ἁφὴ ἐν ὁμοιομερεῖ ἐγγίγνεται μέρει, οἷον ἐν σαρκὶ ἢ τοιούτῳ τινί, καὶ ὅλως ἐν τοῖς αἱματικοῖς, ὅσα ἔχει αἷμα· τοῖς δ' ἐν τῷ ἀνάλογον, πᾶσι δ' ἐν τοῖς ὁμοιομερέσιν. αἱ δὲ ποιητικαὶ δυνάμεις ἐν τοῖς ἀνομοιομερέσιν, οἷον ἡ τῆς τροφῆς ἐργασία ἐν στόματι καὶ ἡ τῆς κινήσεως τῆς κατὰ τόπον ἐν ποσὶν ἢ πτέρυξιν ἢ τοῖς ἀνάλογον.

Πρὸς δὲ τούτοις τὰ μὲν ἔναιμα τυγχάνει ὄντα, οἷον ἄνθρωπος καὶ ἵππος καὶ πάνθ' ὅσα ἢ ἄποδά ἐστι τέλεα ὄντα ἢ δίποδα ἢ τετράποδα, τὰ δ' ἄναιμα, οἷον μέλιττα καὶ σφὴξ καὶ τῶν θαλαττίων σηπία καὶ κάραβος καὶ πάνθ' ὅσα πλείους πόδας ἔχει τεττάρων.

V Καὶ τὰ μὲν ζῳοτόκα τὰ δ' ᾠοτόκα τὰ δὲ σκωλη-

---

¹ ταῦτα A.-W. : τοῦτο codd.   ² τὰ Dt. : τὸ vulg.

---

[a] Here *ichor* (serum) and fibre are said to be, in bloodless animals, the counterparts respectively of blood and bloodvessel in blooded animals (Aristotle allows the name "blood" to be applied only to red blood). We find a similar usage at 511 b 4; and *cf.* 515 b 27. On the other hand, in *P.A.* Aristotle several times mentions "the counterpart of blood" without naming it (*e.g.*, 645 b 10, 648 a 5 (*cf.* a 1), 650 a 35), and at *P.A.* 678 a 9 he explicitly says that what corresponds to blood in the bloodless animals "has no name." This may be due (*a*) to the fact that ἰχώρ had been the name used for the fluid in the veins of the gods (*e.g.*, *Iliad* v. 340); and (*b*) to the fact that ἰχώρ was the term applied to the "watery part of blood" (τὸ ὑδατῶδες τοῦ αἵματος; *P.A.* 651 a 17), its watery character being due either (1) to its not yet having undergone concoction into blood, which is the final form assumed by the nourishment (ἐσχάτη τροφή), or (2) to its having undergone corruption (*cf. H.A.* 521 a 13 ff.: "exces-

HISTORIA ANIMALIUM, I. iv–v

part. Now these parts, respectively, in some animals are blood and blood-vessel; in others, parts analogous to these; but the latter are imperfect, *e.g.*, fibre and serum.[a]

<small>exhibited by it and the parts in which it occurs.</small>

Now touch is brought about in some "uniform" part, *e.g.*, the flesh, or something similar to it, and generally, in those animals which have blood, it occurs in the parts which are supplied with blood. In others, it occurs in the part which is analogous; but in all animals it occurs in the uniform parts. The active faculties, on the other hand, occur in the non-uniform parts, *e.g.*, the mastication of food occurs in the mouth, and the faculty of locomotion in the feet, or wings, or parts analogous to these.

In addition to this, some animals are blooded, *e.g.*, man, horse, and all animals which when full-grown are footless, two-footed, or four-footed; others are bloodless, *e.g.*, the bee, the wasp, and the cuttlefish and crayfish (marine animals, these two), and all animals which have more than four feet.

Again, some animals are viviparous, some ovi- V

<small>Differences in methods</small>

sive liquefaction of the blood causes disease: it becomes *ichor*-like"). At 521 a 17 we read that blood is produced by the concoction of *ichor*, and at 521 a 33 that in the very young the blood is *ichor*-like. It is therefore to some extent confusing to apply the term *ichor* to a fluid in bloodless animals which never will be and never has been blood, and for this reason Aristotle may prefer to avoid doing so.—" Fibres " are defined at 515 b 27 as being " between sinew and bloodvessel "; some of them contain " the liquid of *ichor*," and are clearly there thought of as occurring in blooded animals. The same term is also used, as Aristotle points out immediately after, of the fibres *in* the blood, which enable it to coagulate: *cf.* 520 b 25. At *P.A.* 650 b 14 ff. blood is said to consist of a " more watery part " and an " earthy " part, *viz.*, the fibres; and at *P.A.* 650 b 33 ff. the fibres are said to be " solid and earthy."

23

ARISTOTLE

κοτόκα, ζωοτόκα μὲν οἷον ἄνθρωπος καὶ ἵππος καὶ φώκη καὶ τὰ ἄλλα ὅσα ἔχει τρίχας, καὶ τῶν ἐνύδρων τὰ κητώδη, οἷον δελφίς, καὶ τὰ καλούμενα σελάχη. τούτων δὲ τὰ μὲν αὐλὸν ἔχει, βράγχια δ' οὐκ ἔχει, οἷον δελφὶς καὶ φάλαινα (ἔχει δ' ὁ μὲν δελφὶς τὸν αὐλὸν διὰ τοῦ νώτου, ἡ δὲ φάλαινα ἐν τῷ μετώπῳ), τὰ δ' ἀκάλυπτα βράγχια, οἷον τὰ σελάχη, γαλεοί τε καὶ βάτοι.

Καλεῖται δ' ᾠὸν μὲν τῶν κυημάτων τῶν τελείων, ἐξ οὗ γίγνεται τὸ γιγνόμενον ζῷον, ἐκ μορίου τὴν ἀρχήν, τὸ δ' ἄλλο τροφὴ τῷ γιγνομένῳ ἐστίν· σκώληξ δ' ἐστὶν ἐξ οὗ ὅλου ὅλον γίγνεται τὸ ζῷον, διαρθρουμένου καὶ αὐξανομένου τοῦ κυήματος.

Τὰ μὲν οὖν ἐν αὐτοῖς ᾠοτοκεῖ τῶν ζῳοτόκων, οἷον τὰ σελάχη, τὰ δὲ ζῳοτοκεῖ ἐν αὐτοῖς, οἷον ἄνθρωπος καὶ ἵππος· εἰς δὲ τὸ φανερὸν τῶν μὲν τελεωθέντος τοῦ κυήματος ζῷον ἐξέρχεται, τῶν δ' ᾠόν, τῶν δὲ σκώληξ. τῶν δ' ᾠῶν τὰ μὲν ὀστρακόδερμά ἐστι καὶ δίχροα, οἷον τὰ τῶν ὀρνίθων, τὰ δὲ μαλακόδερμα καὶ μονόχροα, οἷον τὰ τῶν σελαχῶν. καὶ τῶν σκωλήκων οἱ μὲν εὐθὺς κινητικοὶ οἱ δ' ἀκίνητοι. ἀλλὰ περὶ μὲν τούτων ἐν τοῖς περὶ γενέσεως δι' ἀκριβείας ὕστερον ἐροῦμεν.

---

[a] The cartilaginous fishes, *i.e.*, dogfishes, sharks and rays.
[b] See Notes, § 16.
[c] *Cf. G.A.* 732 a 29, where the same difference between an egg and a larva is stated : an egg is something from *part* of which the new creature is formed, while the remainder is nourishment for it, whereas in the case of the larva the *whole* of it is used up to form the whole of the offspring ; *cf.* also *G.A.* 758 b 10 ff. Aristotle further specifies two grades of nourishment (*G.A.* 744 b 33 ff., where see note in the Loeb edition) : the first-grade nourishment, which is described as " nutritive " and " seminal," provides the whole animal and its parts with " being " ; the second-grade nourishment is de-

parous, and some larviparous. Examples of vivipa- of repro-
rous animals are man, horse, seal, and all other hairy duction.
animals ; of water-animals, the cetacea (*e.g.*, the
dolphin) and the Selachia as they are called. Some
of these sea-animals have a blow-hole but no gills
(*e.g.*, the dolphin and the whale : the dolphin's tube
goes through its back, the whale's is in its forehead) ;
some have uncovered gills, *e.g.*, the Selachia [a] (the
dogfishes and rays).

Egg is the name given to that class of perfected
fetation [b] out of which the forming animal comes into
being : in its initial stage the embryo develops from
part of the egg, and the rest serves as nourishment
for the creature while it is forming.[c] A larva is that
out of the *whole* of which the whole of the animal
comes into being, as the fetation becomes articulated
and grows.

Some of the viviparous animals are internally oviparous, *e.g.*, the Selachia ; others are internally viviparous, for example, man and horse. That which is
brought forth when the fetation has reached completion is in some cases a living animal, in others an
egg, in others a larva. With regard to eggs, some
have shelly exteriors and their contents are of two
colours, for example, the eggs of birds ; others are
soft-skinned, and the contents are of a single colour,
for example, the eggs of the Selachia. As for larvae,
some are able to move about straight away, others are
not. However, we shall speak of these matters in
detail later in the treatise on Generation.

scribed as " growth-promoting," and causes increase of bulk.
In the embryo, it is the leavings of the first-grade nourishment, left over after the " supreme " parts (flesh and the other
sense-organs) have been provided for, which are used to
form the bones and sinews.

ARISTOTLE

489 b

Ἔτι δὲ τῶν ζῴων τὰ μὲν ἔχει πόδας τὰ δ' ἄποδα, καὶ τῶν ἐχόντων τὰ μὲν δύο πόδας ἔχει, οἷον ἄνθρωπος καὶ ὄρνις μόνα, τὰ δὲ τέτταρας, οἷον σαύρα καὶ κύων, τὰ δὲ πλείους, οἷον σκολόπενδρα καὶ μέλιττα· πάντα δ' ἀρτίους ἔχει πόδας.

Τῶν δὲ νευστικῶν ὅσα ἄποδα, τὰ μὲν πτερύγια ἔχει, ὥσπερ ἰχθύς, καὶ τούτων οἱ μὲν τέτταρα πτερύγια, δύο μὲν ἄνω ἐν τοῖς πρανέσι, δύο δὲ κάτω ἐν τοῖς ὑπτίοις, οἷον χρύσοφρυς καὶ λάβραξ, τὰ δὲ δύο μόνον, ὅσα προμήκη καὶ λεῖα, οἷον ἔγχελυς καὶ γόγγρος· τὰ δ' ὅλως οὐκ ἔχει, οἷον σμύραινα καὶ ὅσα ἄλλα χρῆται τῇ θαλάττῃ ὥσπερ οἱ ὄφεις τῇ γῇ, ⟨οἳ⟩[1] καὶ ἐν τῷ ὑγρῷ ὁμοίως νέουσιν. τῶν δὲ σελαχῶν ἔνια μὲν οὐκ ἔχει πτερύγια, οἷον τὰ πλατέα καὶ κερκοφόρα, ὥσπερ βάτος καὶ τρυγών, ἀλλ' αὐτοῖς νεῖ τοῖς πλάτεσιν[2]· βάτραχος δ' ἔχει καὶ ὅσα τὸ πλάτος μὴ ἔχει ἀπολελεπτυσμένον. ὅσα δὲ δοκεῖ πόδας ἔχειν, καθάπερ τὰ μαλάκια, τούτοις νεῖ καὶ τοῖς πτερυγίοις, καὶ θᾶττον ἐπὶ κύτος, οἷον σηπία καὶ τευθὶς [καὶ πολύπους][3]· βαδίζει δὲ τούτων οὐδέτερον, ὥσπερ πολύπους.

490 a

Τὰ δὲ σκληρόδερμα, οἷον κάραβος, τοῖς οὐραίοις νεῖ, τάχιστα δ' ἐπὶ τὴν κέρκον τοῖς ἐν ἐκείνῃ πτερυγίοις· καὶ ὁ κορδύλος τοῖς ποσὶ καὶ τῷ οὐραίῳ· ἔχει δ' ὅμοιον γλάνει τὸ οὐραῖον, ὡς μικρὸν εἰκάσαι μεγάλῳ.

Τῶν δὲ πτηνῶν τὰ μὲν πτερωτά ἐστιν, οἷον ἀετός

[1] ⟨οἳ⟩ Pi. [2] πλατέσιν codd. [3] secl. Scal., A.-W.

---

[a] Angler-fish, *Lophius piscatorius*.   [b] *Cf.* 524 a 13.
[c] *Parasilurus aristotelis*, a large representative of the widely-spread family of the Catfishes. See 568 b 13, n.

Further, some animals have feet, some are footless; some footed animals have two feet, for example, man and bird (these are the only ones); some have four, for example, the lizard and the dog; some have more, for example the millipede and the bee; but all of them have an even number. *Differences in means of locomotion.*

Some swimmers are footless; and of these, some have fins, for example, fishes, and of these again some have four fins, two above on the back, and two below on the belly, for example, the gilthead and the seabasse; while some—the very long and smooth fishes—have two only, for example, the eel and the conger. Some have none at all, for example, the muraena, and all others which move in the sea just as serpents do on the land: as a matter of fact serpents swim in water in precisely the same way. As for the Selachia, some have no fins, for example, those that are broad and long-tailed, like the ray and the sting-ray: they swim by means of their broad bodies merely. The fishing-frog,[a] however, has fins, and so have all those whose broad bodies do not taper off to a very thin edge. Some swimmers appear to have feet, for example, the Cephalopods: they use these and also their fins for swimming; and they swim more quickly in the direction of their trunk,[b] for example, the cuttlefish and the calamary. Neither of these is able to walk as the octopus can.

As for the hard-skinned animals, for example, the crayfish, they swim with their tail-parts, and most quickly tail first, by means of the fins they have on it. The newt uses feet and tail-parts alike; the latter are similar to those of the *glanis*,[c] if we may compare small with great.

Some of the fliers have feathered wings, for ex-

ARISTOTLE

καὶ ἱέραξ, τὰ δὲ πτιλωτά, οἷον μέλιττα καὶ μηλολόνθη, τὰ δὲ δερμόπτερα, οἷον ἀλώπηξ καὶ νυκτερίς. πτερωτὰ μὲν οὖν ἐστιν ὅσα ἔναιμα, καὶ δερμόπτερα ὡσαύτως· πτιλωτὰ δ' ὅσα ἄναιμα, οἷον
10 τὰ ἔντομα. ἔστι δὲ τὰ μὲν πτερωτὰ καὶ δερμόπτερα δίποδα πάντα ἢ ἄποδα· λέγονται γὰρ εἶναί τινες ὄφεις τοιοῦτοι περὶ Αἰθιοπίαν.

Τὸ μὲν οὖν πτερωτὸν γένος τῶν ζῴων ὄρνις καλεῖται, τὰ δὲ λοιπὰ δύο ἀνώνυμα ἑνὶ ὀνόματι.

Τῶν δὲ πτηνῶν μὲν ἀναίμων δὲ τὰ μὲν κολεόπτερά ἐστιν (ἔχει γὰρ ἐν ἐλύτρῳ τὰ πτερά, οἷον αἱ
15 μηλολόνθαι καὶ οἱ κάνθαροι), τὰ δ' ἀνέλυτρα, καὶ τούτων τὰ μὲν δίπτερα τὰ δὲ τετράπτερα, τετράπτερα μὲν ὅσα μέγεθος ἔχει ἢ ὅσα ὀπισθόκεντρά ἐστι, δίπτερα δ' ὅσα ἢ μέγεθος μὴ ἔχει ἢ ἐμπροσθόκεντρά ἐστιν. τῶν δὲ κολεοπτέρων οὐδὲν ἔχει κέν-
20 τρον. τὰ δὲ δίπτερα ἔμπροσθεν ἔχει τὰ κέντρα, οἷον μυῖα καὶ μύωψ καὶ οἶστρος καὶ ἐμπίς.

Πάντα δὲ τὰ ἄναιμα ἐλάττω τὰ μεγέθη ἐστὶ τῶν ἐναίμων ζῴων· πλὴν ὀλίγα ἐν τῇ θαλάττῃ μείζονα ἄναιμά ἐστιν, οἷον τῶν μαλακίων ἔνια. μέγιστα δὲ γίγνεται ταῦτα τὰ γένη αὐτῶν ἐν τοῖς ἀλεεινοτάτοις,
25 καὶ ἐν τῇ θαλάττῃ μᾶλλον ἢ ἐν τῇ γῇ καὶ ἐν τοῖς γλυκέσιν ὕδασιν.

Κινεῖται δὲ τὰ κινούμενα πάντα τέτταρσι σημείοις ἢ πλείοσι, τὰ μὲν ἔναιμα τέτταρσι μόνον, οἷον ἄν-

---

[a] See note on 487 b 22.
[b] A large bat; not the animal now called flying-fox.
[c] It is now known that there is a close connexion between the efficiency of animals' respiratory mechanism (which is dependent on the blood) and the limits of their possible size. Cf. J. B. S. Haldane, *Possible Worlds* (*On being the right*

ample, the eagle and the hawk, some have membranous wings, for example, the bee and the cockchafer, some have dermatous [a] wings, for example, the flying-fox [b] and the bat. If a flying animal is blooded it has either feathered or dermatous wings; the bloodless ones have membranous wings, for example, the insects. The animals that have feathered or dermatous wings are all two-footed or footless : I add footless, because there are alleged to be certain flying serpents in Ethiopia.

The name given to the group of animals which have feathered wings is Birds ; there is no single name for either of the remaining two winged groups.

Some of the animals which though fliers are bloodless have sheathed wings (*i.e.* they have their wings in a sheath or shard), for example, the cockchafer and the dung-beetle ; others have no sheath, and of these some have two, some four, wings : four-winged are those which are of some size or have a sting at the rear ; two-winged are those which either are smallish or have their stings in front. No animal whose wings are sheathed has a sting. The two-winged ones have their sting in front, for example, the fly, the horse-fly, the gadfly, and the gnat.

All the bloodless animals are smaller in size than the blooded,[c] except that a few bloodless animals found in the sea are larger, for example, certain of the Cephalopods. The biggest of the bloodless kinds are those which are found in the sunniest climates, and in the sea rather than on land or in fresh water.

Now all animals which can move about do so with four motion-points or more, the blooded ones with

size), pp. 21 f. Aristotle did not apply the name " blood " to the *blue* blood of the Cephalopods.

ARISTOTLE

θρωπος μὲν χερσὶ δυσὶ καὶ ποσὶ δυσίν, ὄρνις δὲ πτέρυξι δυσὶ καὶ ποσὶ δυσί, τὰ δὲ τετράποδα καὶ ἰχθύες τὰ μὲν τέτταρσι ποσίν, οἱ δὲ τέτταρσι πτερυγίοις. ὅσα δὲ δύο ἔχει πτερύγια ἢ ὅλως μή, οἷον ὄφις, τέτταρσι σημείοις οὐδὲν ἧττον· αἱ γὰρ καμπαὶ τέτταρες, ἢ δύο σὺν τοῖς πτερυγίοις. ὅσα δ' ἄναιμα ὄντα πλείους πόδας ἔχει, εἴτε πτηνὰ εἴτε πεζά, σημείοις κινεῖται πλείοσιν, οἷον τὸ καλούμενον ζῷον ἐφήμερον τέτταρσι καὶ ποσὶ καὶ πτεροῖς· τούτῳ γὰρ οὐ μόνον κατὰ τὸν βίον συμβαίνει τὸ ἴδιον, ὅθεν καὶ τὴν ἐπωνυμίαν ἔχει, ἀλλ' ὅτι καὶ πτηνόν ἐστι τετράπουν ὄν. πάντα δὲ κινεῖται ὁμοίως, τὰ τετράποδα καὶ πολύποδα· κατὰ διάμετρον γὰρ κινεῖται. τὰ μὲν οὖν ἄλλα ζῷα δύο τοὺς ἡγεμόνας ἔχει πόδας, ὁ δὲ καρκίνος μόνος τῶν ζῴων τέτταρας.

VI. Γένη δὲ μέγιστα τῶν ζῴων, εἰς ἃ διήρηται τἆλλα ζῷα, τάδ' ἐστίν, ἓν μὲν ὀρνίθων, ἓν δ' ἰχθύων, ἄλλο δὲ κήτους. ταῦτα μὲν οὖν πάντα ἔναιμά ἐστιν. ἄλλο δὲ γένος ἐστὶ τὸ τῶν ὀστρακοδέρμων, ὃ καλεῖται ὄστρεον· ἄλλο τὸ τῶν μαλακοστράκων, ἀνώνυμον ἑνὶ ὀνόματι, οἷον κάραβοι καὶ γένη τινὰ καρκίνων καὶ ἀστακῶν· ἄλλο τὸ τῶν μαλακίων, οἷον τευθίδες τε καὶ τεῦθοι καὶ σηπίαι· ἕτερον τὸ τῶν ἐντόμων.

---

[a] *Cf.* I.A. 712 a 20 ff. and H.A. 498 b 5 ff.

[b] In this passage I have translated Aristotle's terms for the 4th, 5th and 6th groups literally, as is required by the passage itself; elsewhere I have employed the conventional translations Testacea for ὀστρακοδέρμα, Crustacea for μαλακόστρακα (soft-shelled), and Cephalopods for μαλάκια (softies). The English word "shellfish," here used to translate ὄστρεον, may be misleading, but must here be understood to exclude soft-shelled fish such as crabs, which are included in Aristotle's next (5th) group. Also, although most Testacea are water-animals, a few (*e.g.*, the snails, as Aristotle points out at

four only, for example, man, who uses two hands and two feet, and birds, which use two wings and two feet; whereas quadrupeds use four feet and fishes four fins. Animals which have two fins, or none (*e.g.*, serpents), use four motion-points nevertheless: we find there are four bends in their bodies, or two bends in addition to their fins. Those bloodless animals which have a large number of feet, whether they are winged or footed, use numerous motion-points, for example, the day-fly as it is called uses four feet and four wings: this animal is exceptional, not only in the shortness of its existence (whence its name) but also in its being winged as well as having four feet. All animals move in the same manner, whether they are four-footed or many-footed, in other words, they all move diagonally.[a] And whereas all other animals have two feet leading, the crab alone has four.

The main groups of animals, into which the various animals are divided, are these: (1) that of Birds; (2) of Fishes; (3) that of the Cetacea. All these are blooded. A further group (4) is that of the potsherd-skinned animals,[b] popularly known as shellfish; another (5) is that of the soft-shelled animals,[c] which has no single name, *e.g.*, crayfish, and various kinds of crabs and lobsters; another (6) is that of the "softies," *e.g.*, *teuthides* and *teuthoi*[d] and the cuttle-fish; and yet another (7) is that of the "insects."

VI The main groups of animals:

G.A. 761 a 21 ff.) live on land. "Shellfish," however, is a popular term, thus matching ὄστρεον, and is less misleading than "oysters" would be here. "Bivalves" will not do here, since Aristotle includes under Testacea univalves and even sea-urchins.

[c] τὰ μαλακόστρακα is not a "name," but a descriptive epithet. The same is true of the others, including τὰ ἔντομα, *i.e.*, "the insected animals." *Cf.* 487 a 33, and Notes, § 11.

[d] Two kinds of calamary.

ARISTOTLE

ταῦτα δὲ πάντα μέν ἐστιν ἄναιμα, ὅσα δὲ πόδας
ἔχει, πολύποδα· τῶν δ' ἐντόμων ἔνια καὶ πτηνά
ἐστιν. τῶν δὲ λοιπῶν ζῴων οὐκ ἔστι[1] τὰ γένη
μεγάλα· οὐ γὰρ περιέχει πολλὰ εἴδη ἓν εἶδος, ἀλλὰ
τὸ μέν ἐστιν ἁπλοῦν αὐτὸ οὐκ ἔχον διαφορὰν τὸ
εἶδος, οἷον ἄνθρωπος, τὰ δ' ἔχει μέν, ἀλλ' ἀνώνυμα
τὰ εἴδη. ἔστι γὰρ τὰ τετράποδα καὶ μὴ πτερωτὰ
ἔναιμα μὲν πάντα, ἀλλὰ τὰ μὲν ζῳοτόκα τὰ δ'
ᾠοτόκα αὐτῶν. ὅσα μὲν οὖν ζῳοτόκα, πάντα[2]
τρίχας ἔχει, ὅσα δ' ᾠοτόκα, φολίδας[3]· ἔστι δ' ἡ
φολὶς ὅμοιον χώρᾳ λεπίδος.

Ἄπουν δὲ φύσει ἐστὶν ἔναιμον πεζὸν τὸ τῶν ὄ-
φεων γένος· ἔστι δὲ τοῦτο φολιδωτόν. ἀλλ' οἱ μὲν
ἄλλοι ᾠοτοκοῦσιν ὄφεις, ἡ δ' ἔχιδνα μόνον ζῳοτοκεῖ. τὰ μὲν γὰρ ζῳοτοκοῦντα οὐ πάντα τρίχας
ἔχει· καὶ γὰρ τῶν ἰχθύων τινὲς ζῳοτοκοῦσιν· ὅσα
μέντοι ἔχει τρίχας, πάντα ζῳοτοκεῖ. τριχῶν γάρ
τι εἶδος θετέον καὶ τὰς ἀκανθώδεις τρίχας, οἵας οἱ
χερσαῖοι ἔχουσιν ἐχῖνοι καὶ οἱ ὕστριχες· τριχὸς γὰρ
χρείαν παρέχουσιν, ἀλλ' οὐ ποδῶν, ὥσπερ αἱ[4] τῶν
θαλαττίων.

Τοῦ δὲ γένους τοῦ τῶν τετραπόδων ζῴων καὶ
ζῳοτόκων εἴδη μέν ἐστι πολλά, ἀνώνυμα δέ· ἀλλὰ
καθ' ἕκαστον αὐτῶν ὡς εἰπεῖν, ὥσπερ ἄνθρωπος,
εἴρηται[5] λέων, ἔλαφος, ἵππος, κύων καὶ τἆλλα τοῦ-

---

[1] οὐκ ἔστι A : οὐκέτι C, vulg. : οὐκέτι ἐστὶ PD.
[2] οὐ πάντα PDA², vulg. : οὐ om. A¹CΣ.
[3] φολίδας ἔχει PD, vulg. : φολίδα AC.
[4] αἱ PD : οἱ AC, vulg.
[5] sic Dt., post εἴρηται interpungit vulg.

---

[a] It is not clear why Aristotle does not reckon the Vivi-

HISTORIA ANIMALIUM, I. vi

All these last four are bloodless; and if they have feet, the feet are numerous. Some of the insects have wings as well. Among the remaining animals we find no large groups; we do not find one kind comprising many other kinds: either a kind will be simple, showing no differentiation, for example, man; or if it does show differentiation, the kinds which it comprises are unnamed. Thus, all the wingless quadrupeds are blooded, though some are viviparous and others oviparous.[a] The viviparous ones are all coated with hair; the oviparous ones have horny scales; these scales correspond in position with the scales of fishes. *but precise and exhaustive classification is impossible.*

A class of animals which are by nature footless, though blooded and land-dwelling, is that of the serpents; and these have horny scales. All serpents are oviparous except the viper, which is viviparous; for not all viviparous animals are covered with hair, and indeed there are certain fishes too which are viviparous. Nevertheless, all animals which have hair are viviparous (and we must include here as a kind of hair the spiny hairs of hedgehogs and porcupines: these spines serve the purpose of hair, and not of feet as do the spines of the sea-urchin).

There are many kinds included in the group of viviparous quadrupeds; but they are unnamed. Each constituent, we may say, has been named individually, as man has, *e.g.*, lion, deer, horse, dog, and the rest of

parous Quadrupeds among the "main groups" here (as he does at 505 b 28, together with the Oviparous Quadrupeds), since he says below (490 b 32) that they are a group which includes many kinds, whereas both here and in the later passage he reckons the Cetacea as a "main group," although the kinds he cites are not numerically impressive (the same is true, to some extent, of his Cephalopods and Crustacea). See further, Notes, § 11.

ARISTOTLE

τον τὸν τρόπον (ἐπεὶ ἔστιν ἔν τι γένος καὶ[1] ἐπὶ τοῖς λοφούροις καλούμενον,[2] οἷον ἵππῳ καὶ ὄνῳ καὶ ὀρεῖ καὶ γίννῳ [καὶ ἴννῳ][3] καὶ ταῖς ἐν Συρίᾳ καλουμέναις ἡμιόνοις, αἳ καλοῦνται ἡμίονοι δι' ὁμοιότητα, οὐκ οὖσαι ἁπλῶς τὸ αὐτὸ εἶδος· καὶ γὰρ ὀχεύονται καὶ γεννῶνται ἐξ ἀλλήλων). διὸ καὶ χωρὶς λαμβάνοντας ἀνάγκη θεωρεῖν ἑκάστου τὴν φύσιν αὐτῶν.[4]

Ταῦτα μὲν οὖν τοῦτον τὸν τρόπον εἴρηται νῦν ὡς τύπῳ,[5] γεύματος χάριν περὶ ὅσων καὶ ὅσα θεωρητέον (δι' ἀκριβείας δ' ὕστερον ἐροῦμεν) ἵνα πρῶτον τὰς ὑπαρχούσας διαφορὰς καὶ τὰ συμβεβηκότα πᾶσι λάβωμεν. μετὰ δὲ τοῦτο τὰς αἰτίας τούτων πειρατέον εὑρεῖν. οὕτω γὰρ κατὰ φύσιν ἐστὶ ποιεῖσθαι τὴν μέθοδον, ὑπαρχούσης τῆς ἱστορίας τῆς περὶ ἕκαστον· περὶ ὧν τε γὰρ καὶ ἐξ ὧν εἶναι δεῖ τὴν ἀπόδειξιν, ἐκ τούτων γίγνεται φανερόν.

Ληπτέον δὴ[6] πρῶτον τὰ μέρη τῶν ζῴων ἐξ ὧν

---

[1] ἐπεί ἐστιν ἔν τι γένος καὶ vulg. : μόνον loco καὶ coni. Dt. : ἔπεστι δ' ἔν τι ὄνομα A.-W., "bene, sed iusto audacius" (Dt.) ; credo equidem γένος retinendum esse (cf. 505 b 31), et mox καλούμενον scribendum, nusquam enim alibi locutio τὰ λόφουρα καλούμενα invenitur.

[2] καλούμενον scripsi : καλουμένοις vulg., edd. ; vid. et adnot. supra scriptam.

[3] καὶ ἴννῳ secl. A.-W.

[4] καὶ γίννῳ . . . . αὐτῶν non vertit Σ, sed eorum loco *volatilis plumosa et pluma assimilatur capillis.* fortasse plurima hic secludenda.

[5] ἐν τύπῳ A, vulg.

[6] δὴ PAC, δὲ vulg.

---

[a] Lit., tufty- or bushy-tailed animals. *Cf.* 495 a 4.

[b] At 577 b 21 ff. a *ginnos* is said to be the offspring of a mare by a mule. See add. note, p. 237.

[c] These remarks about the half-asses are almost certainly an interpolation.

HISTORIA ANIMALIUM, I. vi

them in the same way (though in fact there is a single named group in the case of the *lophoura*,[a] such as horse, ass, mule, *ginnos*,[b] and the animals known as half-asses in Syria—so called because they resemble mules, though actually they are not the same type, as is shown by their mating and breeding with one another).[c] This, too, is why we are obliged to take them separately and examine the nature of each individual type.[d]

What has just been said has been stated thus by way of outline,[e] so as to give a foretaste of the matters and subjects which we have to examine; detailed statements will follow later; our object being to determine first of all the differences that exist and the actual facts in the case of all of them. Having done this, we must attempt to discover the causes. And, after all, this is the natural method of procedure—to do this only after we have before us the ascertained facts about each item, for this will give us a clear indication of the subjects with which our exposition is to be concerned and the principles upon which it must be based.

METHOD AND PURPOSE OF THE UNDERTAKING.

So first of all we must consider the parts of animals

A. Differences in the PARTS of animals:

[d] In this paragraph (*cf.* 490 b 16 ff. above) three levels are envisaged: (1) the group (here, Viviparous Quadrupeds); (2) the sub-groups; (3) the individual types. The sub-groups (2) are all unnamed, except the *lophoura*; otherwise we find no names until the third level, *i.e.*, the names of the simple types of animals, such as "lion": there is in Greek no name corresponding to *Felidae*. To this extent the situation is parallel to that of "Man," which is the name of a simple or individual type (ἁπλοῦν εἶδος, 490 b 18; *cf.* 505 b 31), though in this case there is no main group or sub-group under which Man falls, and hence (though for a different reason) there is no group name. *Cf.* 505 b 30 ff. See further, Notes, §§ 9 ff.

[e] As promised at 487 a 13.

35

συνέστηκεν. κατὰ γὰρ ταῦτα μάλιστα καὶ πρῶτα
διαφέρει καὶ τὰ ὅλα, ἢ τῷ τὰ μὲν ἔχειν τὰ δὲ μὴ
ἔχειν ἢ τῇ θέσει καὶ τῇ τάξει, ἢ καὶ κατὰ τὰς
εἰρημένας πρότερον διαφοράς, εἴδει καὶ ὑπεροχῇ
καὶ ἀναλογίᾳ καὶ τῶν παθημάτων ἐναντιότητι.

20 Πρῶτον δὲ τὰ τοῦ ἀνθρώπου μέρη ληπτέον· ὥσπερ
γὰρ τὰ νομίσματα πρὸς τὸ αὑτοῖς ἕκαστοι γνωριμώτατον δοκιμάζουσιν, οὕτω δὴ καὶ ἐν τοῖς ἄλλοις·
ὁ δ' ἄνθρωπος τῶν ζῴων γνωριμώτατον ἡμῖν ἐξ
ἀνάγκης ἐστίν.

Τῇ μὲν οὖν αἰσθήσει οὐκ ἄδηλα τὰ μόρια· ὅμως
δ' ἕνεκεν τοῦ μὴ παραλιπεῖν τε τὸ ἐφεξῆς καὶ τοῦ
25 λόγον ἔχειν μετὰ τῆς αἰσθήσεως, λεκτέον τὰ μέρη
πρῶτον μὲν τὰ ὀργανικά, εἶτα τὰ ὁμοιομερῆ.

VII Μέγιστα μὲν οὖν ἐστι τάδε τῶν μερῶν εἰς ἃ
διαιρεῖται τὸ σῶμα τὸ σύνολον, κεφαλή, αὐχήν,
θώραξ, βραχίονες δύο, σκέλη δύο, [τὸ ἀπ' αὐχένος
30 μέχρι αἰδοίων κύτος, ὃ καλεῖται θώραξ].[1]

Κεφαλῆς μὲν οὖν μέρη τὸ μὲν τριχωτὸν κρανίον
καλεῖται. τούτου δὲ μέρη τὸ μὲν πρόσθιον βρέγμα,
ὑστερογενές (τελευταῖον γὰρ τῶν ἐν τῷ σώματι πήγνυται ὀστῶν), τὸ δ' ὀπίσθιον ἰνίον, μέσον δ' ἰνίου
καὶ βρέγματος κορυφή. ὑπὸ μὲν οὖν τὸ βρέγμα ὁ
ἐγκέφαλός ἐστιν, τὸ δ' ἰνίον κενόν. ἔστι δὲ τὸ
κρανίον ἅπαν ἀραιὸν ὀστοῦν, στρογγύλον, ἀσάρκῳ
δέρματι περιεχόμενον. ἔχει δὲ ῥαφὰς τῶν μὲν γυ-

---

[1] secl. Pi.: habet Σ.

[a] See note on 486 b 5; and cf. 486 b 15 ff.
[b] Probably an interpolated explanation.

—the parts of which they are composed; for it is in respect of its parts first and foremost that any animal as a whole differs from another: it may be that one animal has parts which another lacks, or that the parts vary in their position or their arrangement, or the differences may be those which we have already mentioned, *viz.*, those of species, excess or defect, analogy, and oppositions of secondary characteristics.[a]

And first, we should consider the parts of the human body. Every nation reckons currency with reference to the standard most familiar to itself; and we must do the same in other fields: man is, of necessity, the animal most familiar to us.

1. INSTRUMENTAL PARTS OF BLOODED ANIMALS: (i) Vivipara (chiefly man):

So far as sense-observation takes us, the parts are clear enough. Still, in order to make sure that we do not neglect the proper sequence, *i.e.*, so as to include the logical aspect as well as the merely observable, we must begin with the instrumental parts, and then go on to the uniform parts.

Chief among the parts into which the body as a whole is divided are these: the head, the neck, the trunk, the two arms, the two legs, [the bulk extending from the neck as far as the privy parts, which is called the trunk].[b]

VII (a) External parts:

Parts of the head. The part covered with hair is called the skull. The front portion of this is the *bregma* (sinciput), which reaches its formation late in the process—it is the last of all the bones in the body to solidify; the back portion is the occiput, and the part between the *bregma* and the occiput is the crown. Underneath the *bregma* is the brain; the occiput is empty. The whole of the skull is of thin bone, rounded in shape, and surrounded by skin but no flesh. In women, the skull has one circular suture,

The head.

37

491 b

ναικῶν μίαν κύκλῳ, τῶν δ' ἀνδρῶν τρεῖς εἰς ἓν συναπτούσας ὡς ἐπὶ τὸ πολύ· ἤδη δ' ὠμμένη ἐστὶ κεφαλὴ ἀνδρὸς οὐδεμίαν ἔχουσα ῥαφήν. τοῦ δὲ κρανίου κορυφὴ καλεῖται τὸ μέσον λίσσωμα τῶν τριχῶν. τοῦτο δ' ἐνίοις διπλοῦν ἐστιν· γίγνονται γάρ τινες δικόρυφοι, οὐ τῷ ὀστῷ ἀλλὰ τῇ τῶν τριχῶν λισσώσει.

VIII Τὸ δ' ὑπὸ τὸ κρανίον ὀνομάζεται πρόσωπον ἐπὶ μόνου τῶν ἄλλων ζῴων ἀνθρώπου· ἰχθύος γὰρ καὶ βοὸς οὐ λέγεται πρόσωπον· προσώπου δὲ τὸ μὲν ὑπὸ τὸ βρέγμα μεταξὺ τῶν ὀμμάτων μέτωπον. τοῦτο δ' οἷς μὲν μέγα, βραδύτεροι, οἷς δὲ μικρόν, εὐκίνητοι· καὶ οἷς μὲν πλατύ, ἐκστατικοί, οἷς δὲ περιφερές, θυμικοί.[1]

IX Ὑπὸ δὲ τῷ μετώπῳ ὀφρύες διφυεῖς· ὧν αἱ μὲν εὐθεῖαι μαλακοῦ ἤθους σημεῖον, αἱ δὲ πρὸς τὴν ῥῖνα τὴν καμπυλότητ' ἔχουσαι στρυφνοῦ, αἱ δὲ πρὸς τοὺς κροτάφους μωκοῦ καὶ εἴρωνος.[2] ὑφ' αἷς ὀφθαλμοί. οὗτοι κατὰ φύσιν δύο. τούτων μέρη ἑκατέρου βλέφαρον τὸ ἄνω καὶ κάτω. τούτου τρίχες αἱ ἔσχαται βλεφαρίδες. τὸ δ' ἐντὸς τοῦ ὀφθαλμοῦ, τὸ μὲν ὑγρόν, ᾧ βλέπει, κόρη, τὸ δὲ περὶ τοῦτο μέλαν, τὸ δ' ἐκτὸς τούτου λευκόν. κοινὸν δὲ τῆς βλεφαρίδος μέρος τῆς ἄνω καὶ κάτω κανθοὶ δύο, ὁ μὲν πρὸς τῇ ῥινί, ὁ δὲ πρὸς τοῖς κροτάφοις· οἳ ἂν μὲν ὦσι μακροί, κακοηθείας σημεῖον, ἐὰν δ' οἷον οἱ

[1] θυμικοί PD, edd.: εὔκοι Λ: εὔϊκοι C: εὐήκοοι A.-W., coll. Physiog. 811 b 31: significat iracundiam Σ.
[2] huc αἱ δὲ κατεσπασμέναι φθόνου, ex v. 34 mutata, transferunt vulg., edd.

---

[a] Nevertheless, there is mention of the face of a baboon at

in men usually three sutures, which meet at one point, though a male skull has been observed having no suture at all. The middle line of the skull, where the parting of the hair is, is called the crown. In some people this parting is double : some men are double-crowned, not of course in respect of the bone itself, but in respect of the parting of the hair.

The part below the skull is named the face, but only in man, and in no other animal; we do not speak of the face of a fish or of an ox.[a] That part of the face which is below the *bregma* and between the eyes is the forehead. Persons who have a large forehead are sluggish, those who have a small one are fickle; those who have a broad one are excitable, those who have a bulging one, quick-tempered.

VIII The face.

Below the forehead are the eyebrows. Straight ones are a sign of a soft disposition, those which bend in towards the nose, a sign of harshness; those which bend out towards the temples, of a mocking and dissimulating disposition.[b] Below the eyebrows are the eyes, of which the natural number is two. Parts of them are as follows : each eye has an upper and lower eyelid; the hairs on their edges are eyelashes. In the inner part of the eye we have (a) a fluid part, the instrument of sight, called the pupil; surrounding this, (b), the "black," and outside this, (c), the "white." Common to the upper and lower eyelid we have a part known as the nick—there are two of these to each eye, one towards the nose, one towards the temples. If these are long, they are a sign of a malicious disposition; if they have the part towards

IX The eyebrows and the eyes.

---

[a] 502 a 20, and of the face of a chamaeleon (compared to that of a pig-faced baboon) at 503 a 18.

[b] Here some editors insert a sentence from the end of the next paragraph.

ARISTOTLE

ἰκτῖνες[1] κρεῶδες ἔχωσι τὸ πρὸς τῷ μυκτῆρι, πονηρίας.

Τὰ μὲν οὖν ἄλλα γένη πάντα τῶν ζῴων πλὴν τῶν ὀστρακοδέρμων καὶ εἴ τι ἄλλο ἀτελές, ἔχει ὀφθαλμούς· τὰ δὲ ζῳοτόκα πάντα πλὴν ἀσπάλακος. τοῦτον δὲ τρόπον μέν τιν' ἔχειν ἄν θείη τις, ὅλως δ' οὐκ ἔχειν. ὅλως μὲν γὰρ οὔθ' ὁρᾷ οὔτ' ἔχει εἰς τὸ φανερὸν δήλους ὀφθαλμούς· ἀφαιρεθέντος δὲ τοῦ δέρματος ἔχει τήν τε χώραν τῶν ὀμμάτων καὶ τῶν ὀφθαλμῶν τὰ μέλανα κατὰ τὸν τόπον καὶ τὴν χώραν τὴν φύσει τοῖς ὀφθαλμοῖς ὑπάρχουσαν ἐν τῷ ἐκτός, ὡς ἐν τῇ γενέσει πηρουμένων καὶ ἐπιφυομένου τοῦ δέρματος. [αἱ δ' ὀφρύες αἱ κατεσπασμέναι φθόνου.][2]

X Ὀφθαλμοῦ δὲ τὸ μὲν λευκὸν ὅμοιον ὡς ἐπὶ τὸ πολὺ πᾶσιν, τὸ δὲ καλούμενον μέλαν διαφέρει· τοῖς μὲν γάρ ἐστι μέλαν, τοῖς δὲ σφόδρα γλαυκόν, τοῖς δὲ χαροπόν, ἐνίοις δ' αἰγωπόν, ὃ ἤθους βελτίστου σημεῖον καὶ πρὸς ὀξύτητα ὄψεως κράτιστον. μόνον δ' ἢ μάλιστα τῶν ζῴων ἄνθρωπος πολύχρους τὰ ὄμματά ἐστιν· τῶν δ' ἄλλων ἓν εἶδος· ἵπποι δὲ γίγνονται γλαυκοὶ ἔνιοι. τῶν δ' ὀφθαλμῶν οἱ μὲν μεγάλοι, οἱ δὲ μικροί, οἱ δὲ μέσοι· οἱ μέσοι βέλτιστοι. καὶ ἢ ἐκτὸς σφόδρα ἢ ἐντὸς ἢ μέσως· τούτων οἱ ἐντὸς μάλιστα ὀξυωπέστατοι ἐπὶ παντὸς ζῴου, τὸ δὲ μέσον ἤθους βελτίστου σημεῖον. καὶ

---

[1] ita scripsi, Buss., Pi., Canisianum secutus : *sicut accidit oculis milvi* Σ : οἷον οἱ κτένες vulg. : οἱονεὶ κτένας κρεώδεις ἔχωσιν οἱ πρὸς velint A.-W.

[2] mutant et post εἴρωνος 491 b 17 transferunt edd. ; delent A.-W., Dt.

---

[a] But at *P.A.* 640 a 20 Aristotle criticizes Empedocles for accounting for the vertebrae of the backbone by alleging that they are caused by an accident in the process of formation :

HISTORIA ANIMALIUM, I. ix-x

the nose fleshy, as the kites do, it is a sign of dishonesty.

All other kinds of animals have eyes, apart from the Testacea and any other imperfect animals ; and all Vivipara have eyes except the mole, though one might consider that in a way it has eyes, yet not in the full sense. The fact is that it cannot see, and it has no eyes which can be detected externally ; but if the skin is removed we find it has the place for the eyes and the " black " parts of the eye where they should be, and the position which is naturally provided externally for eyes, which suggests that the eyes get stunted in the process of formation[a] and that the skin grows over. [Eyebrows which are drawn down ⟨are a sign⟩ of enviousness.][b]

The white of the eye is very much the same in all X animals, but the black as it is called shows differences. In some animals it really is black, in others, quite blue, in some, greyish-blue, in some, yellow[c] : the last is a sign of the finest disposition, and is best of all for sharpness of sight. Man is the only animal in which a variety of colours is found, or, at any rate, such variety is found most extensively in man : in other kinds of animals all individuals have one and the same colour, though a few horses have blue eyes. Eyes may be large, or small, or medium-sized (these are the best). They may protrude, or be deep-set, or intermediate : those which are deep-set are in all animals the keenest : the intermediate are a sign of the finest disposition. Again, eyes may tend to

[a] " the fetus gets twisted and so the backbone is broken into pieces."

[b] This sentence is transferred by some editors to the end of the first sentence of the previous paragraph.

[c] Or perhaps " greenish " (lit., goat-eyed).

ἢ σκαρδαμυκτικοὶ ἢ ἀτενεῖς ἢ μέσοι· βελτίστου δ' ἤθους οἱ μέσοι, ἐκείνων δ' ὁ μὲν ἀναιδὴς ὁ δ' ἀβέβαιος.

XI Ἔτι δὲ κεφαλῆς μόριον, δι' οὗ ἀκούει, ἄπνουν, τὸ οὖς· Ἀλκμαίων γὰρ οὐκ ἀληθῆ λέγει, φάμενος ἀναπνεῖν τὰς αἶγας κατὰ τὰ ὦτα. ὠτὸς δὲ μέρος τὸ μὲν ἀνώνυμον, τὸ δὲ λοβός. ὅλον δ' ἐκ χόνδρου καὶ σαρκὸς σύγκειται. εἴσω δὲ τὴν μὲν φύσιν ἔχει οἷον οἱ στρόμβοι, τὸ δ' ἔσχατον ὀστοῦν ὅμοιον τῷ ὠτί, εἰς ὃ ὥσπερ ἀγγεῖον ἔσχατον ἀφικνεῖται ὁ ψόφος. τοῦτο δ' εἰς μὲν τὸν ἐγκέφαλον οὐκ ἔχει πόρον, εἰς δὲ τὸν τοῦ στόματος οὐρανόν· καὶ ἐκ τοῦ ἐγκεφάλου φλὲψ τείνει εἰς αὐτό. (περαίνουσι δὲ καὶ οἱ ὀφθαλμοὶ εἰς τὸν ἐγκέφαλον, καὶ κεῖται ἐπὶ φλεβίου ἑκάτερος.)

Ἀκίνητον δὲ τὸ οὖς ἄνθρωπος ἔχει μόνος τῶν ἐχόντων τοῦτο τὸ μόριον. τῶν γὰρ ἐχόντων ἀκοὴν τὰ μὲν ἔχει ὦτα, τὰ δ' οὐκ ἔχει, ἀλλὰ τὸν πόρον φανερόν, οἷον ὅσα πτερωτὰ ἢ φολιδωτά. ὅσα δὲ ζῳοτοκεῖ, ἔξω φώκης τε¹ καὶ δελφῖνος καὶ τῶν ἄλλων ὅσα τοιαῦτα² κητώδη, πάντα ἔχει ὦτα· ζῳοτοκεῖ γὰρ καὶ τὰ σελάχη· [ἀλλὰ μόνον ἄνθρωπος οὐ κινεῖ.]³ (ἡ μὲν οὖν φώκη πόρους ἔχει φανεροὺς ᾗ ἀκούει· ὁ δὲ δελφὶς ἀκούει μέν, οὐκ ἔχει δ' ὦτα.) [τὰ δ' ἄλλα κινεῖ πάντα.]³ κεῖται δὲ τὰ ὦτα ἐπὶ τῆς αὐτῆς περιφερείας τοῖς ὀφθαλμοῖς, καὶ οὐχ

¹ τε om. PD, vulg.
² τοιαῦτα scripsi, cf. *P.A.* 697 a 17 : οὕτω vulg. : οὕτω τε AC.
³ haec tantum seclusi ; sed locus ita corruptus ut vera vix distingui possint.

---

[a] This is the first recorded reference to the Eustachian tubes, rediscovered by Bartolomeo Eustachi (1520–1574), whose work was published in 1714.

blink, or to remain unblinking, or exhibit no extreme in either direction : the last-named show the finest disposition, the first indicates instability, the second impudence.

Furthermore, there is a part of the head through which the animal hears : it is incapable of breathing, and it is called the ear. Incapable of breathing, yes ; Alcmeon is incorrect in saying that goats breathe through their ears. One part of the ear has no special name, the other is called the lobe ; the whole consists of gristle and flesh. The natural structure of the interior of the ear is like the spiral-shells' : the innermost part is a bone similar to the ear, and into this ultimately the sound penetrates, as into a vessel. There is no passage from this to the brain, but there is a passage to the roof of the mouth, and a blood-vessel passes to it from the brain.[a] (The eyes too are connected with the brain, and each eye is situated upon a small blood-vessel.)

XI
The ears.

Of all animals which possess ears, man is the only one which is unable to move them. While some animals which have the power of hearing possess ears, others do not, though they have the passage, which can be plainly seen, *e.g.*, the animals which are feathered or covered with horny scales. All Vivipara have ears, except the seal, the dolphin, and other such Cetacea. The Selachia, also, of course are viviparous. [But man alone cannot move his ears.] The seal, sure enough, has passages, clearly visible, through which it hears ; whereas the dolphin has the power of hearing, but no ears. [All other animals can move their ears.] The ears lie on a level with the eyes, and not higher up, as is the case with some

ARISTOTLE

ὥσπερ ἐνίοις τῶν τετραπόδων ἄνωθεν. ὤτων δὲ τὰ μὲν ψιλά, τὰ δὲ δασέα, τὰ δὲ μέσα· βέλτιστα δὲ τὰ μέσα πρὸς ἀκοήν, ἦθος δ' οὐδὲν σημαίνει. καὶ ἢ μεγάλα ἢ μικρὰ ἢ μέσα. ⟨καὶ⟩ ἢ[1] ἐπανεστηκότα σφόδρα ἢ οὐδὲν ἢ μέσον· τὰ δὲ μέσα βελτίστου ἤθους σημεῖον, τὰ δὲ μεγάλα καὶ ἐπανεστηκότα μωρολογίας καὶ ἀδολεσχίας. τὸ δὲ μεταξὺ ὀφθαλμοῦ καὶ ὠτὸς καὶ κορυφῆς καλεῖται κρόταφος.

Ἔτι προσώπου μέρος τὸ μὲν ὂν τῷ πνεύματι πόρος ῥίς· καὶ γὰρ ἀναπνεῖ καὶ ἐκπνεῖ ταύτῃ, καὶ ὁ πταρμὸς διὰ ταύτης γίγνεται, πνεύματος ἀθρόου ἔξοδος, σημεῖον οἰωνιστικὸν καὶ ἱερὸν μόνον τῶν πνευμάτων. ἅμα δ' ἡ ἀνάπνευσις καὶ ἔκπνευσις γίγνεται εἰς τὸ στῆθος, καὶ ἀδύνατον χωρὶς τοῖς μυκτῆρσιν ἀναπνεῦσαι ἢ ἐκπνεῦσαι, διὰ τὸ ἐκ τοῦ στήθους εἶναι τὴν ἀναπνοὴν καὶ ἐκπνοὴν κατὰ τὸν γαργαρεῶνα, καὶ μὴ ἐκ τῆς κεφαλῆς τινὶ μέρει· ἐνδέχεται δὲ καὶ μὴ χρώμενον ταύτῃ ζῆν.

Ἡ δ' ὄσφρησις γίγνεται διὰ τούτου τοῦ μέρους· αὕτη δ' ἐστὶν ἡ αἴσθησις ὀσμῆς. εὐκίνητος δ' ὁ μυκτήρ, καὶ οὐχ ὥσπερ τὸ οὖς ἀκίνητον κατ' ἰδίαν. μέρος δ' αὐτοῦ τὸ μὲν διάφραγμα χόνδρος, τὸ δ' ὀχέτευμα κενόν· ἔστι γὰρ ὁ μυκτὴρ διχότομος. τοῖς δ' ἐλέφασιν ὁ μυκτὴρ γίγνεται μακρὸς καὶ ἰσχυρός, καὶ χρῆται αὐτῷ ὥσπερ χειρί· προσάγεταί τε γὰρ καὶ λαμβάνει τούτῳ καὶ εἰς τὸ στόμα προσφέρεται τὴν τροφήν, καὶ τὴν ὑγρὰν καὶ τὴν ξηράν, μόνον τῶν ζῴων.

[1] καὶ ἢ Pi.: καὶ A.-W.: ἢ vulg.

of the quadrupeds. Ears may be smooth, or hairy, or intermediate: the last are the best for hearing, but are no sign of disposition. They may be large, or small, or of medium size. They may stand well out, or not stand out at all, or intermediately. The last are a sign of the finest disposition; large, projecting ears are a sign of senseless talk and chatter. The part between the eye, the ear, and the crown is called the temple.

Further, the part of the face which is a passage for the breath is the nose. Through this we breathe in and out; and sneezing too, which is the exit of a collected volume of breath, takes place through the nose. Sneezing is the only sort of breath which has divinatory significance and is supernatural. Both inhalation and exhalation continue as far as the chest, nor is it possible otherwise, with the nostrils alone, to breathe in or out, because these actions take place from the chest along the windpipe, and not from the head by means of some part of that: in fact, it is possible to dispense with breathing through the nose and yet continue to live. *The nose.*

Smelling, too, takes place through the nose— smelling, which is the sensual perception of odour. The nostril can easily be moved, herein differing from the ear, which cannot be moved on its own. Since the nostril is double in structure, there is a part of it known as the septum, or partition, which consists of gristle; the remainder is an open channel. In elephants the nostril is elongated and strong, and the animal uses it as a hand: it pulls objects and seizes them with it and by it conveys food to the mouth, both solid and liquid, and it is the only animal which does this.

ARISTOTLE

492 b

Ἔτι δὲ σιαγόνες δύο· τούτων τὸ πρόσθιον γένειον, τὸ δ' ὀπίσθιον γένυς. κινεῖ δὲ πάντα τὰ ζῷα τὴν κάτωθεν σιαγόνα,[1] πλὴν τοῦ ποταμίου κροκοδείλου· οὗτος δὲ τὴν ἄνω μόνον· μετὰ δὲ τὴν ῥῖνα χείλη δύο, σὰρξ εὐκίνητος. τὸ δ' ἐντὸς σιαγόνων καὶ χειλῶν στόμα.[2] τούτου μέρη τὸ μὲν ὑπερῷα τὸ δὲ φάρυγξ.

Τὸ δ' αἰσθητικὸν χυμοῦ γλῶττα· ἡ δ' αἴσθησις ἐν τῷ ἄκρῳ· ἐὰν δὲ[3] ἐπὶ τὸ πλατὺ ἐπιτεθῇ, ἧττον. αἰσθάνεται δὲ καὶ ὧν ἡ ἄλλη σὰρξ πάντων, οἷον σκληροῦ θερμοῦ καὶ ψυχροῦ, καθ' ὁτιοῦν μέρος [ὥσπερ καὶ χυμοῦ].[4] αὕτη δ' ἢ πλατεῖα ἢ στενὴ ἢ μέση· ἡ μέση δὲ βελτίστη καὶ σαφεστάτη. καὶ ἢ λελυμένη ἢ καταδεδεμένη, ὥσπερ τοῖς ψελλοῖς καὶ τοῖς τραυλοῖς. ἔστι δ' ἡ γλῶττα σὰρξ μανὴ καὶ σομφή. ταύτης τι μέρος ἐπιγλωττίς.

493 a

Καὶ τὸ μὲν διφυὲς τοῦ στόματος παρίσθμιον, τὸ δὲ πολυφυὲς οὖλον· σάρκινα δὲ ταῦτα. ἐντὸς δ' ὀδόντες ὀστέϊνοι. εἴσω[5] δ' ἄλλο μόριον σταφυλοφόρον, κίων ἐπίφλεβος[6]· ὃς ἐὰν ἐξυγρανθεὶς φλεγμήνῃ, σταφυλὴ καλεῖται καὶ πνίγει.[7]

XII   Αὐχὴν δὲ τὸ μεταξὺ προσώπου καὶ θώρακος. [καὶ τούτου τὸ μὲν πρόσθιον λάρυγξ, τὸ δ' ὀπίσθιον στόμαχος.][8] τούτου δὲ τὸ μὲν χονδρῶδες καὶ πρό-

---

[1] σιαγόνα PD : γένυν AC, vulg.
[2] ita edd. : στόμα post ἐντὸς codd. : om. D.
[3] δέ ⟨τι⟩ Dt. : δέ ⟨τι⟩ τῷ πλατεῖ coniecerat Pi.
[4] secl. A.-W.
[5] εἴσω A : ἔσω vulg.
[6] ἐπίφλεβος AC, vulg. : ἐπὶ φλεβός PD : *est posita supra venam* Σ (num igitur κείμενον loco κίων?)

HISTORIA ANIMALIUM, I. xi–xii

Further, there are two jaws : the front part is the chin, the rear is the cheek. All animals can move the lower jaw except the river crocodile, which moves the upper jaw only. After the nose are placed the two lips, consisting of flesh, which is mobile. The interior of the jaws and the lips is the mouth. The mouth has these parts : the palate and the pharynx. *The jaws, lips, and mouth.*

The part which can perceive taste is the tongue. Its power of sensation is located in the tip, and if it is placed so that the flat wide part of it makes contact, the sensation is less vivid. The tongue can also perceive all the things that other flesh can, *e.g.*, it can perceive hard, warm, cold, in any part of itself, [just as it can perceive taste]. Now the tongue can be broad, or narrow, or intermediate ; and the last is the best and gives the clearest perception. Further, the tongue may be loosely fastened, or tightly, as occurs in those who mumble and lisp. The tongue consists of flesh, which is soft and spongy. The epiglottis is a part of the tongue. *The tongue etc.*

Further, that region of the mouth which is divided into two is the tonsils, and that which is divided into many portions is the gums : these all are fleshy. Inside are the teeth, which are bony. Right inside is another part, resembling a bunch of grapes, a pillar with prominent veins ; if this gets moistened unduly and inflamed, it is called *staphyle* (uvula) and tends to cause suffocation.

The neck is the part between the face and the trunk. [The front of this is the larynx, and the back the gullet.] The front part of this is of gristle, and *XII The neck.*

---

[7] add. Σ *et sub radice lingue in lateribus colli domestici sunt due amygdale.* καὶ τὸ . . . πνίγει damnant A.-W.

[8] seclusi ; τὸ δ' . . . στόμαχος secl. Dt.

σθιον, δι' οὗ ἡ φωνὴ καὶ ἡ ἀναπνοή, ἀρτηρία· τὸ δὲ σαρκῶδες στόμαχος, ἐντὸς πρὸ τῆς ῥάχεως. τὸ δ' ὀπίσθιον αὐχένος μέρος[1] ἐπωμίς.

Ταῦτα μὲν οὖν τὰ μόρια μέχρι τοῦ θώρακος.

Θώρακος δὲ μέρη τὰ μὲν πρόσθια τὰ δ' ὀπίσθια. πρῶτον μὲν μετὰ τὸν αὐχένα ἐν τοῖς προσθίοις στῆθος διφυὲς μαστοῖς. τούτων ἡ θηλὴ διφυής, δι' ἧς τοῖς θήλεσι τὸ γάλα διηθεῖται· ὁ δὲ μαστὸς μανός. ἐγγίγνεται δὲ καὶ ἐν[2] τοῖς ἄρρεσι γάλα· ἀλλὰ πυκνὴ ἡ σὰρξ τοῖς ἄρρεσι, ταῖς δὲ γυναιξὶ σομφὴ καὶ πόρων μεστή.

XIII Μετὰ δὲ τὸν θώρακα ἐν τοῖς προσθίοις γαστήρ, καὶ ταύτης ῥίζα ὀμφαλός. ὑπόρριζον δὲ τὸ μὲν διφυὲς λαγών, τὸ δὲ μονοφυὲς τὸ μὲν ὑπὸ τὸν ὀμφαλὸν ἦτρον (τούτου δὲ τὸ ἔσχατον ἐπίσιον), τὸ δ' ὑπὲρ τὸν ὀμφαλὸν ὑποχόνδριον, τὸ δὲ ⟨κοῖλον⟩ κοινὸν[3] ὑποχονδρίου καὶ λαγόνος χολάς. τῶν δ' ὄπισθεν διάζωμα μὲν ἡ ὀσφύς, ὅθεν καὶ τοὔνομ' ἔχει (δοκεῖ γὰρ εἶναι ἰσοφυές), τοῦ δὲ διεξοδικοῦ τὸ μὲν οἷον ἐφέδρανον γλουτός, τὸ δ' ἐν ᾧ στρέφεται ὁ μηρός, κοτυληδών.

Τοῦ δὲ θήλεος ἴδιον μέρος ὑστέρα, καὶ τοῦ ἄρρενος αἰδοῖον, ἔξωθεν ἐπὶ τῷ τέλει τοῦ θώρακος, διμερές, τὸ μὲν ἄκρον σαρκῶδες καὶ ἀεὶ[4] ὡς εἰπεῖν ἴσον, ὃ καλεῖται βάλανος, τὸ δὲ περὶ αὐτὴν ἀνώνυμον δέρμα, ὃ ἐὰν διακοπῇ, οὐ συμφύεται, οὐδὲ γνάθος οὐδὲ βλεφαρίς. κοινὸν δὲ τούτου καὶ τῆς βαλάνου ἀκροποσθία. τὸ δὲ λοιπὸν μέρος χονδρῶ-

---

[1] μέρος AC : μόριον PD, vulg.
[2] ἐν AC : om. PD, vulg.
[3] ⟨κοῖλον⟩ κοινὸν A.-W. e Gaza : κοῖλον PD : κοινὸν vulg.
[4] ἀεὶ A.-W. : ἀεὶ λεῖον AC : λεῖον vulg.

through it speech and respiration take place ; it is known as the windpipe. The fleshy part is the gullet (oesophagus) ; it is inside, in front of the backbone. The rear part of the neck is the *epomis*.

This completes our list of parts down to the trunk.

The trunk has a back and a front. First of all, in front, after the neck there is the chest, twofold in structure, with a breast each side. Upon each breast is a nipple, through which in females the milk percolates. The breast is loose in texture. Milk is found to occur in males as well as females ; but in the male the flesh is firm, whereas in woman it is spongy and full of passages. The trunk: front parts.

Next after the trunk, in front, is the belly, and its root is the navel. Underneath this root is a dual part, the flank ; and the single part (*a*) below the navel is the abdomen, the extreme part of which is the pubic region, (*b*) above the navel is the hypochondrium ; the hollowed part common to this and the flank is the gut-cavity. As a brace for the rear parts is the pelvis —indeed this circumstance provides its name *osphys* : as we can see, it is symmetrical (*isophyes*). The part of the fundament which forms as it were a seat is the rump, and the part on which the thigh pivots is the hip-socket. XIII

A part peculiar to the female is the womb, while the penis is peculiar to the male. The penis is situated externally, at the base of the trunk, and is of two parts : the extremity is fleshy, and always, one may say, equal in size ; it is called the glans. The skin round it (which has no special name) if cut does not grow together (nor does the jaw or the eyelid). Common to this part and the glans is the *acroposthia* (**frenum**). The remaining part consists of gristle ; it

493 a

δες, εὐαυξές, καὶ ἐξέρχεται καὶ εἰσέρχεται ἐναντίως
ἢ τοῖς αἰλούροις.[1] τοῦ δ' αἰδοίου ὑποκάτω ὄρχεις
δύο. τὸ δὲ πέριξ δέρμα, ὃ καλεῖται ὄσχεος. οἱ
δ' ὄρχεις οὔτε ταὐτὸ σαρκὶ οὔτε πόρρω σαρκός· ὃν

493 b

τρόπον δ' ἔχουσιν, ὕστερον δι' ἀκριβείας λεχθήσεται καθόλου περὶ πάντων τῶν τοιούτων μορίων.

XIV Τὸ δὲ τῆς γυναικὸς αἰδοῖον ἐξ ἐναντίας τῷ τῶν
ἀρρένων· κοῖλον γὰρ τὸ ὑπὸ τὴν ἥβην καὶ οὐχ
ὥσπερ τὸ τοῦ ἄρρενος ἐξεστηκός. καὶ οὐρήθρα ἔξω
5 τῶν ὑστερῶν, δίοδος τῷ σπέρματι τῷ τοῦ ἄρρενος,
τοῦ δ' ὑγροῦ περιττώματος ἀμφοῖν ἔξοδος.

Κοινὸν δὲ μέρος αὐχένος καὶ στήθους σφαγή,
πλευρᾶς δὲ καὶ βραχίονος καὶ ὤμου μασχάλη, μηροῦ δὲ καὶ ἤτρου βουβών. μηροῦ δὲ καὶ γλουτοῦ
10 τὸ ἐντὸς περίνεος, μηροῦ δὲ καὶ γλουτοῦ τὸ ἔξω
ὑπογλουτίς.

Θώρακος δὲ περὶ μὲν τῶν ἔμπροσθεν εἴρηται, τοῦ
XV δὲ στήθους τὸ ὄπισθεν νῶτον. νώτου δὲ μέρη
ὠμοπλάται δύο καὶ ῥάχις, ὑποκάτωθεν δὲ κατὰ τὴν
γαστέρα τοῦ θώρακος ὀσφύς. κοινὸν δὲ τοῦ ἄνω
καὶ κάτω πλευραί, ἑκατέρωθεν ὀκτώ· περὶ γὰρ
15 Λιγύων τῶν καλουμένων ἑπταπλεύρων οὐδενός πω
ἀξιοπίστου ἀκηκόαμεν.

Ἔχει δ' ὁ ἄνθρωπος καὶ τὸ ἄνω καὶ τὸ κάτω,
καὶ πρόσθια καὶ ὀπίσθια, καὶ δεξιὰ καὶ ἀριστερά.
τὰ μὲν οὖν δεξιὰ καὶ ἀριστερὰ ὅμοια σχεδὸν ἐν τοῖς
20 μέρεσι καὶ ταὐτὰ πάντα, πλὴν ἀσθενέστερα τὰ ἀρι-

[1] αἰλούροις Th. : λοφούροις vulg.

enlarges easily, and it protrudes and retracts in the reverse way to that which occurs in cats. Below the penis are the two testicles, and the skin which surrounds them is the scrotum. The testicles are not the same thing as flesh, yet they do not substantially differ from it. We shall speak in detail fully later about the nature of them, and generally of all such parts.

The privy part of the female is opposite in structure XIV to that of the male : the part below the pubes is receding, and does not protrude as in the male. Further, there is an " urethra " outside the womb ; it serves as a passage for the semen of the male : in both sexes the urethra serves as an outlet for the liquid residue.

The part common to the neck and the chest is the throat ; the part common to the side, the arm, and the shoulder is the armpit ; the part common to the thigh and abdomen is the groin. The part inside the thigh and the buttocks is the perineum ; the part exterior to them is the *hypoglutis*.

We have now mentioned the front parts of the trunk. The part in the rear corresponding to the chest is the back. Parts of the back are : shoulder-blades (two) ; the backbone ; and below, on a level with where the belly is in the trunk, is the loin. Common to the upper and lower trunk are the ribs, eight on each side. (We have received no reliable evidence about the alleged seven-ribbed Ligurians.) _{The trunk: rear parts. XV}

Thus man has an upper and lower part of the body, a front and a rear, a right side and a left. Now the right and left sides are practically alike so far as concerns their parts, and the same in every way, except that the left side is weaker ; but the rear parts are _{Right and left alike, but not}

ARISTOTLE

493 b

στερά· τὰ δ' ὀπίσθια τοῖς προσθίοις ἀνόμοια, καὶ τὰ κάτω τοῖς ἄνω, πλὴν ὧδε ὅμοια· τὰ κάτω τοῦ ἤτρου οἷον τὸ πρόσωπον εὐσαρκίᾳ καὶ ἀσαρκίᾳ καὶ τὰ σκέλη πρὸς τοὺς βραχίονας ἀντίκειται· καὶ οἷς βραχεῖς οἱ ἀγκῶνες, καὶ οἱ μηροὶ ὡς ἐπὶ τὸ πολύ, 25 καὶ οἷς οἱ πόδες μικροί, καὶ αἱ χεῖρες.

Κώλων[1] δὲ τὸ μὲν διφυὲς βραχίων· βραχίονος δ' ὦμος, ἀγκών, ὠλέκρανον, πῆχυς, χείρ· χειρὸς δὲ θέναρ, δάκτυλοι πέντε· δακτύλου δὲ τὸ μὲν καμπ- τικὸν κόνδυλος, τὸ δ' ἄκαμπτον φάλαγξ. δάκτυλος 30 δ' ὁ μὲν μέγας μονοκόνδυλος, οἱ δ' ἄλλοι δικόνδυλοι. ἡ δὲ κάμψις καὶ τῷ βραχίονι καὶ τῷ δακτύλῳ εἴσω πᾶσιν· κάμπτεται δ' ὁ βραχίων κατὰ τὸ ὠλέκρανον. χειρὸς δὲ τὸ μὲν ἐντὸς θέναρ, σαρκῶδες καὶ διηρη- μένον ἄρθροις, τοῖς μὲν μακροβίοις ἑνὶ ἢ δυσὶ δι'
494 a ὅλου, τοῖς δὲ βραχυβίοις δυσὶ καὶ οὐ δι' ὅλου. ἄρθρον[2] δὲ χειρὸς καὶ βραχίονος καρπός. τὸ δὲ ἔξω τῆς χειρὸς νευρῶδες καὶ ἀνώνυμον.

Κώλων[3] δὲ διμερὲς ἄλλο σκέλος. σκέλους δὲ τὸ 5 μὲν ἀμφικέφαλον μηρός, τὸ δὲ πλανησίεδρον μύλη, τὸ δὲ διόστεον κνήμη, καὶ ταύτης τὸ μὲν πρόσθιον ἀντικνήμιον, τὸ δ' ὀπίσθιον γαστροκνημία, σάρξ νευρώδης καὶ[4] φλεβώδης, τοῖς μὲν ἀνεσπασμένη ἄνω πρὸς τὴν ἰγνύν, ὅσοι μεγάλα τὰ ἰσχία ἔχουσι, τοῖς δ' ἐναντίως κατεσπασμένη· τὸ δ' ἔσχατον ἀντι- 10 κνημίου σφυρόν, διφυὲς ἐν ἑκατέρῳ τῷ σκέλει. τὸ δὲ πολύοστεον τοῦ σκέλους πούς. τούτου δὲ τὸ μὲν ὀπίσθιον μέρος πτέρνα, τὸ δ' ἐμπρόσθιον τοῦ ποδὸς τὸ μὲν ἐσχισμένον δάκτυλοι πέντε, τὸ δὲ σαρκῶδες κάτωθεν στῆθος, τὸ δ' ἄνωθεν ἐν τοῖς πρανέσι

---

[1] κώλων Dt. : κώλου vulg.   [2] ἄρθρον PD : ἄρθρα A, vulg.
[3] κώλων Dt. : κώλου vulg.   [4] καὶ Scal. : ἢ vulg.

HISTORIA ANIMALIUM, I. xv

not similar to the front parts, nor the lower to the upper, except in the following way : if the face is well or poorly covered with flesh, so are the lower parts of the abdomen, and the legs correspond to the arms ; and those who have short upper arms generally have short thighs, and people with small feet have small hands as well. *[front and rear, or upper and lower.]*

One pair of limbs is the arms. Parts of the arm are : shoulder, upper arm, elbow, forearm, and hand ; of the hand, the parts are the palm and the five fingers. The part of the finger which bends is the knuckle, the inflexible part is the phalanx. The thumb is single-jointed, the other fingers have two joints. Both arm and finger bend inwards in all instances, the arm bending at the elbow. The inner surface of the hand is the palm, which is fleshy and divided by lines ; in long-lived persons by one or two lines, which go right across, in short-lived by two which do not go right across. The joint connecting hand and arm is the wrist. The outer surface of the hand is sinewy and has no special name. *[The arms and their parts.]*

Another pair of limbs is the legs. The double-headed part of the leg is the thigh-bone, the sliding part is the knee-cap, the double-boned part is the shank, of which the front is the shin and the back is the calf, in which the flesh is sinewy and contains blood-vessels : in some persons (those with large hips) it is drawn upwards towards the ham, in others it is drawn in the contrary direction. At the extremity of the shin is the ankle, double in either leg. The many-boned part of the leg is the foot. The back part of the foot is the heel : the front part is divided, making the five toes ; the fleshy part underneath is the ball of the foot ; the upper part at the back is *[The legs and their parts.]*

53

ARISTOTLE

νευρῶδες καὶ ἀνώνυμον. δακτύλου δὲ τὸ μὲν ὄνυξ, τὸ δὲ καμπή· πάντων δ' ὁ ὄνυξ ἐπ' ἄκρῳ· μονόκαμπτοι δὲ πάντες οἱ κάτω δάκτυλοι. τοῦ δὲ ποδὸς ὅσοις τὸ ἐντὸς παχὺ καὶ μὴ κοῖλον, ἀλλὰ βαίνουσιν ὅλῳ, πανοῦργοι. κοινὸν δὲ μηροῦ καὶ κνήμης γόνυ, καμπή.

Ταῦτα μὲν οὖν τὰ μέρη κοινὰ καὶ ἄρρενος καὶ θήλεος. ἡ δὲ θέσις τῶν μερῶν πρὸς τὸ ἄνω καὶ κάτω καὶ πρόσθιον καὶ ὀπίσθιον καὶ δεξιὸν καὶ ἀριστερὸν ὡς ἔχει, φανερὰ μὲν ἂν εἶναι δόξειε τὰ ἔξωθεν κατὰ τὴν αἴσθησιν, οὐ μὴν ἀλλὰ διὰ τὴν αὐτὴν αἰτίαν λεκτέον δι' ἥνπερ καὶ τὰ πρότερον εἰρήκαμεν, ἵνα περαίνηται τὸ ἐφεξῆς, καὶ καταριθμουμένων ὅπως ἧττον λανθάνῃ τὰ μὴ τὸν αὐτὸν ἔχοντα τρόπον ἐπί τε τῶν ἄλλων ζῴων καὶ ἐπὶ τῶν ἀνθρώπων.

Μάλιστα δ' ἔχει διωρισμένα πρὸς τοὺς κατὰ φύσιν τόπους τὰ ἄνω καὶ κάτω ἄνθρωπος τῶν ἄλλων ζῴων· τά τε γὰρ ἄνω καὶ κάτω πρὸς τὰ τοῦ παντὸς ἄνω καὶ κάτω τέτακται. τὸν αὐτὸν τρόπον καὶ τὰ πρόσθια καὶ τὰ ὀπίσθια καὶ τὰ δεξιὰ καὶ τὰ ἀριστερὰ κατὰ φύσιν ἔχει. τῶν δ' ἄλλων ζῴων τὰ μὲν οὐκ ἔχει, τὰ δ' ἔχει μὲν συγκεχυμένα δ' ἔχει μᾶλλον. ἡ μὲν οὖν κεφαλὴ πᾶσιν ἄνω πρὸς τὸ σῶμα τὸ ἑαυτῶν· ὁ δ' ἄνθρωπος μόνος, ὥσπερ εἴρηται, πρὸς τὸ τοῦ ὅλου τελειωθεὶς ἔχει τοῦτο τὸ μόριον.

Μετὰ δὲ τὴν κεφαλήν ἐστιν ὁ αὐχήν, εἶτα στῆθος

sinewy and has no special name. Parts of the toe are the nail, and the joint; the nail is always at the end of the toe, and all toes have one joint only. People who have the sole of the foot thick and not arched, so that they walk on the whole of it, are liable to be mischievous. Common to thigh and shank is the knee, which is a joint.

I have now mentioned the parts which are common to both sexes. As for the position of the parts in respect of up and down, front and back, and right and left, all this, so far as the external ones are concerned, might be considered clear enough to mere perception. Nevertheless, we must deal with them for the same reason on account of which we have given our previous description—*viz.*, so that we may provide a complete list in proper sequence, and by enumerating them make it less likely that we fail to observe those differences which exist between the parts of man and of other animals.

In man more than in any other animal the upper and the lower parts of the body are determined in accordance with what is naturally upper and lower: in other words, upper and lower in man correspond with upper and lower in the universe itself. Similarly, in man, front and rear, right and left as applied to his parts, have their proper natural meaning. In some of the other animals this is not so at all; in some the distinctions exist but in a somewhat confused manner. Of course, in all animals the head is up above with regard to the creature's own body; but, as I have said, man is the only animal which, when fully developed, has the head up above in the sense in which " up " is applied to the universe.

After the head is placed the neck, then the chest

<small>Man's unique (and " natural ") posture.</small>

καὶ νῶτον, τὸ μὲν ἐκ τοῦ πρόσθεν τὸ δ' ἐκ τοῦ ὄπισθεν. καὶ ἐχόμενα τούτων γαστὴρ καὶ ὀσφὺς καὶ αἰδοῖον καὶ ἰσχίον, εἶτα μηρὸς καὶ κνήμη, τελευταῖον δὲ πόδες. εἰς τὸ πρόσθεν δὲ καὶ τὰ σκέλη τὴν κάμψιν ἔχει, ἐφ' ὃ καὶ ἡ πορεία, καὶ τῶν ποδῶν τὸ κινητικώτερον μέρος καὶ ἡ κάμψις· ἡ δὲ πτέρνα ἐκ τοῦ ὄπισθεν· τῶν δὲ σφυρῶν ἑκάτερον κατὰ τὸ οὖς. ἐκ δὲ τῶν πλαγίων τῶν δεξιῶν καὶ τῶν ἀριστερῶν οἱ βραχίονες, τὴν κάμψιν ἔχοντες εἰς τὸ ἐντός, ὥστε τὰ κυρτὰ τῶν σκελῶν καὶ τῶν βραχιόνων πρὸς ἄλληλα εἶναι ἐπ' ἀνθρώπου μάλιστα.

Τὰς δ' αἰσθήσεις καὶ τὰ αἰσθητήρια, ὀφθαλμοὺς καὶ μυκτῆρας καὶ γλῶτταν, ἐπὶ ταὐτὸ καὶ εἰς τὸ πρόσθιον ἔχει· τὴν δ' ἀκοὴν καὶ τὸ αἰσθητήριον αὐτῆς,[1] τὰ ὦτα, ἐκ τοῦ πλαγίου μέν, ἐπὶ τῆς αὐτῆς δὲ περιφερείας τοῖς ὄμμασιν. τὰ δ' ὄμματα ἐλάχιστον κατὰ μέγεθος διέστηκεν ἀνθρώπῳ τῶν ζῴων. ἔχει δ' ἀκριβεστάτην ἄνθρωπος τῶν αἰσθήσεων τὴν ἁφήν, δευτέραν δὲ τὴν γεῦσιν· ἐν δὲ ταῖς ἄλλαις λείπεται πολλῶν.

XVI Τὰ μὲν οὖν μόρια τὰ πρὸς τὴν ἔξω ἐπιφάνειαν τοῦτον τέτακται τὸν τρόπον, καὶ καθάπερ ἐλέχθη, διωνόμασταί τε μάλιστα καὶ γνώριμα διὰ τὴν συνήθειάν ἐστιν· τὰ δ' ἐντὸς τοὐναντίον. ἄγνωστα γάρ ἐστι μάλιστα τὰ τῶν ἀνθρώπων, ὥστε δεῖ πρὸς τὰ τῶν ἄλλων μόρια ζῴων ἀνάγοντας σκοπεῖν, οἷς ἔχει παραπλησίαν τὴν φύσιν.

Πρῶτον μὲν οὖν τῆς κεφαλῆς κεῖται τὴν θέσιν ἐν τῷ πρόσθεν ἔχων ὁ ἐγκέφαλος. ὁμοίως δὲ καὶ τοῖς

---

[1] add. καὶ C, vulg. : om. PD.

HISTORIA ANIMALIUM, I. xv-xvi

and the back, the former in front, the latter behind. Next after these are the belly, the loin, the generative parts, and the haunch; then the thigh and the shank, and last of all the feet. The bending of the legs is in a forward direction, *i.e.*, the same direction as that in which the animal advances, and the more movable part of the foot is forward too, and so is its bending; the heel is at the back, and each of the two ankles earwise.[a] The arms are placed at the sides to right and left, and bend inwards. Consequently, the convexities of the bent legs and bent arms most nearly face towards each other in man. {Flexion of the limbs.}

With regard to the senses and their organs: Eyes, nostrils, tongue, all face the same way, *viz.*, front; the sense of hearing, and its organ, the ear, is, sure enough, placed sideways, but at the same height horizontally as the eyes. For his size, man's eyes are less far apart than those of any other animal. Man's sense of touch is more accurate than any other creature's, and his sense of taste comes second; in the other senses he falls short of many animals. {Sense-organs.}

I have now described the arrangement of the parts which are on the visible surface; and, as I said, they mostly have their own proper names and are well known through their familiarity. With the inner parts the reverse is true. They are for the most part unknown—at least, those of man are, and hence we have to refer to those of other animals, the natural structure of whose parts those of man resemble, and examine them. {XVI INSTRUMENTAL PARTS OF BLOODED ANIMALS: (i) Vivipara: (b) Internal parts:}

First of all, then, the brain has its place in the front portion of the head. This is the case also with all {The brain, etc.}

---

[a] The meaning of this is not clear, and the text may be corrupt.

ἄλλοις ζῴοις, ὅσα ἔχει τοῦτο τὸ μόριον· ἔχει δ᾽ ἅπαντα ὅσα ἔχει αἷμα, καὶ ἔτι τὰ μαλάκια· κατὰ μέγεθος δ᾽ ἔχει[1] ἄνθρωπος πλεῖστον ἐγκέφαλον καὶ ὑγρότατον. ὑμένες δ᾽ αὐτὸν δύο περιέχουσιν, ὁ μὲν περὶ τὸ ὀστοῦν ἰσχυρότερος, ὁ δὲ περὶ αὐτὸν τὸν ἐγκέφαλον ἥττων ἐκείνου. διφυὴς δ᾽ ἐν πᾶσίν ἐστιν ὁ ἐγκέφαλος. καὶ ἐπὶ τούτου ἡ καλουμένη παρεγκεφαλὶς ἔσχατον, ἑτέραν ἔχουσα τὴν μορφὴν καὶ κατὰ τὴν ἁφὴν καὶ κατὰ τὴν ὄψιν.

Τὸ δ᾽ ὄπισθεν τῆς κεφαλῆς κενὸν καὶ κοῖλον πᾶσιν, ὡς ἑκάστοις ὑπάρχει μεγέθους. ἔνια μὲν γὰρ μεγάλην ἔχει τὴν κεφαλήν, τὸ δ᾽ ὑποκείμενον τοῦ προσώπου μόριον ἔλαττον, ὅσα στρογγυλοπρόσωπα· τὰ δὲ τὴν μὲν κεφαλὴν μικράν, τὰς δὲ σιαγόνας μακράς, οἷον τὸ τῶν λοφούρων γένος πᾶν.

Ἄναιμος δ᾽ ὁ ἐγκέφαλος ἅπασι, καὶ οὐδεμίαν ἔχων ἐν αὑτῷ φλέβα, καὶ θιγγανόμενος κατὰ φύσιν ψυχρός. ἔχει δ᾽ ἐν τῷ μέσῳ ὁ τῶν πλείστων [πᾶς][2] κοῖλόν τι μικρόν. ἡ δὲ περὶ αὐτὸν μῆνιγξ φλεβώδης· ἔστι δ᾽ ἡ μῆνιγξ ὑμὴν δερματικὸς[3] ὁ περιέχων τὸν ἐγκέφαλον. ὑπὲρ δὲ τοῦ ἐγκεφάλου λεπτότατον ὀστοῦν καὶ ἀσθενέστατον τῆς κεφαλῆς ἐστιν, ὃ καλεῖται βρέγμα.

Φέρουσι δ᾽ ἐκ τοῦ ὀφθαλμοῦ τρεῖς πόροι εἰς τὸν ἐγκέφαλον, ὁ μὲν μέγιστος καὶ ὁ μέσος εἰς τὴν παρεγκεφαλίδα, ὁ δ᾽ ἐλάχιστος εἰς αὐτὸν τὸν ἐγκέφαλον· ἐλάχιστος δ᾽ ἐστὶν ὁ πρὸς τῷ μυκτῆρι μάλιστα. οἱ μὲν οὖν μέγιστοι παράλληλοί[4] εἰσι καὶ οὐ συμπίπτουσιν, οἱ δὲ μέσοι συμπίπτουσι (δῆλον δὲ τοῦτο μάλιστα ἐπὶ τῶν ἰχθύων)· καὶ γὰρ ἐγγύτε-

---

[1] ὁμοίως ἔχει AC, vulg.: ἔχει D: ἑαυτοῦ ἔχει Dt. (ἑαυτοῦ supra μέγεθος C). [2] πᾶς secl. Sn.

other animals which possess a brain. All blooded animals have one, and so have the Cephalopods; but for his size man has the largest brain and the most fluid one. The brain is surrounded by two membranes: the one round the bone [a] is the stronger, the one round the brain itself [b] less so. In all animals the brain is double. Beyond this, at the far end, is the cerebellum as it is called; its form is different from that of the brain, as can be both felt and seen.

In all animals the back of the head is empty and hollow, whatever its size in each animal may be. Thus some have a large head, while the lower part of the face is smaller than might be expected; this is so in the round-faced animals; others have a small head, but long jaws, *e.g.*, the whole class of bushy-tailed animals.[c]

In all animals the brain is bloodless; there is not a single blood-vessel in it, and it feels cold to the touch. In most animals it has a small hollow in the middle. The membrane which surrounds it is patterned with blood-vessels: this is the skin-like one which surrounds the brain. Above the brain is the lightest and weakest bone in the head, called the *bregma*.

From the eye three passages lead to the brain: the largest and second-largest to the cerebellum, the smallest to the brain itself: this last is the one nearest to the nostril. So the two largest run side by side and do not coalesce; the two second-largest do coalesce (as can be very clearly seen in fishes), for in

---

[a] The *dura mater*.   [b] The *pia mater*.
[c] See note on 491 a 1.

---

[3] hoc ordine (sed ὁ μ.) Λ, ἔστι δ' ὑ. δ. ἡ μ. vulg.
[4] AC: παρ' ἀλλήλους PD, vulg.

ARISTOTLE

ρον οὗτοι τοῦ ἐγκεφάλου ἢ οἱ μεγάλοι· οἱ δ' ἐλάχιστοι πλεῖστόν τ' ἀπήρτηνται ἀλλήλων καὶ οὐ συμπίπτουσιν.

Ἐντὸς δὲ τοῦ αὐχένος ὅ τ' οἰσοφάγος καλούμενός ἐστιν (ἔχων τὴν ἐπωνυμίαν ἀπὸ τοῦ μήκους καὶ τῆς στενότητος), καὶ ἡ ἀρτηρία. πρότερον δὲ τῇ θέσει ἡ ἀρτηρία κεῖται τοῦ οἰσοφάγου ἐν πᾶσι τοῖς ἔχουσιν αὐτήν· ἔχει δὲ ταύτην πάντα ὅσαπερ πνεύμονα ἔχει. ἔστι δ' ἡ μὲν ἀρτηρία χονδρώδης τὴν φύσιν καὶ ὀλίγαιμος, πολλοῖς λεπτοῖς φλεβίοις περιεχομένη, κεῖται δ' ἐπὶ μὲν τὰ ἄνω πρὸς τὸ στόμα κατὰ τὴν ἐκ τῶν μυκτήρων σύντρησιν εἰς τὸ στόμα, ᾗ καὶ ὅταν πίνοντες ἀνασπάσωσί τι τοῦ ποτοῦ, χωρεῖ ἐκ τοῦ στόματος διὰ τῶν μυκτήρων ἔξω. μεταξὺ δ' ἔχει τῶν τρήσεων τὴν ἐπιγλωττίδα καλουμένην, ἐπιπτύσσεσθαι δυναμένην ἐπὶ τὸ τῆς ἀρτηρίας τρῆμα τὸ εἰς τὸ στόμα τεῖνον. ταύτῃ δὲ τὸ πέρας συνήρτηται τῆς γλώττης. ἐπὶ δὲ θάτερα καθήκει εἰς τὸ μεταξὺ τοῦ πνεύμονος, εἶτ' ἀπὸ τούτου σχίζεται εἰς ἑκάτερον τῶν μερῶν τοῦ πνεύμονος. θέλει γὰρ εἶναι διμερὴς ὁ πνεύμων ἐν ἅπασι τοῖς ἔχουσιν αὐτόν· ἀλλ' ἐν μὲν τοῖς ζῳοτόκοις οὐχ ὁμοίως ἡ διάστασις φανερά, ἥκιστα δ' ἐν ἀνθρώπῳ. ἔστι δ' οὐ πολυσχιδὴς ὁ τοῦ ἀνθρώπου, ὥσπερ ἐνίων ζῳοτόκων, οὐδὲ λεῖος, ἀλλ' ἔχει ἀνωμαλίαν. ἐν δὲ τοῖς ᾠοτόκοις, οἷον ὄρνισι καὶ τῶν τετραπόδων ὅσα ᾠοτόκα, πολὺ τὸ μέρος ἑκάτερον ἀπ' ἀλλήλων ἔσχισται, ὥστε δοκεῖν δύο ἔχειν πνεύμονας· καὶ ἀπὸ μιᾶς δύο ἐστὶ μόρια τῆς ἀρτηρίας, εἰς ἑκάτερον τὸ μέρος τείνοντα τοῦ πνεύμονος. συνήρτηται δὲ καὶ

---

[a] *Stomachos*: the suggested derivation seems to be from *stenos* and *mekos*.

any case they are much nearer the brain than the large ones are. The smallest ones are widely spaced from each other and do not coalesce.

Within the neck is what is called the oesophagus (it gets its popular name *a* from its length and its narrowness), and the windpipe. The position of the windpipe in all animals where it is present is in front of the oesophagus, and all which have a lung have a windpipe. The windpipe consists of gristle, and contains little blood ; it is covered with a large number of very small blood-vessels. Its upper portion is situated near the mouth, where the aperture from the nostrils comes into the mouth—this is the place where any liquid which you may draw in too quickly when drinking makes its way out of the mouth through the nostrils. In between the apertures is what is called the epiglottis, which can be spread over the aperture of the windpipe which leads to the mouth. To it the extremity of the tongue is attached. At the other end the windpipe extends to the region between the lungs, and therefrom branches into two, into each of the two parts of the lung. The lung, of course, in all animals where it is present has a tendency to be double ; but in viviparous animals this duplication is not very plainly discernible, and least so in man ; though in man it is not divided into numerous parts, as in some Vivipara, nor is it smooth, but it exhibits some unevenness. In the Ovipara, on the other hand, *e.g.*, the Birds, and the oviparous quadrupeds, the two parts of the lung are quite separate and well removed from each other, so that they appear to have two lungs ; and from the windpipe, originally single, two parts branch off, one going to each of the two parts of the lung. It is also attached to the

τῇ μεγάλῃ φλεβὶ καὶ τῇ ἀορτῇ καλουμένῃ. φυσωμένης δὲ τῆς ἀρτηρίας διαδιδῶσιν εἰς τὰ κοῖλα μέρη τοῦ πνεύμονος τὸ πνεῦμα. ταῦτα δὲ διαφύσεις ἔχει χονδρώδεις εἰς ὀξὺ συνηκούσας· ἐκ δὲ τῶν διαφύσεων τρήματα διὰ παντός ἐστι τοῦ πνεύμονος, ἀεὶ ἐκ μειζόνων εἰς ἐλάττω διαδιδόμενα. συνήρτηται δὲ καὶ ἡ καρδία τῇ ἀρτηρίᾳ πιμελώδεσι καὶ χονδρώδεσι καὶ ἰνώδεσι δεσμοῖς· ᾗ δὲ συνήρτηται, κοῖλόν ἐστιν. φυσωμένης δὲ τῆς ἀρτηρίας ἐν ἐνίοις μὲν οὐ κατάδηλον ποιεῖ, ἐν δὲ τοῖς μείζοσι τῶν ζῴων δῆλον ὅτι εἰσέρχεται τὸ πνεῦμα εἰς αὐτήν. ἡ μὲν οὖν ἀρτηρία τοῦτον ἔχει τὸν τρόπον, καὶ δέχεται μόνον τὸ πνεῦμα καὶ ἀφίησιν, ἄλλο δ' οὐδὲν οὔτε ξηρὸν οὔθ' ὑγρόν, ἢ πόνον παρέχει, ἕως ἂν ἐκβήξῃ τὸ κατελθόν.

Ὁ δὲ στόμαχος ἤρτηται μὲν ἄνωθεν ἀπὸ τοῦ στόματος, ἐχόμενος τῆς ἀρτηρίας, συνεχὴς ὢν πρός τε τὴν ῥάχιν καὶ τὴν ἀρτηρίαν ὑμενώδεσι δεσμοῖς, τελευτᾷ δὲ διὰ τοῦ διαζώματος εἰς τὴν κοιλίαν, σαρκοειδὴς ὢν τὴν φύσιν, καὶ τάσιν ἔχων καὶ ἐπὶ μῆκος καὶ ἐπὶ πλάτος. ἡ δὲ κοιλία ἡ τοῦ ἀνθρώπου ὁμοία τῇ κυνείᾳ ἐστίν· οὐ πολλῷ γὰρ τοῦ ἐντέρου μείζων, ἀλλ' ἐοικυῖα οἱονεὶ ἐντέρῳ τινὶ[1] εὖρος ἔχοντι· εἶτα ἔντερον ἁπλοῦν, εἱλιγμένον,[2] ἐπιεικῶς πλατύ. ἡ δὲ κάτω κοιλία ὁμοία τῇ ὑείᾳ· πλατεῖά τε γάρ ἐστι, καὶ τὸ ἀπὸ ταύτης πρὸς τὴν ἕδραν παχὺ καὶ βραχύ. τὸ δ' ἐπίπλοον ἀπὸ μέσης τῆς κοιλίας ἤρτηται, ἔστι δὲ τὴν φύσιν ὑμὴν πιμελώδης, ὥσπερ καὶ τοῖς ἄλλοις τοῖς μονοκοιλίοις καὶ ἀμφώδουσιν.

[1] τινὶ A¹ : ἐνὶ C, om. vulg.
[2] εἶτα ἔντερον add. vulg. ex PD, qui etiam εὖρος ἔχον addunt.

Great Blood-vessel and to the so-called Aorta. When the windpipe is inflated with air, the breath gets distributed to the hollow parts of the lung. These parts have divisions, consisting of gristle, which meet in a point; and from the divisions there are apertures running through the whole of the lung, breaking up into smaller and smaller ones. The heart also is attached to the windpipe by means of ligatures consisting of fat, gristle, and sinew, and there is a hollow at the place of attachment. When the windpipe is inflated with air, in some animals the entry of the breath into the heart is not clearly perceptible, though it is clear enough in the larger ones. Such then is the arrangement of the windpipe: it does nothing more than take in the breath and emit it. It admits nothing else whatever, solid or liquid—or if it does, it causes discomfort until the intruding object has been coughed up.

The oesophagus at the top is attached to the mouth, close beside the windpipe, and is contiguous with the backbone and the windpipe, being attached by membranous ligatures; and having passed through the midriff it terminates at the belly. Its substance is fleshlike, and it can stretch both lengthways and sideways. Man's stomach resembles the dog's: it is not much larger than the gut, but rather resembles a somewhat unusually wide gut. Next after it is the gut, which is single, twisted, and fairly wide. The lower stomach is like the pig's: it is wide, and the part from it to the buttocks is thick and short. The Omentum is attached to the middle of the stomach, and its substance is that of a fatty membrane, as indeed it is in all other animals which have only one stomach and have teeth in both jaws.

ARISTOTLE

Ὑπὲρ δὲ τῶν ἐντέρων τὸ μεσεντέριόν ἐστιν· ὑμενῶδες δ' ἐστὶ καὶ τοῦτο καὶ πλατύ, καὶ πῖον γίγνεται. ἐξήρτηται δ' ἐκ τῆς μεγάλης φλεβὸς καὶ τῆς ἀορτῆς, καὶ δι' αὐτοῦ φλέβες πολλαὶ καὶ πυκναί, κατατείνουσαι πρὸς τὴν τῶν ἐντέρων θέσιν, ἄνωθεν ἀρξάμεναι μέχρι κάτω.

Τὰ μὲν οὖν περὶ τὸν στόμαχον καὶ τὴν ἀρτηρίαν οὕτως ἔχει, καὶ τὰ περὶ τὴν κοιλίαν.

XVII. Ἡ δὲ καρδία ἔχει μὲν τρεῖς κοιλίας, κεῖται δ' ἀνωτέρω τοῦ πνεύμονος κατὰ τὴν σχίσιν τῆς ἀρτηρίας, ἔχει δ' ὑμένα πιμελώδη καὶ παχύν, ᾗ προσπέφυκε τῇ φλεβὶ τῇ μεγάλῃ καὶ τῇ ἀορτῇ. κεῖται δ' ἐπὶ τῇ ἀορτῇ κατὰ τὰ ὀξέα. κεῖται δὲ τὰ ὀξέα κατὰ τὸ στῆθος ὁμοίως ἁπάντων τῶν ζῴων, ὅσα ἔχει στῆθος. πᾶσι δ' ὁμοίως καὶ τοῖς ἔχουσι καὶ τοῖς μὴ ἔχουσι τοῦτο τὸ μόριον εἰς τὸ πρόσθεν ἔχει ἡ καρδία τὸ ὀξύ· λάθοι δ' ἂν πολλάκις διὰ τὸ μεταπίπτειν διαιρουμένων. τὸ δὲ κυρτὸν αὐτῆς ἐστιν ἄνω. ἔχει δὲ τὸ ὀξὺ σαρκῶδες ἐπὶ πολὺ καὶ πυκνόν, καὶ ἐν τοῖς κοίλοις αὐτῆς νεῦρα ἔνεστιν. κεῖται δὲ τὴν θέσιν ἐν μὲν τοῖς ἄλλοις κατὰ μέσον τὸ στῆθος, ὅσα ἔχει στῆθος, τοῖς δ' ἀνθρώποις ἐν τοῖς ἀριστεροῖς μᾶλλον, μικρὸν τῆς διαιρέσεως τῶν μαστῶν ἐγκλίνουσα εἰς τὸν ἀριστερὸν μαστὸν ἐν τῷ ἄνω μέρει τοῦ στήθους. καὶ οὔτε μεγάλη, τό θ' ὅλον αὐτῆς εἶδος οὐ πρόμηκές ἐστιν ἀλλὰ στρογγυλώτερον· πλὴν τὸ ἄκρον εἰς ὀξὺ συνῆκται. ἔχει δὲ κοιλίας τρεῖς, ὥσπερ εἴρηται, μεγίστην μὲν τὴν ἐν τοῖς δεξιοῖς, ἐλαχίστην δὲ τὴν ἐν τοῖς ἀριστεροῖς, μέσην δὲ μεγέθει τὴν ἀνὰ μέσον. ἁπάσας δ' ἔχει, καὶ τὰς δύο μικράς, εἰς τὸν πνεύμονα τετρημένας, κατάδηλον δὲ κατὰ μίαν τῶν κοιλιῶν. κάτωθεν δ'

Above the guts is the mesentery; this too is membranous and wide, and becomes fatty. It is attached to the Great Blood-vessel and the Aorta, and through it there run numerous blood-vessels packed close together, which extend to the region of the guts, starting from above and continuing to down below. *(The mesentery.)*

Such then is the arrangement of the oesophagus, the windpipe and the stomach and its related parts.

Now the heart has three cavities, and it lies above the lung at the point where the windpipe divides into two, and has a fat, thick membrane at the place where it is attached to the Great Blood-vessel and the Aorta. And it lies with its pointed end upon the Aorta. This end of it lies towards the chest in all animals which have a chest. And in all animals, whether they have a chest or not, the pointed end of the heart is always forwards, though this fact may very likely escape observation owing to some change in position while dissection is in progress. The rounded end of the heart is at the top. The pointed end is very largely fleshy and firm in texture, and there are sinews in its cavities. In animals other than man which have a chest, its position is in the middle of the chest, but in man it is somewhat over towards the left, inclining slightly from the division of the breasts towards the left breast in the upper part of the chest. Further, the heart is not large, and its shape as a whole is not elongated but roundish, except of course that it is pointed at the end. As I have already said, it has three cavities, the largest being on the right hand side, the smallest on the left, and the medium-sized one in the middle. All of them, even the two small ones, have a connexion with the lung, and this is quite clearly visible in respect of *(XVII The heart.)*

ARISTOTLE

ἐκ τῆς προσφύσεως κατὰ μὲν τὴν μεγίστην κοιλίαν ἐξήρτηται τῇ μεγάλῃ φλεβί [πρὸς ἣν καὶ τὸ μεσεντέριόν ἐστι],[1] κατὰ δὲ τὴν μέσην τῇ ἀορτῇ.

Φέρουσι δὲ καὶ εἰς τὸν πνεύμονα πόροι ἀπὸ τῆς καρδίας, καὶ σχίζονται τὸν αὐτὸν τρόπον ὅνπερ ἡ ἀρτηρία, κατὰ πάντα τὸν πνεύμονα παρακολουθοῦντες τοῖς ἀπὸ τῆς ἀρτηρίας. ἐπάνω δ' εἰσὶν οἱ ἀπὸ τῆς καρδίας πόροι· οὐδεὶς δ' ἐστὶ κοινὸς πόρος, ἀλλὰ διὰ τὴν σύναψιν δέχονται τὸ πνεῦμα καὶ τῇ καρδίᾳ διαπέμπουσιν· φέρει γὰρ ὁ μὲν εἰς τὸ δεξιὸν κοῖλον τῶν πόρων, ὁ δ' εἰς τὸ ἀριστερόν.

Περὶ δὲ τῆς φλεβὸς τῆς μεγάλης καὶ τῆς ἀορτῆς κατ' αὐτὰς κοινῇ περὶ ἀμφοτέρων ἐροῦμεν ὕστερον.

Αἷμα δὲ πλεῖστον μὲν ὁ πνεύμων ἔχει τῶν ἐν τοῖς ζῴοις μορίων τοῖς ἔχουσί τε πνεύμονα καὶ ζωοτοκοῦσιν ἐν αὑτοῖς τε καὶ ἐκτός· ἅπας μὲν γάρ ἐστι σομφός, παρ' ἑκάστην δὲ τὴν σύριγγα πόροι φέρουσι τῆς μεγάλης φλεβός. ἀλλ' οἱ νομίζοντες εἶναι κενὸν διηπάτηνται, θεωροῦντες τοὺς ἐξῃρημένους ἐκ τῶν διαιρουμένων τῶν ζῴων, ὧν εὐθὺς[2] ἐξελήλυθε τὸ αἷμα ἀθρόον.

Τῶν δ' ἄλλων σπλάγχνων ἡ καρδία μόνον ἔχει

---

[1] secl. A.-W., qui et 22 ἁπάσας hucusque correxere. totum 19 ἔχει δὲ ... 27 ἀορτῇ secl. Dt. de codd. singula hic non protuli.
[2] εὐθὺς AC : εὐθέως PD, vulg.

---

[a] This reference to passages leading into the lung from the heart recalls the developed Galenic theory of natural, vital, and animal spirits, which persisted through the Middle Ages and even into the 18th century. According to this theory, blood was formed in the liver and there charged with natural spirits. Thence it was distributed through the veins, part of it entering the right ventricle of the heart : of this the greater part, after purification, was returned to the venous system,

one of them. Below, at the place of attachment, from the largest cavity there is a connexion to the Great Blood-vessel [beside which lies the mesentery], and from the middle cavity there is a connexion to the Aorta.

Passages also lead into the lung from the heart,[a] and they divide off just as the windpipe does, running all over the lung and accompanying those which come from the windpipe. Those from the heart are uppermost. There is no common passage, but in virtue of their contact they receive the breath and convey it to the heart, one passage leading to the right cavity and the other to the left.

I shall deal later with the Great Blood-vessel and the Aorta together, giving details of each, in a paragraph devoted to them only.

Of all the parts in the body, so far as those animals are concerned which have a lung and are both internally and externally viviparous, the lung is the part which contains most blood, for the whole substance of the lung is spongy, and alongside every duct passages lead from the Great Blood-vessel. Those who suppose the lung to be empty are quite misled: they have observed lungs which have been removed from animals while being dissected, and the blood has rushed out from them immediately upon killing.

The lung.

The only other one of the viscera which contains

but a small part of the venous blood trickled through to the left ventricle, where it met *air conveyed from the lung*, and produced a higher type of spirit, the vital spirits: these were distributed through the arteries, some of which went to the brain. Here the blood was changed into animal spirits, which were distributed through the nerves, supposedly hollow (see C. Singer, *A History of Biology*[3], 1959, p. 104). For Aristotle's doctrine of the pneumatization of the blood in the heart see *G.A.*, Loeb edition, Introd. § 63, App. B §§ 31, 32.

αἷμα. καὶ ὁ μὲν πνεύμων οὐκ ἐν αὑτῷ ἀλλ' ἐν ταῖς φλεψίν, ἡ δὲ καρδία ἐν αὑτῇ· ἐν ἑκάστῃ γὰρ ἔχει αἷμα τῶν κοιλιῶν, λεπτότατον δ' ἐστὶ τὸ ἐν τῇ μέσῃ.

Ὑπὸ δὲ τὸν πνεύμονά ἐστι τὸ διάζωμα τὸ τοῦ θώρακος, αἱ καλούμεναι φρένες, πρὸς μὲν τὰ πλευρὰ καὶ τὰ ὑποχόνδρια καὶ τὴν ῥάχιν συνηρτημέναι, ἐν μέσῳ δ' ἔχει τὰ λεπτὰ καὶ ὑμενώδη. ἔχει δὲ δι' αὐτοῦ καὶ φλέβας τεταμένας· εἰσὶ δ' αἱ τοῦ ἀνθρώπου φρένες[1] παχεῖαι ὡς κατὰ λόγον τοῦ σώματος.

Ὑπὸ δὲ τὸ διάζωμα ἐν μὲν τοῖς δεξιοῖς κεῖται τὸ ἧπαρ, ἐν δὲ τοῖς ἀριστεροῖς ὁ σπλήν, ὁμοίως ἐν ἅπασι τοῖς ἔχουσι ταῦτα τὰ μόρια κατὰ φύσιν καὶ μὴ τερατωδῶς· ἤδη γὰρ ὧπται μετηλλαχότα τὴν τάξιν ἔν τισι τῶν τετραπόδων. συνήρτηται δὲ τῇ[2] κοιλίᾳ κατὰ τὸ ἐπίπλοον.

Τὴν δ' ὄψιν ἐστὶν ὁ τοῦ ἀνθρώπου σπλὴν στενὸς καὶ μακρός, ὅμοιος τῷ ὑείῳ. τὸ δ' ἧπαρ ὡς μὲν ἐπὶ τὸ πολὺ καὶ ἐν τοῖς πλείστοις οὐκ ἔχει[3] χολήν, ἐπ' ἐνίοις δ' ἔπεστιν. στρογγύλον δ' ἐστὶ τὸ τοῦ ἀνθρώπου ἧπαρ καὶ ὅμοιον τῷ βοείῳ. συμβαίνει δὲ τοῦτο καὶ ἐν τοῖς ἱερείοις, οἷον ἐν μὲν τόπῳ τινὶ τῆς ἐν Εὐβοίᾳ Χαλκιδικῆς οὐκ ἔχει τὰ πρόβατα χολήν, ἐν δὲ Νάξῳ πάντα σχεδὸν τὰ τετράποδα τοσαύτην ὥστ' ἐκπλήττεσθαι τοὺς θύοντας τῶν ξένων, οἰομένους αὐτῶν ἴδιον εἶναι τὸ σημεῖον, ἀλλ' οὐ φύσιν αὐτῶν εἶναι ταύτην. προσπέφυκε δὲ τῇ

---

[1] C, edd., *paries corporis* Σ: φλέβες APD, vulg.
[2] τῇ κάτω PD, vulg.
[3] ut eis congruant quae alibi (*P.A.* 676 b 16) narrat Arist., οὐκ ante 23 ἔπεστιν transposuit Dt.

blood is the heart. Indeed, the blood which the lung contains is not in the lung itself but in the blood-vessels, whereas the heart has blood in itself : there is blood in each of the cavities, the thinnest being that in the middle cavity.

Below the lung is the thoracic diaphragm, or what is called the midriff, which is attached to the ribs, the hypochondria, and the backbone ; it has a thin membranous centre. There are blood-vessels running through it. In man the midriff is thick in proportion to the size of the body. *The diaphragm.*

Below the diaphragm on the right side is the liver, and on the left side the spleen. This is the natural arrangement of these parts in all animals which have them, monstrosities excluded : cases have been observed in some quadrupeds in which spleen and liver were transposed. These two parts have a connexion with the stomach by way of the Omentum. *The liver, spleen, and gall-bladder.*

To look at, the spleen in man is narrow and long, resembling that of the pig. The liver generally and in most animals contains no gall-bladder, though it is present in some.[a] In man the liver is round in shape, resembling that of the ox. This erratic presence or absence of a gall-bladder comes to notice in sacrificial victims. Thus, in a certain district of the Chalcidian settlement in Euboea the sheep have no gall-bladder, whereas in Naxos practically all the quadrupeds have so large a one that when foreigners are offering sacrifice they get quite a shock, supposing that this is some sign from heaven meant for themselves, instead of a natural phenomenon. The liver

[a] The reverse is stated at *P.A.* 676 b 16 : " the majority of the blooded animals have a gall-bladder." Dittmeyer transposes the negative in the present passage to bring it into agreement with *P.A.*

μεγάλῃ φλεβὶ τὸ ἧπαρ, τῇ δ' ἀορτῇ οὐ κοινωνεῖ· διὰ γὰρ τοῦ ἥπατος διέχει ἡ ἀπὸ τῆς μεγάλης φλεβὸς φλέψ, ᾗ αἱ καλούμεναι πύλαι εἰσὶ τοῦ ἥπατος. συνήρτηται δὲ καὶ ὁ σπλὴν τῇ μεγάλῃ φλεβὶ μόνον· τείνει γὰρ ἀπ' αὐτῆς φλὲψ εἰς τὸν σπλῆνα.

Μετὰ δὲ ταῦτα οἱ νεφροὶ πρὸς αὐτῇ τῇ ῥάχει κεῖνται, ὅμοιοι τὴν φύσιν ὄντες τοῖς βοείοις. ἀνώτερος δ' ὁ δεξιός ἐστιν ἐν πᾶσι τοῖς ζῴοις τοῖς ἔχουσι νεφρούς· καὶ ἐλάττω δὲ πιμελὴν ἔχει τοῦ ἀριστεροῦ καὶ αὐχμηρότερος ὁ δεξιός. ἐν πᾶσι δ' ἔχει ὁμοίως τοῖς ἄλλοις καὶ τοῦτο.

Φέρουσι δ' εἰς αὐτοὺς πόροι ἔκ τε τῆς μεγάλης φλεβὸς καὶ τῆς ἀορτῆς, πλὴν οὐκ εἰς τὸ κοῖλον. ἔχουσι γὰρ οἱ νεφροὶ ἐν μέσῳ κοῖλον, οἱ μὲν μεῖζον οἱ δ' ἔλαττον, πλὴν οἱ τῆς φώκης· οὗτοι δ' ὅμοιοι τοῖς βοείοις ὄντες στερεώτατοι πάντων εἰσίν.

Οἱ δὲ πόροι οἱ τείνοντες εἰς αὐτοὺς εἰς τὸ σῶμα καταναλίσκονται τῶν νεφρῶν· σημεῖον δ' ὅτι οὐ περαίνουσι τὸ μὴ ἔχειν αἷμα[1] μηδὲ πήγνυσθαι ἐν αὐτοῖς. [ἔχουσι δὲ κοιλίαν, ὥσπερ εἴρηται, μικράν.][2] ἐκ δὲ τοῦ κοίλου τῶν νεφρῶν φέρουσιν εἰς τὴν κύστιν πόροι δύο νεανικοί, καὶ ἄλλοι ἐκ τῆς ἀορτῆς ἰσχυροὶ καὶ συνεχεῖς. ἐκ μέσου δὲ τῶν νεφρῶν ἑκατέρου φλὲψ κοίλη καὶ νευρώδης ἐξήρτηται, τείνουσα παρ' αὐτὴν τὴν ῥάχιν διὰ τῶν στενῶν· εἶτα εἰς ἑκάτερον τὸ ἰσχίον ἀφανίζονται, καὶ πάλιν δῆλαι γίγνονται τεταμέναι πρὸς τὸ ἰσχίον. αὗται δ' αἱ ἀποτομαὶ τῶν φλεβίων εἰς τὴν κύστιν καθήκουσιν· τελευταία γὰρ ἡ κύστις κεῖται, τὴν

---

[1] ⟨τὰ κοῖλα⟩ Dt.
[2] ἔχουσι ... μικράν secl. A.-W., om. Σ.

HISTORIA ANIMALIUM, I. xvii

also is attached to the Great Blood-vessel, though it has no connexion with the Aorta : it is the blood-vessel from the Great Blood-vessel which passes through the liver, where are the so-called portals of the liver. The spleen too is connected only with the Great Blood-vessel : a blood-vessel extends from it to the spleen.

After these parts are the kidneys, lying close beside the backbone, and resembling those of oxen. In all animals which have kidneys, the right one is placed higher up than the left; it also contains less fat and is drier—this is so in all other animals too. *The kidneys.*

Passages lead to the kidneys both from the Great Blood-vessel and from the Aorta, though they do not lead into the cavity. I have not so far mentioned this cavity in the middle of the kidney. In some animals it is larger, in others smaller; but there is none in the seal's kidneys. These are similar to those of oxen, and are more solid than those of all other animals.

The passages which lead to the kidneys peter out in the mass of the kidneys themselves. That they do not continue through the kidneys is shown by the fact that they contain no blood, nor does it congeal in them. [They contain, however, as has been said, a small cavity.] From the cavity of the kidneys two sturdy passages lead into the bladder, and other strong continuous ones come from the Aorta. To the middle of each kidney is attached a hollow and sinewy blood-vessel, which extends alongside the backbone through the narrow regions; after that these blood-vessels disappear into either loin, and then reappear extending to the loin. These branches from the small blood-vessels go as far as the bladder, and there stop, for the bladder is the last of the *The bladder.*

ARISTOTLE

μὲν ἐξάρτησιν ἔχουσα τοῖς ἀπὸ τῶν νεφρῶν τεταμένοις πόροις παρὰ τὸν καυλὸν τὸν ἐπὶ τὴν οὐρήθραν τείνοντα, καὶ σχεδὸν πάντῃ κύκλῳ λεπτοῖς καὶ ἰνώδεσιν ὑμενίοις ἐστὶ προσειλημμένη,[1] παραπλησίοις οὖσι τρόπον τινὰ τῷ διαζώματι τοῦ θώρακος. ἔστι δ' ἡ τοῦ ἀνθρώπου κύστις ἐπιεικῶς ἔχουσα μέγεθος.

Πρὸς δὲ τὸν καυλὸν τὸν τῆς κύστεως συνήρτηται τὸ αἰδοῖον—τὸ μὲν ἐξωτάτω τρῆμα συνερρωγὸς εἰς ταὐτό, μικρὸν δ' ὑποκάτω τὸ μὲν[2] εἰς τοὺς ὄρχεις φέρει τῶν τρημάτων, τὸ δὲ εἰς τὴν κύστιν—νευρῶδες καὶ χονδρῶδες ὄν.[3] τούτου δ' ἐξήρτηνται οἱ ὄρχεις τοῖς ἄρρεσι, περὶ ὧν ἐν τοῖς κοινῇ λεγομένοις διορισθήσεται πῶς ἔχουσιν.

Τὸν αὐτὸν δὲ τρόπον καὶ ἐν τῷ θήλει πάντα πέφυκεν· διαφέρει γὰρ οὐδενὶ τῶν ἔσω πλὴν ταῖς ὑστέραις, ὧν ἡ μὲν ὄψις θεωρείσθω ἐκ τῆς διαγραφῆς τῆς ἐν ταῖς ἀνατομαῖς, ἡ δὲ θέσις ἐστὶν ἐπὶ τοῖς ἐντέροις· ἐπὶ δὲ τῆς ὑστέρας ἡ κύστις. λεκτέον δὲ καὶ περὶ ὑστερῶν κοινῇ πασῶν ἐν τοῖς ἑπομένοις· οὔτε γὰρ ὅμοιαι πᾶσιν οὔθ' ὁμοίως ἔχουσιν.

Τὰ μὲν οὖν μόρια καὶ τὰ ἐντὸς καὶ τὰ ἐκτὸς τοῦ ἀνθρώπου ταῦτα καὶ τοιαῦτα, καὶ τοῦτον ἔχει τὸν τρόπον.

---

[1] fortasse προσειλημένη scribendum.
[2] add. οὖν vulg., om. AC.
[3] sic interpunx. Buss., A.-W.; ν. καὶ χ. ὄν post v. 25 αἰδοῖον Sn.

HISTORIA ANIMALIUM, I. xvii

series, and is attached by the passages which extend from the kidneys along the stalk which extends to the urethra, and nearly all round it is fastened by thin sinewy membranes, in some ways not unlike the thoracic diaphragm. In man, the bladder is of considerable size.

To the stalk of the bladder is attached the privy member: the endmost part of its passage is a single united orifice, but a little further along there are two passages, one leading to the testicles, the other to the bladder; the member itself is sinewy and gristly. To it in males are attached the testicles, the arrangement of which we shall deal with in our general remarks on this subject.

All the parts mentioned are similar in the female as well; there is no difference so far as the internal parts are concerned, except for the uterus, the appearance of which should be studied in the diagram in the *Dissections*[a]; its position is over the bowels, and the bladder is over the uterus. However, we shall have to treat generally later on of the uterus of all animals: it is not the same in all of them, nor has it a similar arrangement.

We have now given a description of the parts, both internal and external, of man, indicating what they are and what is their character.

[a] The *Dissections*, in seven Books, is no longer extant. Aristotle several times refers to " the diagrams in the *Dissections* " (*e.g.*, 525 a 8, *G.A.* 746 a 14; *cf.* 509 b 23, 565 a 13). Jaeger (*Aristotle*, Eng. transl., p. 336), following V. Rose, describes it as " an illustrated, atlas-like work." See also Jaeger, *Diokles von Karystos*, pp. 165-167.

B

497 b

I Τῶν δ' ἄλλων ζῴων τὰ μόρια τὰ μὲν κοινὰ πάντων ἐστίν, ὥσπερ εἴρηται πρότερον, τὰ δὲ γενῶν τινων. ταυτὰ δὲ καὶ ἕτερά ἐστιν ἀλλήλων τὸν ἤδη πολλάκις εἰρημένον τρόπον. σχεδὸν γὰρ ὅσα γ'
10 ἐστὶ γένει ἕτερα τῶν ζῴων, καὶ τὰ πλεῖστα τῶν μερῶν ἔχει ἕτερα τῷ εἴδει, καὶ τὰ μὲν κατ' ἀναλογίαν ἀδιάφορα μόνον, τῷ γένει δ' ἕτερα, τὰ δὲ τῷ γένει μὲν ταυτὰ τῷ εἴδει δ' ἕτερα· πολλὰ δὲ τοῖς μὲν ὑπάρχει, τοῖς δ' οὐχ ὑπάρχει.

Τὰ μὲν οὖν τετράποδα καὶ ζῳοτόκα κεφαλὴν μὲν
15 ἔχει καὶ αὐχένα καὶ τὰ ἐν τῇ κεφαλῇ μόρια ἅπαντα, διαφέρει δὲ τὰς μορφὰς τῶν μορίων ἕκαστον. καὶ ὅ γε λέων τὸ τοῦ αὐχένος ἔχει ἓν ὀστοῦν, σφονδύλους δ' οὐκ ἔχει· τὰ δ' ἐντὸς ἀνοιχθεὶς ὅμοια πάντ' ἔχει κυνί.

Ἔχει δὲ τὰ τετράποδα ζῷα καὶ ζῳοτόκα ἀντὶ τῶν βραχιόνων σκέλη πρόσθια, πάντα μὲν τὰ τετράποδα,
20 μάλιστα δ' ἀνάλογον ταῖς χερσὶ τὰ πολυσχιδῆ αὐτῶν· χρῆται γὰρ πρὸς πολλὰ ὡς χερσίν (καὶ τὰ ἀριστερὰ δ' ἧττον ἔχει ἀπολελυμένα τῶν ἀνθρώ-

BOOK II

Now with regard to animals generally, some of the parts are common to all, as I have said before, whereas others are found in certain classes only. And I have repeatedly stated the manner in which parts are identical or different. Thus, it is practically true to say that those animals which are generically different have the majority of their parts as well different in form; and of the remaining parts some are identical only by analogy,[a] for generically they are different; and others, although identical generically, are different specifically. There are, too, many parts which occur in some animals, but not in others.

I Recapitulation of modes of difference.

To give examples: the viviparous quadrupeds all have a head and a neck and the parts pertaining to the head; but each differs in respect of the shape of these parts. The lion's neck consists of one single bone,[b] and has no vertebrae; but when it is opened up we find that all its internal parts resemble those of the dog.

Examples of differences and correspondences: various.

The viviparous quadrupeds also have forelegs instead of arms: all the quadrupeds, I say, though the polydactylous ones have something very analogous to hands: at any rate, they use them as hands for many purposes—and we must remember that the limbs on their left side are less independent than in

---

[a] *Cf.* 486 a 15 ff., especially 486 b 18 ff., and Notes, §§ 3 ff.
[b] This is true of the whale, but not of the lion.

πων),[1] πλὴν ἐλέφαντος· οὗτος δὲ τά τε περὶ τοὺς δακτύλους ἀδιαρθρωτότερα ἔχει τῶν ποδῶν, καὶ τὰ πρόσθια σκέλη πολλῷ μείζω. ἔστι δὲ πενταδά-
κτυλον, καὶ πρὸς τοῖς ὀπισθίοις σκέλεσι σφυρὰ ἔχει βραχέα. ἔχει δὲ μυκτῆρα τοιοῦτον καὶ τηλικοῦτον ὥστε ἀντὶ χειρῶν ἔχειν αὐτόν· πίνει γὰρ καὶ ἐσθίει ὀρέγων τούτῳ εἰς τὸ στόμα, καὶ τῷ ἐλεφαντιστῇ [ἀνορέγει ἄνω][2] . . . . τούτῳ καὶ δένδρα ἀνασπᾷ, καὶ διὰ τοῦ ὕδατος βαδίζων τούτῳ ἀναφυσᾷ. τῷ δ' ἄκρῳ ἐγκλίνει, οὐ κάμπτεται δέ· χονδρῶδες γὰρ ἔχει.

Μόνον δὲ καὶ ἀμφιδέξιον γίγνεται τῶν ἄλλων ζῴων ἄνθρωπος.

Τῷ δὲ στήθει τῷ τοῦ ἀνθρώπου πάντα τὰ ζῷα ἀνάλογον ἔχει τοῦτο τὸ μόριον, ἀλλ' οὐχ ὅμοιον· ὁ μὲν γὰρ πλατὺ τὸ στῆθος, τὰ δ' ἄλλα στενόν. μαστοὺς δ' οὐκ ἔχει οὐδὲν ἐν τῷ πρόσθεν ἀλλ' ἢ ἄνθρωπος· ὁ δ' ἐλέφας ἔχει μὲν μαστοὺς δύο, ἀλλ' οὐκ ἐν τῷ στήθει, ἀλλὰ πρὸς τῷ στήθει.

Τὰς δὲ κάμψεις τῶν κώλων καὶ τῶν ἔμπροσθεν καὶ τῶν ὄπισθεν ὑπεναντίας ἔχουσι καὶ ἑαυταῖς καὶ

---

[1] seclus. Wiegmann, edd. nonnulli.
[2] seclusi: ἂν ὀρέγει ἄνω τούτῳ C: ἄνω ὀρέγει τούτῳ A: ἀνορέγει ἄνω τοῦτο P: ἀνορέγει τοῦτο ἄνω D: ὀρέγει ἄνω τούτῳ Aldus, A.-W.: πᾶν ὀρέγει ἄνω τούτῳ Dt.: *et per ipsum etiam domino suo cum voluerit eradicat arbores* Σ. lacunam statui; fortasse voc. ὀργάνῳ hic latet [Warmington]. credo equidem καὶ τῷ ἐλεφαντιστῇ . . . ἔχει secludenda.

---

[a] *Cf.* I.A. 706 a 18 ff., where it is said that of all animals man has his left limbs most independent (ἀπολελυμένα μάλιστα) because his construction is most in accord with nature; and by nature the right is better than the left and separate from it. Hence in man the right limbs are *most* right. And since

man *a*—with the exception of the elephant: this animal has somewhat indistinctly articulated toes, and its forelegs are much larger than the hind ones. Still, it has five toes, and on its hind legs it has short ankles. Its nose, however, is of such a kind and of such a size that it can be used instead of hands: its method of eating and drinking is to reach with this organ into its mouth, and for its driver .... ; it even pulls up trees with it, and, when passing through the water, it blows upwards with it.*b* The nose can coil at the tip, but does not bend as a joint, because it contains gristly substance.

Man is the only animal which can actually become ambidextrous.

All animals have a part which is analogous to the chest in man, though dissimilar: in man the chest is broad, in the other animals it is narrow. Only man has breasts in front: the elephant has two, which are near, but not on, the chest.

Apart from the elephant, animals have the flexions of their hind limbs as well as their fore limbs opposite

Flexion of limbs.

the right limbs are distinct (διωρισμένα), it is reasonable that the left limbs should be less mobile, and independent (ἀπολελυμένα) most of all in man. This is perhaps elucidated to some extent by the statement (705 b 20 ff.) that the origin of movement is from the right [side]: the nature of the right is to initiate movement, of the left to be set in movement; hence we step out with the left foot. Aristotle's view is that the better, *i.e.*, the more *natural*, the animal, the more distinct the right is from the left, and hence the more independent (in one sense) the left is from the right. Thus too, in man (706 a 24 ff.) " upper " and " front " are most in accord with nature, *i.e.*, most distinct (διωρισμένα) from " lower " and " rear" respectively. Aristotle holds a similar view about the sexes (see especially *G.A*. 732 a 3 ff.): the higher the animal, the more clearly the sexes are distinguished: in plants they are indistinct.   *b* And breathes; *cf.* 630 b 28.

ARISTOTLE

ταῖς τοῦ ἀνθρώπου καμπαῖς, πλὴν ἐλέφαντος. τοῖς μὲν γὰρ ζῳοτόκοις τῶν τετραπόδων κάμπτεται τὰ μὲν πρόσθια εἰς τὸ πρόσθεν τὰ δ' ὀπίσθια εἰς τοὔπισθεν, καὶ ἔχουσι τὰ κοῖλα τῆς περιφερείας πρὸς ἄλληλα ἀντεστραμμένα· ὁ δ' ἐλέφας οὐχ οὕτως ὥσπερ ἔλεγόν τινες, ἀλλὰ συγκαθίζει καὶ κάμπτει τὰ σκέλη, πλὴν οὐ δύναται διὰ τὸ βάρος ἐπ' ἀμφότερα ἅμα, ἀλλ' ἀνακλίνεται ἢ ἐπὶ τὰ εὐώνυμα ἢ ἐπὶ τὰ δεξιά, καὶ καθεύδει ἐν τούτῳ τῷ σχήματι, κάμπτει δὲ τὰ ὀπίσθια σκέλη ὥσπερ ἄνθρωπος.

Τοῖς ᾠοτόκοις δέ, ὥσπερ κροκοδείλῳ καὶ σαύρᾳ καὶ τοῖς ἄλλοις τοῖς τοιούτοις ἅπασιν, ἀμφότερα τὰ σκέλη καὶ τὰ πρόσθια καὶ τὰ ὀπίσθια εἰς τὸ πρόσθεν κάμπτεται, μικρὸν εἰς τὸ πλάγιον παρεγκλίνοντα. ὁμοίως δὲ καὶ τοῖς ἄλλοις τοῖς πολύποσιν· πλὴν τὰ μεταξὺ τῶν ἐσχάτων ἀεὶ ἐπαμφοτερίζει καὶ τὴν κάμψιν ἔχει εἰς τὸ πλάγιον μᾶλλον. ὁ δ' ἄνθρωπος ἄμφω τὰς καμπὰς τῶν κώλων ἐπὶ ταὐτὸ ἔχει καὶ ἐξ ἐναντίας· τοὺς μὲν γὰρ βραχίονας εἰς τοὔπισθεν κάμπτει, πλὴν μικρὸν βεβλαίσωται[1] ἐπὶ τὰ πλάγια τὰ ἐντός,[2] τὰ δὲ σκέλη εἰς τοὔμπροσθεν. εἰς δὲ τὸ ὄπισθεν τά τε πρόσθια καὶ τὰ ὀπίσθια οὐδὲν κάμπτει τῶν ζῴων. ἐναντίως δὲ τοῖς ἀγκῶσι καὶ τοῖς προσθίοις σκέλεσιν ἡ τῶν ὤμων ἔχει καμπὴ πᾶσι, καὶ ⟨τῇ⟩[3] τῶν ὄπισθεν γονάτων ἡ τῶν ἰσχίων, ὥστ' ἐπεὶ ὁ ἄνθρωπος τοῖς ἄλλοις[4] ἐναντίως κάμπτει, ⟨κάμπτει⟩ καὶ τοὺς ἀγκῶνας[5] ἐναντίως.

Παραπλησίους δὲ τὰς καμπὰς ἔχει καὶ ὁ ὄρνις

to one another in direction, and to the corresponding flexions in man. Thus, in the viviparous quadrupeds the front limbs bend forwards and the hind ones backwards, and the concavities of the curves are therefore turned to face each other. The elephant does not behave as some used to allege, but settles down and bends its legs, though it cannot on account of its weight settle down on both sides simultaneously, but reclines either on to the left or on to the right, and in that posture goes to sleep. Its hind legs it bends just as a human being does.

In the Ovipara, *e.g.*, the crocodile and the lizard and all such creatures, both fore and hind legs bend forwards, with a slight sideways deviation. A similar thing occurs in the many-footed animals, except that the legs between the extreme ones always dualize: in other words, they bend more sideways. Man bends both his pairs of limbs towards the same direction,[a] which means in opposite ways: his arms he bends backwards (except that there is a slight splaying of them inwards) and his legs forwards. No animal at all bends both its forelegs and its hind legs backwards. But in all of them the flexion of the shoulders is opposed to that of the elbows or the foreleg joints, and the flexion of the hips is opposed to that of the knees of the hind legs. Therefore, since man's flexions are contrary to those of the other animals, the flexion of his elbows is contrary too.

The flexion of birds' limbs is comparable to that of

[a] *i.e.*, the flexions are towards the trunk.

[1] ἐβλαίσωται PD, vulg.   [2] τὰ ἑ.] καὶ ἐκτός tent. A.-W.
[3] ⟨τῇ⟩ Th.   [4] ἄλλοις AC : πολλοῖς D, vulg.
[5] ita correxi : οἱ τὰ (τὰ om. P) τοιαῦτ' ἔχοντες D, vulg.: οἱ ταῦτ' ἔχ. AC : τὰ ταῦτ' ἔχ. coni. Dt.

ARISTOTLE

τοῖς τετράποσι ζῴοις· δίπους γὰρ ὢν τὰ μὲν σκέλη εἰς τὸ ὄπισθεν κάμπτει, ἀντὶ δὲ βραχιόνων καὶ σκελῶν τῶν ἔμπροσθεν πτέρυγας ἔχει, ὧν ἡ κάμψις ἐστὶν εἰς τὸ πρόσθεν.

Ἡ δὲ φώκη ὥσπερ πεπηρωμένον ἐστὶ τετράπουν· εὐθὺς γὰρ ἔχει μετὰ τὴν ὠμοπλάτην τοὺς πόδας ὁμοίους χερσίν, ὥσπερ καὶ οἱ τῆς ἄρκτου· πενταδάκτυλοι γάρ εἰσι, καὶ ἕκαστος τῶν δακτύλων καμπὰς ἔχει τρεῖς καὶ ὄνυχα οὐ μέγαν· οἱ δ' ὀπίσθιοι πόδες πενταδάκτυλοι μέν εἰσι, καὶ τὰς καμπὰς καὶ τοὺς ὄνυχας ὁμοίους ἔχουσι τοῖς προσθίοις, τῷ δὲ σχήματι παραπλήσιοι ταῖς τῶν ἰχθύων οὐραῖς εἰσιν.

Αἱ δὲ κινήσεις τῶν ζῴων τῶν μὲν τετραπόδων καὶ πολυπόδων κατὰ διάμετρόν εἰσι, καὶ ἑστᾶσιν οὕτως· ἡ δ' ἀρχὴ ἀπὸ τῶν δεξιῶν πᾶσιν. κατὰ σκέλος δὲ βαδίζουσιν ὅ τε λέων καὶ αἱ κάμηλοι ἀμφότεραι, αἵ τε Βακτριαναὶ καὶ αἱ Ἀράβιαι. τὸ

---

[a] The definition of movement κατὰ διάμετρον given at *I.A.* 712 a 25 ff. is : the right foreleg is moved, then the left hind leg, then the left foreleg, then the right hind leg. (*Cf.* also *I.A.* 712 b 7, where it is stated that horses and similar animals stand with their legs advanced diagonally, and not with both right or both left legs advanced together.) The two methods of progression described there as impossible are (1) first both forelegs together, and (2) first both right legs together. It seems, however, to be established that the camel does in fact walk by putting forward both legs on the same side simultaneously (*i.e.*, the second of the " impossible " methods of *I.A.*) ; and as κατὰ σκέλος in the present passage is clearly intended to be contrasted with κατὰ διάμετρον, no doubt it is

the quadrupeds': a bird has two legs only, yet it bends its legs backwards, and its wings, which replace arms or forelegs, it bends forwards.

The seal is a sort of stunted quadruped. Immediately behind its shoulder-blades are its front feet, similar to hands, like those of the bear, for each has five toes, and each toe has three flexions and a smallish nail. The hind feet also have five toes, and flexions and nails similar to those of the front feet, but in shape they are comparable to the tail of a fish.

The movements of animals, both quadruped and many-footed, are diagonal, and when standing still their stance is diagonal.[a] In all of them the origin of movement is from the right side. The lion, however, and the two species of camel, Bactrian and Arabian, walk laterally; and in lateral walking sometimes the

intended to describe this method of walking. (There is no mention of the camel or of any such exception in *I.A.*; nor is the phrase κατὰ σκέλος there used.) In view of this contrast with κατὰ διάμετρον, I have translated κατὰ σκέλος " laterally," *i.e.*, with both legs moving forward one side at a time (*cf.* the use of σκέλος in ἰσοσκελής). The translation " amble " would fail to bring out the particular contrast between the two Greek phrases, both of which appear to be of geometrical reference.

It seems to be clearly implied in the present passage that in moving off animals normally step out *on* the right foot (or feet, in the case of the camel), although the phrase ἡ ἀρχὴ ἀπὸ τῶν δεξιῶν could, in itself, be interpreted to mean that the right side is the *source* of movement, *i.e.*, the side from which the animal " pushes off," the first step being taken on the left foot. This is the view expressed in *I.A.* 705 b 20 ff., quoted in the note on 497 b 22 above. *Cf.* also *I.A.* 706 a 7 : human beings all step out with the left foot, and when standing they tend to have the left foot in front. Nevertheless, the definition of movement κατὰ διάμετρον given at *I.A.* 712 a 25 (quoted at the beginning of this note) seems to assume that the right limb is put forward first.

δὲ κατὰ σκέλος ἔστιν[1] ὅτε οὐ προβαίνει τῷ ἀριστερῷ τὸ δεξιόν, ἀλλ' ἐπακολουθεῖ.

Ἔχουσι δὲ τὰ τετράποδα ζῷα, ὅσα μὲν ὁ ἄνθρωπος μόρια ἔχει ἐν τῷ πρόσθεν, κάτω ἐν τοῖς ὑπτίοις, τὰ δ' ὀπίσθια ἐν τοῖς πρανέσιν. ἔτι δὲ τὰ πλεῖστα κέρκον ἔχει· καὶ γὰρ ἡ φώκη μικρὰν ἔχει, ὁμοίαν τῇ τοῦ ἐλάφου. περὶ δὲ τῶν πιθηκοειδῶν ζώων ὕστερον διορισθήσεται.

Πάντα δ' ὅσα τετράποδα καὶ ζωοτόκα, δασέα ὡς εἰπεῖν ἐστι, καὶ οὐχ ὥσπερ ὁ ἄνθρωπος, ⟨ὃς⟩[2] ὀλιγότριχον καὶ μικρότριχον πλὴν τῆς κεφαλῆς, τὴν δὲ κεφαλὴν δασύτατον τῶν ζώων. ἔτι δὲ τῶν μὲν ἄλλων ζώων τῶν ἐχόντων τρίχας τὰ πρανῆ δασύτερα, τὰ δ' ὕπτια ἢ λεῖα πάμπαν ἢ ἧττον δασέα· ὁ δ' ἄνθρωπος τοὐναντίον. καὶ βλεφαρίδας ὁ μὲν ἄνθρωπος ἐπ' ἄμφω ἔχει, καὶ ἐν μασχάλαις ἔχει τρίχας καὶ ἐπὶ τῆς ἥβης· τῶν δ' ἄλλων οὐδὲν οὔτε τούτων οὐδέτερον οὔτε τὴν κάτω βλεφαρίδα, ἀλλὰ κάτωθεν τοῦ βλεφάρου ἐνίοις μαναὶ τρίχες πεφύκασιν. αὐτῶν δὲ τῶν τετραπόδων καὶ τρίχας ἐχόντων τῶν μὲν ἅπαν τὸ σῶμα δασύ, καθάπερ ὑὸς καὶ ἄρκτου καὶ κυνός· τὰ δὲ δασύτερα τὸν αὐχένα ὁμοίως πάντῃ, οἷον ὅσα χαίτην ἔχει, ὥσπερ λέων· τὰ δ' ἐπὶ τῷ πρανεῖ τοῦ αὐχένος ἀπὸ τῆς κεφαλῆς μέχρι τῆς ἀκρωμίας, οἷον ὅσα λοφιὰν ἔχει, ὥσπερ ἵππος καὶ ὀρεὺς καὶ τῶν ἀγρίων καὶ κερατοφόρων βόνασος.

---

[1] ἔστιν Warmington : ἐστὶν vulg., edd.
[2] ⟨ὃς⟩ Dt.

---

[a] I am indebted to my friend Professor E. H. Warmington for the solution of this sentence, a solution which consists in

right foot is not advanced before the left, but follows it.[a]

Whatever parts man has in the forefront of the body, the animals have on the underpart, *i.e.*, on the belly; and man's rear parts they have on their dorsal sides. Most of them have a tail: even the seal has a small one, similar to that of the deer. As for the apelike animals, we will distinguish their characteristics later.

All viviparous quadrupeds are, we may say, covered with hair, unlike man, who has only a few short hairs except on his head, though in respect of his head he has a thicker coat of hair than any animal. Furthermore, animals which are coated with hair have a thicker coat on their backs than on their bellies, which are either completely hairless or less hairy. With man the situation is reversed. Man also has eyelashes, upper and lower, and hair under the armpits and on the pubes. No other animal has either of the latter two, or the lower eyelashes, though some animals have some sparse hairs growing under the eyelid. Some of the hairy-coated quadrupeds have thick hair all over the body, *e.g.*, the pig, the bear and the dog; others are especially hairy all round the neck, *e.g.*, those which have a shaggy mane, like the lion; others are especially hairy on the upper surface of the neck from the head as far as the withers, *e.g.*, those which have a crested mane, like the horse, the mule and the bison (an example from the wild horned animals).

INSTRU-
MENTAL
PARTS OF
BLOODED
ANIMALS:
(i) Vivi-
para:
(c) Exter-
nal parts
(resumed):
Hair.

substituting ἔστιν for ἐστὶν, thereby removing the necessity for regarding the sentence as a definition of κατὰ σκέλος, which is how previous interpreters have attempted to take it, without much success. προβαίνειν is used elsewhere with the accusative of the foot moved, and I have so taken it here; the dative τῷ ἀριστερῷ is however somewhat strange.

ARISTOTLE

Ἔχει δὲ καὶ ὁ ἱππέλαφος καλούμενος ἐπὶ τῇ ἀκρωμίᾳ χαίτην καὶ τὸ θηρίον τὸ πάρδιον ὀνομαζόμενον· ἀπὸ δὲ τῆς κεφαλῆς ἐπὶ τὴν ἀκρωμίαν λεπτὴν ἑκάτερον· ἰδίᾳ δ' ὁ ἱππέλαφος πώγωνα ἔχει κατὰ τὸν λάρυγγα. ἔστι δ' ἀμφότερα κερατοφόρα καὶ διχαλά· ἡ δὲ θήλεια ἱππέλαφος οὐκ ἔχει κέρατα. τὸ δὲ μέγεθός ἐστι τούτου τοῦ ζῴου ἐλάφῳ προσεμφερές. γίγνονται δ' οἱ ἱππέλαφοι ἐν Ἀραχώταις, οὗπερ καὶ οἱ βόες οἱ ἄγριοι. διαφέρουσι δ' οἱ ἄγριοι τῶν ἡμέρων ὅσον περ οἱ ὗες οἱ ἄγριοι πρὸς τοὺς ἡμέρους· μέλανές τε γάρ εἰσι καὶ ἰσχυροὶ τῷ εἴδει καὶ ἐπίγρυποι, τὰ δὲ κέρατα ἐξυπτιάζοντα ἔχουσι μᾶλλον· τὰ δὲ τῶν ἱππελάφων κέρατα παραπλήσια τοῖς τῆς δορκάδος ἐστίν. ὁ δ' ἐλέφας ἥκιστα δασύς ἐστι τῶν τετραπόδων. ἀκολουθοῦσι δὲ κατὰ τὸ σῶμα καὶ αἱ κέρκοι δασύτητι καὶ ψιλότητι, ὅσων αἱ κέρκοι μέγεθος ἔχουσιν· ἔνια γὰρ μικρὰν ἔχει πάμπαν.

Αἱ δὲ κάμηλοι ἴδιον ἔχουσι παρὰ τἆλλα τετράποδα τὸν καλούμενον ὕβον ἐπὶ τῷ νώτῳ. διαφέρουσι δ' αἱ Βάκτριαι τῶν Ἀραβίων· αἱ μὲν γὰρ δύο ἔχουσιν ὕβους, αἱ δ' ἕνα μόνον· ἄλλον δ' ἔχουσιν ὕβον τοιοῦτον οἷον ἄνω ἐν τοῖς κάτω, ἐφ' οὗ, ὅταν κατακλιθῇ εἰς γόνατα, ἐστήρικται τὸ ἄλλο σῶμα. θηλὰς δ' ἔχει τέτταρας ἡ κάμηλος ὥσπερ βοῦς, καὶ κέρκον ὁμοίαν ὄνῳ· τὸ δ' αἰδοῖον ὄπισθεν. καὶ γόνυ δ' ἔχει ἐν ἑκάστῳ τῷ σκέλει ἕν, καὶ τὰς καμπὰς οὐ πλείους, ὥσπερ λέγουσί τινες, ἀλλὰ φαίνεται διὰ τὴν ὑπόσταλσιν[1] τῆς κοιλίας· καὶ ἀστράγαλον

---

[1] ὑπόσταλσιν Sn. : ὑπόστασιν codd., vulg.

---

[a] Lit., "horse-deer," probably the nylghau (= " blue

HISTORIA ANIMALIUM, II. 1

The hippelaphus [a] (as it is called) also has a mane on its withers, and so has the beast known as the pardion [b] : in each of these the mane is a thin one extending from the head to the withers. The hippelaphus is peculiar in having a beard beside the larynx. Both these animals have horns and are cloven-footed, though the female hippelaphus is hornless. In size this animal is comparable to the deer. It is found in Arachotae,[c] where also the wild oxen are found. Wild oxen [d] differ from domesticated ones just as the wild boar differs from the domesticated. They are black, of strong physique, with a hooked muzzle, and their horns tend rather to tilt backwards. The horns of the hippelaphus resemble those of the gazelle. The elephant is the least hairy of the quadrupeds. In those animals which have tails of considerable size, the tail tends to correspond to the rest of the body in respect of the thickness and thinness of its hair. This does not apply to those which have very small tails.

Camels have a peculiarity which distinguishes them from all other quadrupeds : this is what is known as the " hump " on their backs. The Bactrian camel differs from the Arabian in having two humps as against the latter's one, though this has a sort of hump below like the one above, on which it takes the weight of the whole body when it kneels down. The camel has four teats, like the cow, and a tail similar to the ass's ; its privy member is directed backwards. On each leg it has one knee : the limb has not several flexions, as some allege, although there appear to be several owing to the contracted formation of the

Peculiarities of the camel.

---

bull "), *Boselaphus tragocamelus*, a large short-horned Indian antelope. *Cf.* add. n., p. 237.   [b] Not otherwise known.
[c] Baluchistan, now included in Pakistan.   [d] Buffaloes.

499 a

ὅμοιον μὲν βοΐ, ἰσχνὸν δὲ καὶ μικρὸν ὡς κατὰ τὸ μέγεθος. ἔστι δὲ διχαλὸν καὶ οὐκ ἄμφωδον, διχαλὸν δ' ὧδε. ἐκ μὲν τοῦ ὄπισθεν μικρὸν ἔσχισται μέχρι τῆς δευτέρας καμπῆς τῶν δακτύλων· τὸ δ' ἔμπροσθεν ἔσχισται μακράν,[1] ὅσον ἄχρι τῆς πρώτης καμπῆς τῶν δακτύλων, ἐπ' ἄκρῳ τέτταρα[2]· καὶ ἔστι τι καὶ διὰ μέσου τῶν σχισμάτων, ὥσπερ τοῖς χησίν. ὁ δὲ πούς ἐστι κάτωθεν σαρκώδης, ὥσπερ καὶ οἱ τῶν[3] ἄρκτων· διὸ καὶ τὰς εἰς πόλεμον ἰούσας ὑποδοῦσι καρβατίναις, ὅταν ἀλγήσωσιν.

Πάντα δὲ τὰ τετράποδα ὀστώδη τὰ σκέλη ἔχει καὶ νευρώδη καὶ ἄσαρκα· ὅλως δὲ καὶ τἆλλα ζῷα

499 b

ἅπαντα, ὅσα ἔχει πόδας, ἐκτὸς ἀνθρώπου. ἔστι δὲ καὶ ἀνίσχια· καὶ γὰρ οἱ ὄρνιθες ἔτι μᾶλλον τοῦτο πεπόνθασιν. ὁ δ' ἄνθρωπος τοὐναντίον· σαρκώδη γὰρ ἔχει σχεδὸν μάλιστα τοῦ σώματος τὰ ἰσχία καὶ τοὺς μηροὺς καὶ τὰς κνήμας· αἱ γὰρ καλούμεναι γαστροκνημίαι ἐν ταῖς κνήμαις εἰσὶ σαρκώδεις.

Τῶν δὲ τετραπόδων καὶ ἐναίμων καὶ ζῳοτόκων τὰ μέν ἐστι πολυσχιδῆ, ὥσπερ αἱ τοῦ ἀνθρώπου χεῖρες καὶ οἱ πόδες (πολυδάκτυλα γὰρ ἔνιά ἐστιν, οἷον κύων, λέων, πάρδαλις), τὰ δὲ δισχιδῆ, καὶ ἀντὶ τῶν ὀνύχων χηλὰς ἔχει, ὥσπερ πρόβατον καὶ αἲξ καὶ ἔλαφος καὶ ἵππος ὁ ποτάμιος· τὰ δ' ἀσχιδῆ, οἷον τὰ μώνυχα, ὥσπερ ἵππος καὶ ὀρεύς. τὸ δὲ τῶν ὑῶν γένος ἐπαμφοτερίζει· εἰσὶ γὰρ καὶ ἐν Ἰλλυριοῖς καὶ ἐν Παιονίᾳ καὶ ἄλλοθι μώνυχες ὕες.

---

[1] μακράν Dt.: μικρά vulg.
[2] ἄκρων PD; δέ loco τέτταρα Dt.; τὸ δ' ἔμπροσθεν ἔχει μικρὰ ὀνύχια τῆς πρώτης καμπῆς τῶν δακτύλων ἐπ' ἀκροτάτῳ A.-W.: μέχρι τῆς δευτέρας ... τέτταρα om. Σ. [3] τῶν A: om. vulg.

belly. It has a huckle-bone similar to that of the ox, but small and paltry for the animal's bulk. It is cloven-footed, and not ambidentate.[a] It is cloven-footed in the following way : at the rear there is a small cleft reaching as far as the second joint of the toes ; in front there is a cleft for a long distance, right along as far as the first joint of the toes [b] ; and there is something stretched across the clefts, like the webbing on a goose's foot. The foot is fleshy underneath, like that of the bear ; and that is why when camels are taken on war service and get footsore, their keepers strap leather shoes on their feet.

All quadrupeds have legs which are bony, sinewy, and fleshless. So too have all footed animals, apart from man. Also, they lack buttocks : this absence is even more marked in birds. With man the reverse is true : his buttocks, thighs and calves are almost the fleshiest part of the whole body—the so-called *gastroknemiai* on the calves being fleshy.

Some among the blooded viviparous quadrupeds Feet. have many-cloven feet, like the human foot and hand (certain animals being polydactylous, *e.g.*, the dog, the lion, and the leopard) ; others of them have feet cloven into two, and have hooves instead of nails, *e.g.*, the sheep, the goat, the deer, and the hippopotamus. Some have uncloven feet, examples are the solid-hoofed animals, such as the horse and the mule. Swine as a tribe are dualizers [c] : ⟨some are cloven-hoofed, whereas⟩ there are solid-hoofed swine in Il-

---

[a] *i.e.*, has no front teeth in the upper jaw.

[b] I have omitted from the translation the words ἐπ' ἄκρῳ (or ἄκρων) τέτταρα, which seem unintelligible. A.-W. rewrite the phrase as follows : " The front has small hooves at the very tip of the first joint of the toes."

[c] See note on 488 a 2, and Notes, §§ 28 ff.

τὰ μὲν οὖν διχαλὰ δύο ἔχει σχίσεις ὄπισθεν· τοῖς δὲ μώνυξι τοῦτ' ἐστὶ συνεχές.

Ἔστι δὲ καὶ τὰ μὲν κερατοφόρα τῶν ζώων τὰ δ' ἄκερα. τὰ μὲν οὖν πλεῖστα τῶν ἐχόντων κέρατα διχαλὰ κατὰ φύσιν ἐστίν, οἷον βοῦς καὶ ἔλαφος καὶ αἴξ· μώνυχον δὲ καὶ δίκερων οὐδὲν ἡμῖν ὦπται. μονοκέρατα δὲ καὶ μώνυχα ὀλίγα, οἷον ὁ Ἰνδικὸς ὄνος. μονόκερων δὲ καὶ διχαλὸν ὄρυξ. καὶ ἀστράγαλον δ' ὁ Ἰνδικὸς ὄνος ἔχει τῶν μωνύχων μόνον· ἡ γὰρ ὗς, ὥσπερ ἐλέχθη πρότερον, ἐπαμφοτερίζει, διὸ καὶ οὐ καλλιαστράγαλόν ἐστιν. τῶν δὲ διχαλῶν πολλὰ ἔχει ἀστράγαλον. πολυσχιδὲς δ' οὐδὲν ὦπται τοιοῦτον ἔχον ἀστράγαλον, ὥσπερ οὐδ' ἄνθρωπος, ἀλλ' ἡ μὲν λὺγξ ὅμοιον ἡμιαστραγαλίῳ, ὁ δὲ λέων, οἷόν περ πλάττουσι, λαβυρινθώδη. πάντα δὲ τὰ ἔχοντα ἀστράγαλον ἐν τοῖς ὄπισθεν ἔχει σκέλεσιν. ἔχει δ' ὀρθὸν τὸν ἀστράγαλον ἐν τῇ καμπῇ, τὸ μὲν πρανὲς ἔξω, τὸ δ' ὕπτιον εἴσω, καὶ τὰ μὲν κῷα ἐντὸς ἐστραμμένα πρὸς ἄλληλα, τὰ δὲ χῖα[1] καλούμενα ἔξω, καὶ τὰς κεραίας ἄνω. ἡ μὲν οὖν θέσις τῶν ἀστραγάλων τοῖς ἔχουσι πᾶσι τοῦτον ἔχει τὸν τρόπον.

[1] κῷα et χῖα Iunt. e Gaza (*veneres . . . canes*): κῶλα et ἰσχία codd.

---

[a] Probably the rhinoceros; it and the oryx are mentioned

HISTORIA ANIMALIUM, II. i

lyria, Paeonia and elsewhere. The cloven-hoofed animals have two clefts at the rear of the foot ; in the solid-hoofed this part is continuous.

Further, some animals are horned, some hornless. Most of the horned ones are cloven-hoofed, *e.g.*, the ox, the deer, and the goat ; we have seen no solid-hoofed animal with a pair of horns. But a few, *e.g.*, the Indian ass,[a] have a single horn and are solid-hoofed. The oryx has a single horn and cloven hooves. The only solid-hoofed animal with a huckle-bone is the Indian ass—as we said before, the pig is a dualizer, and therefore it has no proper huckle-bone. The huckle-bone is present in many of the cloven-hoofed animals. No polydactylous animal has been observed to possess a huckle-bone of this sort, any more than man has. Nevertheless, the lynx has one like a half-astragal,[b] and the lion has one like the " labyrinth " used in moulding. All those which have a huckle-bone have it in the hind legs. The huckle-bone is set upright in the joint : the upper part outside and the lower part inside : the sides known as the Coan are inside, turned towards each other, the sides known as the Chian [c] are outside, and the *keraiai* (" horns ") are on top. So this is the position of the huckle-bone in all animals that have one.

Horns and huckle-bones.

together at *P.A.* 663 a 23. The oryx is a North African antelope. See add. note, p. 237.

[b] Lit., half-hucklebone. " Labyrinth fret," or simply " fret," is still known as an architectural term (*O.E.D.*). According to Thompson, the lion's astragalus actually has a spiral twist. On these terms see further, add. note, p. 238.

[c] These are dicing terms, restored by the Juntine edition from Gaza's version. The Coan throw, with the inner side of the hucklebone uppermost, counted highest, the Chian (or " dog ") counted lowest.

Διχαλὰ δ' ἅμα καὶ χαίτην ἔχοντα καὶ κέρατα δύο κεκαμμένα εἰς αὑτά ἐστιν ἔνια τῶν ζῴων, οἷον ὁ βόνασος, ὃς γίγνεται περὶ τὴν Παιονίαν καὶ τὴν Μαιδικήν.[1] πάντα δ' ὅσα κερατοφόρα, τετράποδά ἐστιν, εἰ μή τι κατὰ μεταφορὰν λέγεται ἔχειν κέρας καὶ λόγου χάριν, ὥσπερ τοὺς περὶ Θήβας ὄφεις οἱ Αἰγύπτιοί φασιν, ἔχοντας ἐπανάστασιν ὅσον προφάσεως χάριν. τῶν δ' ἐχόντων κέρας δι' ὅλου μὲν ἔχει στερεὸν μόνον ἔλαφος, τὰ δ' ἄλλα κοῖλα μέχρι τινός, τὸ δ' ἔσχατον στερεόν. τὸ μὲν οὖν κοῖλον ἐκ τοῦ δέρματος πέφυκε μᾶλλον· περὶ ὃ[2] δὲ τοῦτο περιήρμοσται, τὸ στερεόν, ἐκ τῶν ὀστῶν, οἷον τὰ κέρατα τῶν βοῶν. ἀποβάλλει δὲ τὰ κέρατα μόνον ἔλαφος κατ' ἔτος, ἀρξάμενος ἀπὸ διετοῦς, καὶ πάλιν φύει· τὰ δ' ἄλλα συνεχῶς ἔχει, ἐὰν μή τι βίᾳ πηρωθῇ.

Ἔτι δὲ περί τε τοὺς μαστοὺς ὑπεναντίως ἐν τοῖς ἄλλοις ζῴοις ὑπάρχει πρὸς αὐτά[3] τε καὶ πρὸς τὸν ἄνθρωπον, καὶ περὶ τὰ ὄργανα τὰ χρήσιμα πρὸς τὴν ὀχείαν. τὰ μὲν γὰρ ἔμπροσθεν ἔχει τοὺς μαστοὺς ἐν τῷ στήθει ἢ πρὸς τῷ στήθει, καὶ δύο μαστοὺς καὶ δύο θηλάς, ὥσπερ ἄνθρωπος καὶ ἐλέφας, καθάπερ εἴρηται πρότερον. καὶ γὰρ ὁ ἐλέφας ἔχει τοὺς μαστοὺς δύο περὶ τὰς μασχάλας· ἔχει δ' ἡ θήλεια τοὺς μαστοὺς μικροὺς παντελῶς καὶ οὐ κατὰ λόγον τοῦ σώματος, ὥστ' ἐκ τοῦ πλαγίου μὴ πάνυ ὁρᾶν· ἔχουσι δὲ καὶ οἱ ἄρρενες μαστούς, ὥσπερ αἱ θήλειαι, μικροὺς παντελῶς. ἡ δ' ἄρκτος τέτταρας. τὰ δὲ δύο μὲν μαστοὺς ἔχει, ἐν τοῖς μη-

---

[1] Sylb.: μηδικήν codd.
[2] ὃ AC: om. vulg.; cf. *P.A.* 663 b 17 sqq.
[3] αὑτά Dt.: αὐτά vulg.

Certain animals have, simultaneously, cloven hoofs, a mane, and two horns bent inwards towards each other, *e.g.*, the bison, which is found in Paeonia and Maedica.[a] All horned animals are quadrupeds, if we except any which are said metaphorically, by a figure of speech, to be horned : for example, the serpents in the neighbourhood of Thebes are so described by the Egyptians : actually they have a protuberance on the head, enough to excuse the description. The deer is the only horned animal whose horns (antlers) are solid throughout. Other animals' horns are hollow for a certain distance, and solid at the tip. The hollow part grows out rather from the skin, whereas the solid central part round which this is fitted [b] grows out from the bones : an instance is the horns of oxen. The deer is the only animal which sheds its horns ; it does this every year as soon as it is two years old, and then grows them again. Other animals keep their horns continuously, unless they are damaged by force.

Furthermore, in regard to breasts and generative organs, animals differ among themselves and also from man. Some have their breasts in front, on the chest or near it, and these have two breasts and two teats as do man and the elephant, as I have said before. The elephant has its two breasts toward the axillae ; the female has two quite small breasts, quite out of keeping with the size of its frame, indeed from the side they cannot be seen at all. The males also have very small breasts, like the females. The she-bear has four. Again, some animals have two breasts,

Breasts and generative organs.

[a] In Northern Macedonia.
[b] The insertion of ὅ, which Bekker's text omits, makes sense of this statement : *cf.* the passage at *P.A.* 663 b 17 (Loeb ed.), where a somewhat similar correction is required.

ροῖς δ' ἔχει, καὶ τὰς θηλὰς δύο, ὥσπερ πρόβατον· τὰ δὲ τέτταρας θηλάς, ὥσπερ βοῦς. τὰ δ' οὔτ' ἐν τῷ στήθει ἔχει τοὺς μαστοὺς οὔτ' ἐν τοῖς μηροῖς, ἀλλ' ἐν τῇ γαστρί, οἷον κύων καὶ ὗς, καὶ πολλούς, οὐ πάντας δ' ἴσους. τὰ μὲν οὖν ἄλλα πλείους ἔχει, ἡ δὲ πάρδαλις τέτταρας ἐν τῇ γαστρί, ἡ δὲ λέαινα δύο ἐν τῇ γαστρί. ἔχει δὲ καὶ ἡ κάμηλος μαστοὺς δύο καὶ θηλὰς τέτταρας, ὥσπερ¹ βοῦς. τῶν δὲ μωνύχων τὰ ἄρρενα οὐκ ἔχουσι μαστούς, πλὴν ὅσα ἐοίκασι τῇ μητρί, ὅπερ συμβαίνει ἐπὶ τῶν ἵππων.

Τὰ δ' αἰδοῖα τῶν μὲν ἀρρένων τὰ μὲν ἔξω ἔχει, οἷον ἄνθρωπος καὶ ἵππος καὶ ἄλλα πολλά, τὰ δ' ἐντός, ὥσπερ δελφίς· καὶ τῶν ἔξω δ' ἐχόντων τὰ μὲν εἰς τὸ πρόσθεν, ὥσπερ καὶ τὰ εἰρημένα, καὶ τούτων τὰ μὲν ἀπολελυμένα καὶ τὸ αἰδοῖον καὶ τοὺς ὄρχεις, ὥσπερ ἄνθρωπος, τὰ δὲ πρὸς τῇ γαστρὶ καὶ τοὺς ὄρχεις καὶ τὸ αἰδοῖον, καὶ τὰ μὲν μᾶλλον τὰ δ' ἧττον ἀπολελυμένα· οὐ γὰρ ὡσαύτως ἀπολέλυται κάπρῳ καὶ ἵππῳ τοῦτο τὸ μόριον. ἔχει δὲ καὶ ὁ ἐλέφας τὸ αἰδοῖον ὅμοιον μὲν ἵππῳ, μικρὸν δὲ καὶ οὐ κατὰ λόγον τοῦ σώματος, τοὺς δ' ὄρχεις οὐκ ἔξω φανερούς, ἀλλ' ἐντὸς περὶ τοὺς νεφρούς· διὸ καὶ ἐν τῇ ὀχείᾳ ἀπαλλάττεται ταχέως. ἡ δὲ θήλεια τὸ αἰδοῖον ἔχει ἐν ᾧ τόπῳ τὰ οὔθατα τῶν προβάτων ἐστίν· ὅταν δ' ὀργᾷ ὀχεύεσθαι, ἀνασπᾷ ἄνω καὶ ἐκτρέπει πρὸς τὸν ἔξω τόπον, ὥστε ῥᾳδίαν εἶναι τῷ ἄρρενι τὴν ὀχείαν· ἀνέρρωγε δ' ἐπιεικῶς ἐπὶ πολὺ τὸ αἰδοῖον.

Τοῖς μὲν οὖν πλείστοις αὐτῶν τὰ αἰδοῖα τοῦτον

but near the thighs, and two teats, as the sheep; others have four teats, as the cow. Some have their breasts neither on the chest nor near the thighs, but on the belly, for example the dog and the pig, and have many of them, though not all are equal in size. While other animals have numerous ones, the she-leopard has four, and the lioness two, on the belly. The she-camel also has two breasts and four teats, like the cow. The male solid-hoofed animals have no breasts, other than those which take after their dams: this occurs with horses.[a]

In some male animals the generative organs are external (examples are man, the horse, and many others), in others internal (example, the dolphin). In some of those where the organs are external, they are in front (examples as just quoted); and of these, some have both penis and testicles clear of the body (example, man), others have them closely adhering to the belly, some more, some less so: thus, in the wild boar and the horse this part does not stand clear in the same way. The penis of the elephant resembles that of the horse, though it is disproportionately small for the animal's size; the testicles are not visible externally, being placed within near the kidneys, and for this reason the male quickly frees itself in intercourse. The genital organ of the female is situated where the udder is in sheep, and when she is on heat, she draws it back and protrudes it externally, so as to facilitate intercourse for the male; and the organ opens out quite considerably.

In most animals the genital organs are situated as

[a] *Cf. P.A.* 688 b 33.

---

[1] ὁ add. codd., vulg.: om. Dt.

ARISTOTLE

ἔχει τὸν τρόπον· ἔνια δ' ὀπισθουρητικά ἐστιν, οἷον λύγξ καὶ λέων καὶ κάμηλος καὶ δασύπους. τὰ μὲν οὖν ἄρρενα ὑπεναντίως ἔχει ἀλλήλοις, καθάπερ εἴρηται, τὰ δὲ θήλεα πάντα ὀπισθουρητικά ἐστιν· καὶ γὰρ ὁ θῆλυς ἐλέφας, ἔχων[1] τὰ αἰδοῖα ὑπὸ τοῖς μηροῖς, καθάπερ καὶ τἆλλα.

Τῶν δ' αἰδοίων διαφορὰ πολλή ἐστιν. τὰ μὲν γὰρ ἔχει χονδρῶδες τὸ αἰδοῖον καὶ σαρκῶδες, ὥσπερ ἄνθρωπος· τὸ μὲν οὖν σαρκῶδες οὐκ ἐμφυσᾶται, τὸ δὲ χονδρῶδες ἔχει αὔξησιν. τὰ δὲ νευρώδη, οἷον καμήλου καὶ ἐλάφου, τὰ δ' ὀστώδη, ὥσπερ ἀλώπεκος καὶ λύκου καὶ ἴκτιδος καὶ γαλῆς· καὶ γὰρ ἡ γαλῆ ὀστέινον[2] ἔχει τὸ αἰδοῖον.

Πρὸς δὲ τούτοις ὁ μὲν ἄνθρωπος τελεωθεὶς τὰ ἄνω ἔχει ἐλάττω τῶν κάτωθεν, τὰ δ' ἄλλα ζῷα, ὅσα ἔναιμα, τοὐναντίον. λέγομεν δ' ἄνω τὸ ἀπὸ κεφαλῆς μέχρι τοῦ μορίου ᾗ ἡ τοῦ περιττώματός ἐστιν ἔξοδος, κάτω δὲ τὸ ἀπὸ τούτου λοιπόν. τοῖς μὲν οὖν ἔχουσι πόδας τὸ ὀπίσθιόν ἐστι σκέλος τὸ κάτωθεν μέρος πρὸς τὸ μέγεθος, τοῖς δὲ μὴ ἔχουσιν οὐραὶ καὶ κέρκοι καὶ τὰ τοιαῦτα.

Τελεούμενα μὲν οὖν τοιαῦτ' ἐστίν, ἐν δὲ τῇ αὐξήσει διαφέρει· ὁ μὲν γὰρ ἄνθρωπος μείζω τὰ ἄνω ἔχει νέος ὢν ἢ τὰ κάτω, αὐξανόμενος δὲ μεταβάλλει τοὐναντίον (διὸ καὶ μόνον οὐ τὴν αὐτὴν ποιεῖται

---

[1] ἔχων A.-W. (καίπερ ἔχων Pi.): ἔχει vulg.
[2] ὀστέινον scripsi Dt. docente; ὀστοῦν vulg.

---

[a] Aristotle's nomenclature here is due to his belief that man's upright posture is the normal and natural one. *Cf.* 494 a 27 ff., and note on 497 b 23.

[b] I have retained Aristotle's phrase "upper parts," although he clearly means the upper *portion* of the body; and so throughout this paragraph. What he has in mind is the

described. But some animals discharge their urine to the rear, as do the lynx, the lion, the camel, and the hare. Now male animals, as stated already, differ from one another in this respect; but all females are retromingent, even the female elephant, just like the others, although her privy parts are below the thighs.

The male organ shows much diversity. In some it consists of gristle and flesh, as in man; and the fleshy part does not become inflated, while the gristly part becomes enlarged. In some it is sinewy, as in the camel and the deer; in others, bony, as in the fox, the wolf, the marten, and the weasel: it is a fact that the weasel's is bony.

Furthermore, when man has reached his full development the upper parts of his body are less in overall length than the lower, whereas the reverse is true of the other blooded animals. By the "upper parts" I mean everything from the head as far as the part which serves for the exit of the residue, and by the "lower parts" all the remainder from that point. In footed animals the "hind" leg counts as "lower" for this purpose of comparative length; in footless animals the tail, rump, etc.[a]

Relative bulk of parts.

When they reach their full development, then, animals are as just described; but while still growing they show differences. Man, for instance, when young has his upper parts [b] greater than the lower, but as he grows he changes over to the reverse state [c]; and that is why he alone moves about in a different

comparative overall length of the two portions. *Cf.* 502 b 15 below.

[c] Aristotle again refers to the difference in the relative sizes of parts of the body at different stages of development at *G.A.* 741 b 27 ff., 742 b 12 ff. On "heterauxesis" see J. S. Huxley, *Problems of Relative Growth*, London, 1932.

ARISTOTLE

κίνησιν τῆς πορείας νέος ὢν καὶ τελεωθείς, ἀλλὰ τὸ πρῶτον παιδίον ὢν ἕρπει τετραποδίζων,)[1] τὰ δ' ἀνὰ λόγον ἀποδίδωσι τὴν αὔξησιν, οἷον κύων. ἔνια δὲ τὸ πρῶτον ἐλάττω τὰ ἄνω, τὰ δὲ κάτω μείζω ἔχει, αὐξανόμενα δὲ τὰ ἄνω γίγνεται μείζω, ὥσπερ τὰ λόφουρα· τούτων γὰρ οὐδὲν γίγνεται μεῖζον ὕστερον τὸ ἀπὸ τῆς ὁπλῆς μέχρι τοῦ ἰσχίου.

Ἔστι δὲ καὶ περὶ τοὺς ὀδόντας πολλὴ διαφορὰ τοῖς ἄλλοις ζῴοις καὶ πρὸς αὐτὰ καὶ πρὸς τὸν ἄνθρωπον. ἔχει μὲν γὰρ πάντα ὀδόντας ὅσα τετράποδα καὶ ἔναιμα καὶ ζῳοτόκα, ἀλλὰ πρῶτον τὰ μέν ἐστιν ἀμφώδοντα τὰ δ' οὔ. ὅσα μὲν γάρ ἐστι κερατοφόρα, οὐκ ἀμφώδοντά ἐστιν· οὐ γὰρ ἔχει τοὺς προσθίους ὀδόντας ἐπὶ τῆς ἄνω σιαγόνος. ἔστι δ' ἔνια οὐκ ἀμφώδοντα καὶ ἀκέρατα, οἷον κάμηλος. καὶ τὰ μὲν χαυλιόδοντας ἔχει, ὥσπερ οἱ ἄρρενες ὕες, τὰ δ' οὐκ ἔχει. ἔτι δὲ τὰ μέν ἐστι καρχαρόδοντα αὐτῶν, οἷον λέων καὶ πάρδαλις καὶ κύων, τὰ δ' ἀνεπάλλακτα, οἷον ἵππος καὶ βοῦς· καρχαρόδοντα γάρ ἐστιν ὅσα ἐπαλλάττει τοὺς ὀδόντας τοὺς ὀξεῖς. ἅμα δὲ χαυλιόδοντα καὶ κέρας οὐδὲν ἔχει ζῷον, οὐδὲ καρχαρόδουν καὶ τούτων θάτερον. τὰ δὲ πλεῖστα τοὺς προσθίους ἔχει ὀξεῖς, τοὺς δ' ἐντὸς πλατεῖς. ἡ δὲ φώκη καρχαρόδουν ἐστὶ πᾶσι τοῖς ὀδοῦσιν, ὡς ἐπαλλάττουσα τῷ γένει τῶν ἰχθύων· οἱ γὰρ ἰχθύες πάντες σχεδὸν καρχαρόδοντές εἰσιν. διστοίχους δ' ὀδόντας οὐδὲν ἔχει τούτων τῶν γενῶν. ἔστι δέ τι, εἰ δεῖ πιστεῦσαι Κτησίᾳ· ἐκεῖνος γὰρ τὸ ἐν Ἰνδοῖς θηρίον, ᾧ ὄνομα εἶναι μαρτιχόραν, τοῦτ'

HISTORIA ANIMALIUM, II. i

manner when young and when fully developed; as a child to begin with he creeps about on all fours. Some animals, such as the dog, maintain the relative proportions as they grow up. Some to begin with have their upper parts less and the lower greater, and as they grow up the upper parts become greater, *e.g.*, the bushy-tailed animals [a]: in none of these do we find that at a later stage the portion from the hoof to the haunch becomes greater.

Again, with regard to teeth the animals exhibit great differences both from each other and from man. All blooded viviparous quadrupeds have teeth, but to begin with some have teeth in both jaws while others have not. Thus horned animals have not teeth in both jaws: they have no front teeth in the upper jaw; and some, like the camel, lack these front teeth although they have no horns. And some have tusks, *e.g.*, the boar; others have not. Furthermore, some of them are saw-toothed, *e.g.*, the lion, the leopard, and the dog; some have teeth which do not interlock, *e.g.*, the horse and the ox. (By " saw-toothed " is meant those animals whose sharp-pointed teeth interlock.) No animal has both tusks and horns; and no animal which possesses saw-teeth has either tusks or horns. In most animals the front teeth are sharp and the inner ones broad and flat. The seal is completely saw-toothed, being a sort of link with the class of fishes, nearly all of which are saw-toothed. No animal of these classes has double rows of teeth. There is, however, one such, if we are to believe Ktesias. He asserts that the Indian beast which bears

Teeth.

(The *martichoras*.)

---

[a] See note on 491 a 1.

---

[1] Sn.: ὅν ... τετραποδίζον vulg.

ἔχειν ἐπ' ἀμφότερά φησι τριστοίχους τοὺς ὀδόντας· εἶναι δὲ μέγεθος μὲν ἡλίκον λέοντα καὶ δασὺ ὁμοίως, καὶ πόδας ἔχειν ὁμοίους, πρόσωπον δὲ καὶ ὦτα ἀνθρωποειδές, τὸ δ' ὄμμα γλαυκόν, τὸ δὲ χρῶμα κινναβάρινον, τὴν δὲ κέρκον ὁμοίαν τῇ τοῦ σκορπίου τοῦ χερσαίου, ἐν ᾗ κέντρον ἔχειν, καὶ τὰς ἀποφυάδας ἀπακοντίζειν, φθέγγεσθαι δ' ὅμοιον φωνῇ ἅμα σύριγγος καὶ σάλπιγγος, ταχὺ δὲ θεῖν οὐχ ἧττον τῶν ἐλάφων, καὶ εἶναι ἄγριον καὶ ἀνθρωποφάγον.[1]

Ἄνθρωπος μὲν οὖν βάλλει τοὺς ὀδόντας, βάλλει δὲ καὶ ἄλλα τῶν ζῴων, οἷον ἵππος καὶ ὀρεὺς καὶ ὄνος. βάλλει δ' ἄνθρωπος τοὺς προσθίους, τοὺς δὲ γομφίους οὐδὲν βάλλει τῶν ζῴων. ὗς δ' ὅλως οὐδένα βάλλει τῶν ὀδόντων. περὶ δὲ τῶν κυνῶν ἀμφισβητεῖται, καὶ οἱ μὲν ὅλως οὐκ οἴονται βάλλειν οὐδένα αὐτούς, οἱ δὲ τοὺς κυνόδοντας μόνον· ὦπται δ' ὅτι βάλλει καθάπερ καὶ ἄνθρωπος, ἀλλὰ λανθάνει διὰ τὸ μὴ βάλλειν πρότερον πρὶν ὑποφυῶσιν ἐντὸς ἴσοι. ὁμοίως δὲ καὶ ἐπὶ τῶν ἄλλων τῶν ἀγρίων εἰκὸς συμβαίνειν, ἐπεὶ λέγονταί γε τοὺς κυνόδοντας μόνον βάλλειν. τοὺς δὲ κύνας διαγιγνώσκουσι τοὺς νεωτέρους καὶ πρεσβυτέρους ἐκ τῶν ὀδόντων· οἱ μὲν γὰρ νέοι λευκοὺς ἔχουσι καὶ ὀξεῖς τοὺς ὀδόντας, οἱ δὲ πρεσβύτεροι μέλανας καὶ ἀμβλεῖς. ἐναντίως δὲ πρὸς τἆλλα ζῷα καὶ ἐπὶ τῶν ἵππων συμβαίνει· τὰ μὲν γὰρ ἄλλα ζῷα πρεσβύτερα γιγνόμενα μελαντέρους ἔχει τοὺς ὀδόντας, ὁ δ' ἵππος λευκοτέρους.

Ὁρίζουσι δὲ τούς τ' ὀξεῖς καὶ τοὺς πλατεῖς οἱ καλούμενοι κυνόδοντες, ἀμφοτέρων μετέχοντες τῆς

---

[1] 25 ἔστι δέ τι hucusque secl. A.-W.

the name of *martichoras*[a] has a triple row of teeth in both jaws; that it is as big as a lion and just as hairy, with similar feet; that its face and ears resemble a man's, that its eyes are blue and its general colour vermilion; that its tail is like that of the land-scorpion; that it has a sting in its tail, and can shoot off like arrows the spines attached to it; that its voice is like a mixture of a shepherd's pipe and a trumpet; that it can run as fast as a deer; and that it is fierce and man-eating.[b]

Man sheds his teeth, and so do other animals, *e.g.*, the horse, the mule, the ass. Man sheds his front teeth; no animal sheds its molars. The pig sheds none of its teeth at all. There is disagreement about II dogs, whether they shed their teeth; some think they shed none of their teeth at all, others think they shed their canines only, yet observation has shown that they do shed them, as man does, though the fact is concealed because they do not shed them until new ones have grown up within of equal size to take their place. It seems probable that a similar thing occurs with the other wild animals as well, since they are alleged to shed their canine teeth only. People distinguish young dogs from old by their teeth: young dogs have sharp white teeth, old ones black and blunt. In the horse the reverse occurs from what III happens in the other animals, whose teeth, as they themselves grow older, become blacker; the horse's become whiter.

The canines, as they are called, mark the boundary between the sharp teeth and the broad flat ones, as

[a] An Old Persian word meaning man-eater; *cf.* Persian *mard-khor*. The tiger.

[b] The passage about the martichoras is considered by A.-W. and Dittmeyer to be an interpolation. See add. n., p. 238.

ARISTOTLE

μορφῆς· κάτωθεν μὲν γὰρ πλατεῖς, ἄνωθεν δ' εἰσὶν ὀξεῖς.

Ἔχουσι δὲ πλείους οἱ ἄρρενες τῶν θηλειῶν ὀδόντας καὶ ἐν ἀνθρώποις καὶ ἐπὶ προβάτων καὶ αἰγῶν καὶ ὑῶν· ἐπὶ δὲ τῶν ἄλλων οὐ τεθεώρηταί πω. ὅσοι δὲ πλείους ἔχουσι, μακροβιώτεροι ὡς ἐπὶ τὸ πολύ εἰσιν, οἱ δ' ἐλάττους καὶ ἀραιόδοντες ὡς ἐπὶ τὸ πολὺ βραχυβιώτεροι.

IV Φύονται δ' οἱ τελευταῖοι τοῖς ἀνθρώποις γόμφιοι, οὓς καλοῦσι κραντῆρας, περὶ τὰ εἴκοσιν ἔτη καὶ ἀνδράσι καὶ γυναιξίν. ἤδη δέ τισι γυναιξὶ καὶ ὀγδοήκοντα ἐτῶν οὔσαις ἔφυσαν γόμφιοι ἐν τοῖς ἐσχάτοις, πόνον παρασχόντες ἐν τῇ ἀνατολῇ, καὶ ἀνδράσιν ὡσαύτως· τοῦτο δὲ συμβαίνει ὅσοις ἂν μὴ ἐν τῇ ἡλικίᾳ ἀνατείλωσιν οἱ κραντῆρες.

V Ὁ δ' ἐλέφας ὀδόντας μὲν ἔχει τέτταρας ἐφ' ἑκάτερα, οἷς κατεργάζεται τὴν τροφήν (λεαίνει δ' ὥσπερ κρίμνα[1]), χωρὶς δὲ τούτων ἄλλους δύο τοὺς μεγάλους. ὁ μὲν οὖν ἄρρην τούτους ἔχει μεγάλους τε καὶ ἀνασίμους, ἡ δὲ θήλεια μικροὺς καὶ ἐξ ἐναντίας τοῖς ἄρρεσιν· κάτω γὰρ οἱ ὀδόντες βλέπουσιν. ἔχει δ' ὁ ἐλέφας εὐθὺς γενόμενος ὀδόντας, τοὺς μέντοι VI μεγάλους ἀδήλους τὸ πρῶτον. γλῶτταν δ' ἔχει μικράν τε σφόδρα καὶ ἐντός, ὥστε ἔργον ἐστὶν ἰδεῖν.

VII Ἔχουσι δὲ τὰ ζῷα καὶ τὰ μεγέθη διαφέροντα τοῦ στόματος. τῶν μὲν γάρ ἐστι τὰ στόματα ἀνερρωγότα, ὥσπερ κυνὸς καὶ λέοντος καὶ πάντων τῶν καρχαροδόντων, τὰ δὲ μικρόστομα, ὥσπερ ἄνθρωπος, τὰ δὲ μεταξύ, οἷον τὸ τῶν ὑῶν γένος.

Ὁ δ' ἵππος ὁ ποτάμιος ὁ ἐν Αἰγύπτῳ χαίτην

---

[1] Sn.: κριμνά vulg.

they partake of the form of both : they are broad
and flat at the base and sharp at the top.

Males have more numerous teeth than females not
only in men but also in sheep, goats, and pigs. Examination has not yet been fully made of other animals. In general, the more teeth animals have, the
longer they tend to live ; those with fewer and more
widely spaced teeth are on the whole shorter-lived.

IV In man the last teeth to come through are molars,
wisdom-teeth [a] as they call them, which come through
in both sexes about the age of twenty. Instances have
been known of women whose wisdom-teeth have come
through at the age of eighty or more, at the very
extremity of life, causing much pain in so doing ; the
same has occurred in men also. This occurs in persons
whose wisdom-teeth have not come through in youth.

V The elephant has four teeth on either side, which
it uses to masticate its food, grinding it like coarse
barley meal, and apart from these it has two others,
the " great " ones.[b] In the male these are quite big
and turned up at the end ; in the female they are
small and turned in the opposite direction from those
of the male—*i.e.*, they face downwards. The elephant
has teeth when it is born, but the " great " ones [b] are
not visible at first. VI It also has a tongue, quite a small
one, and placed well inside the mouth, so that it is
difficult to see.

VII Again, animals vary in the sizes of their mouths.
Some have mouths which open wide (*e.g.*, the dog, the
lion, and all the saw-toothed animals), others have
small mouths (*e.g.*, man), others have mouths of intermediate size (*e.g.*, the pig tribe).

The Egyptian hippopotamus has a mane like a

[a] Lit., " completers."   [b] *i.e.*, the tusks.

μὲν ἔχει ὥσπερ ἵππος, διχαλὸν δ' ἐστὶν ὥσπερ βοῦς, τὴν δ' ὄψιν σιμός. ἔχει δὲ καὶ ἀστράγαλον ὥσπερ τὰ διχαλά, καὶ χαυλιόδοντας ὑποφαινομένους, κέρκον δ' ὑός, φωνὴν δ' ἵππου· μέγεθος δ' ἐστὶν ἡλίκον ὄνος. τοῦ δὲ δέρματος τὸ πάχος ὥστε δόρατα ποιεῖσθαι ἐξ αὐτοῦ. τὰ δ' ἐντὸς ἔχει ὅμοια ἵππῳ καὶ ὄνῳ.[1]

VIII. Ἔνια δὲ τῶν ζῴων ἐπαμφοτερίζει τὴν φύσιν τῷ τ' ἀνθρώπῳ καὶ τοῖς τετράποσιν, οἷον πίθηκοι καὶ κῆβοι καὶ κυνοκέφαλοι. ἔστι δ' ὁ μὲν κῆβος πίθηκος ἔχων οὐράν. καὶ οἱ κυνοκέφαλοι δὲ τὴν αὐτὴν ἔχουσι μορφὴν τοῖς πιθήκοις, πλὴν μείζονές τ' εἰσὶ καὶ τὰ πρόσωπα ἔχοντες κυνοειδέστερα, ἔτι δ' ἀγριώτερά τε τὰ ἤθη καὶ τοὺς ὀδόντας ἔχουσι κυνοειδεστέρους καὶ ἰσχυροτέρους. οἱ δὲ πίθηκοι δασεῖς μέν εἰσι τὰ πρανῆ ὡς ὄντες τετράποδες, καὶ τὰ ὕπτια δ' ὡσαύτως ὡς ὄντες ἀνθρωποειδεῖς (τοῦτο γὰρ ἐπὶ τῶν ἀνθρώπων ἐναντίως ἔχει καὶ ἐπὶ τῶν τετραπόδων, καθάπερ ἐλέχθη πρότερον)· πλὴν ἥ γε θρὶξ παχεῖα, ὥστε[2] δασεῖς ἐπ' ἀμφότερα σφόδρα εἰσὶν οἱ πίθηκοι. τὸ δὲ πρόσωπον ἔχει πολλὰς ὁμοιότητας τῷ τοῦ ἀνθρώπου· καὶ γὰρ μυκτῆρας καὶ ὦτα παραπλήσια ἔχει, καὶ ὀδόντας ὥσπερ ὁ ἄνθρωπος, καὶ τοὺς προσθίους καὶ τοὺς γομφίους. ἔτι δὲ βλεφαρίδας τῶν ἄλλων τετραπόδων ἐπὶ θάτερα οὐκ ἐχόντων οὗτος ἔχει μὲν λεπτὰς δὲ σφόδρα, καὶ μᾶλλον τὰς κάτω, καὶ μικρὰς πάμπαν· τὰ γὰρ ἄλλα τετράποδα ταύτας οὐκ ἔχει.

Ἔτι δ' ἐν τῷ στήθει δύο θηλὰς μαστῶν μικρῶν.

horse, is cloven-hoofed like an ox, and is snub-nosed. It has a huckle-bone like the cloven-footed animals, tusks which just show through, the tail of a pig, the neigh of a horse, and the size of an ass. Its hide is so thick that spears are made from it.[a] Its internal organs resemble those of the horse and the ass.[b]

Some animals dualize[c] in their nature with man and the quadrupeds, *e.g.*, the ape, the monkey, and the baboon. The monkey is an ape with a tail. The baboon is identical in form with the ape, except that it is bigger and stronger and its face[d] is more like a dog's; also it is of a fiercer disposition and its teeth are more doglike and stronger. Apes are hairy on the back, in virtue of their being quadrupeds, and hairy on their fronts, in virtue of being man-like (that, as I mentioned before, is a point in which the arrangement in man is the reverse of that in the quadrupeds), except that the hair is coarse; hence apes are covered with thick hair both back and front. Its face shows many resemblances to that of man: it has similar nostrils and ears, and teeth, both front and molars, like man's. Furthermore, whereas other quadrupeds have no lashes on one of the two eyelids, the ape has them on both, though they are extremely fine, especially the lower ones, and very small indeed. This is unique, for the other quadrupeds lack these lower eyelashes.

Also, the ape has on its chest two teats, though the breasts on which they are placed are small. It has

VIII
Apes, etc.

---

[a] *Cf.* Herodotus ii. 71.
[b] This paragraph is considered by Schneider, A.-W. and Dittmeyer to be an interpolation.
[c] See Notes, §§ 28 ff.   [d] But see 491 b 10.

---

[1] 9 ὁ δ' ἵππος hucusque secl. edd.   [2] ὥστε Dt. : καὶ vulg.

ARISTOTLE

ἔχει δὲ καὶ βραχίονας ὥσπερ ἄνθρωπος, πλὴν δασεῖς· καὶ κάμπτει καὶ τούτους καὶ τὰ σκέλη ὥσπερ ἄνθρωπος, τὰς περιφερείας πρὸς ἀλλήλας ἀμφοτέρων τῶν κώλων. πρὸς δὲ τούτοις χεῖρας καὶ δακτύλους καὶ ὄνυχας ὁμοίους ἀνθρώπῳ, πλὴν πάντα ταῦτα ἐπὶ τὸ θηριωδέστερον. ἰδίους δὲ τοὺς πόδας· εἰσὶ γὰρ οἷον χεῖρες μεγάλαι, καὶ οἱ δάκτυλοι ὥσπερ οἱ τῶν χειρῶν, ὁ μέσος μακρότατος, καὶ τὸ κάτω τοῦ ποδὸς χειρὶ ὅμοιον, πλὴν ἐπιμηκέστερον[1] τῆς χειρός, ἐπὶ τὰ ἔσχατα τεῖνον, καθάπερ θέναρ· τοῦτο δ' ἐπ' ἄκρου σκληρότερον, κακῶς καὶ ἀμυδρῶς μιμούμενον πτέρνην. κέχρηται δὲ τοῖς ποσὶν ἐπ' ἄμφω, καὶ ὡς χερσὶ καὶ ὡς ποσί, καὶ συγκάμπτει ὥσπερ χεῖρας. ἔχει δὲ τὸν ἀγκῶνα καὶ τὸν μηρὸν βραχεῖς ὡς πρὸς τὸν βραχίονα καὶ τὴν κνήμην. ὀμφαλὸν δ' ἐξέχοντα μὲν οὐκ ἔχει, σκληρὸν δὲ τὸ κατὰ τὸν τόπον τὸν τοῦ[2] ὀμφαλοῦ. τὰ δ' ἄνω τῶν[3] κάτω πολὺ μείζονα ἔχει, ὥσπερ τὰ τετράποδα· σχεδὸν γὰρ ὡς πέντε πρὸς τρία ἐστί. καὶ διά τε ταῦτα καὶ διὰ τὸ τοὺς πόδας ἔχειν ὁμοίους χερσὶ καὶ ὡσπερανεὶ συγκειμένους ἐκ χειρὸς καὶ ποδός—ἐκ μὲν ποδὸς κατὰ τὸ τῆς πτέρνης ἔσχατον, ἐκ δὲ χειρὸς τἆλλα μέρη· καὶ γὰρ οἱ δάκτυλοι ἔχουσι τὸ καλούμενον θέναρ—διατελεῖ[4] τὸν πλείω χρόνον τετράπουν ὂν μᾶλλον ἢ ὀρθόν· καὶ οὔτ' ἰσχία ἔχει ὡς τετράπουν ὂν οὔτε κέρκον ὡς δίπουν, πλὴν μικρὰν τὸ ὅλον, ὅσον σημείου χάριν. ἔχει δὲ καὶ τὸ αἰδοῖον ἡ θήλεια ὅμοιον γυναικί, ὁ δ' ἄρρην κυνωδέστερον ἢ ἄνθρωπος. οἱ δὲ κῆβοι, καθάπερ

[1] Dt. ex coniect. Pi. : ἐπὶ τὸ μῆκος τὸ vulg.

arms like a man, though covered in hair, and it bends these and its legs as a man does, *i.e.*, with the convexities of both limbs towards each other. Besides this, the ape has hands, fingers and nails like a man, except that all these parts tend to be more beastlike. Its feet, however, are a peculiarity. They are like large hands and the toes are like fingers, the middle one being the longest; and the lower part of the foot is like a hand, except that it is more prolonged than a hand, and extends towards the extremities, like a palm; and this part towards the end is harder, in a poor and indistinct way resembling a heel. The ape uses its feet both as feet and hands, and clenches them like fists. The upper arm and thigh are short compared with the lower arm and the calf. It has no protruding navel, but the part where the navel should be is hard. Its upper parts[a] are much larger than the lower, as with quadrupeds: the proportion is about 5 to 3. Because of this, and because it has feet which are similar to hands and consist as it were of hand as well as foot (of foot so far as the extremity of the heel is concerned, and of hand for the remainder, for even the toes have what is called a palm), because of all this the ape spends most of its time on all fours rather than upright. It has neither buttocks (*qua* quadruped)[b] nor tail (*qua* biped)—or only a very small one indeed, just enough to give an indication of one. The generative organs of the female are similar to those of a woman; those of the male are more like a dog's than a man's are. The IX

[a] See above, note on 500 b 34.   [b] See 499 b 1.

² τὸν τοῦ A.-W.: τὸν om. vulg.: τοῦτον τοῦ AC.
³ τῶν Sylb.: τοῦ vulg.
⁴ add. δὲ codd., vulg.: om. Cs., A.-W.

25 εἴρηται πρότερον, ἔχουσι κέρκον. τὰ δ' ἐντὸς διαιρεθέντα ὅμοια ἔχουσιν ἀνθρώπῳ πάντα τὰ τοιαῦτα.

Τὰ μὲν οὖν τῶν εἰς τὸ ἐκτὸς ζῳοτοκούντων μόρια τοῦτον ἔχει τὸν τρόπον.

X Τὰ δὲ τετράποδα μὲν ᾠοτόκα δὲ καὶ ἔναιμα (οὐδὲν δ' ᾠοτοκεῖ χερσαῖον καὶ ἔναιμον μὴ τετρά-
30 πουν ὂν ἢ ἄπουν) κεφαλὴν μὲν ἔχει καὶ αὐχένα καὶ νῶτον καὶ τὰ πρανῆ καὶ τὰ ὕπτια τοῦ σώματος, ἔτι δὲ σκέλη πρόσθια καὶ ὀπίσθια καὶ τὸ ἀνάλογον τῷ στήθει, ὥσπερ τὰ ζῳοτόκα τῶν τετραπόδων, καὶ κέρκον τὰ μὲν πλεῖστα μείζω, ὀλίγα δ' ἐλάττω. πάντα δὲ πολυδάκτυλα καὶ πολυσχιδῆ ἐστι τὰ
35 τοιαῦτα. πρὸς δὲ τούτοις τὰ αἰσθητήρια καὶ γλῶτταν πάντα, πλὴν ὁ ἐν Αἰγύπτῳ κροκόδειλος. οὗτος δὲ παραπλησίως τῶν ἰχθύων τισίν· ὅλως μὲν γὰρ οἱ ἰχθύες ἀκανθώδη καὶ οὐκ ἀπολελυμένην ἔχουσι τὴν γλῶτταν, ἔνιοι δὲ πάμπαν λεῖον καὶ ἀδιάρθρωτον τὸν τόπον μὴ ἐγκλίναντι σφόδρα τὸ χεῖλος. ὦτα δ'
5 οὐκ ἔχουσιν ἀλλὰ τὸν πόρον τῆς ἀκοῆς μόνον πάντα τὰ τοιαῦτα· οὐδὲ μαστούς, οὐδ' αἰδοῖον, οὐδ' ὄρχεις ἔξω φανεροὺς ἀλλ' ἐντός, οὐδὲ τρίχας, ἀλλὰ πάντ' ἐστὶ φολιδωτά. ἔτι δὲ καρχαρόδοντα πάντα.

Οἱ δὲ κροκόδειλοι οἱ ποτάμιοι ἔχουσιν ὀφθαλμοὺς μὲν ὑός, ὀδόντας δὲ μεγάλους καὶ χαυλιόδοντας καὶ
10 ὄνυχας ἰσχυροὺς καὶ δέρμα ἄρρηκτον φολιδωτόν· βλέπουσι δ' ἐν μὲν τῷ ὕδατι φαύλως, ἔξω δ' ὀξύτατον. τὴν μὲν οὖν ἡμέραν ἐν τῇ γῇ τὸ πλεῖστον

---

[a] It appears from this that Aristotle dissected monkeys and other such animals; see also 511 b 21. For Galen's dissection of monkeys see C. Singer, "Galen as a Modern," in *Proc. Roy. Soc. Med.*, 1949, 42, pp. 563-70.

HISTORIA ANIMALIUM, II. ix–x

monkey, as I have mentioned before, has a tail. In all animals of this sort the internal parts, when dissected,[a] resemble those of man.

Such then is the character of the parts of the externally viviparous animals.

The quadrupeds which are blooded and oviparous (and no terrestrial[b] blooded animal is oviparous unless it is a quadruped or footless) have a head, a neck, a back, dorsal and ventral parts, also front and hind legs and a counterpart of the chest, just as the viviparous quadrupeds do, and most of them have a fairly large tail, though a few have a smallish one. All of them are polydactylous, *i.e.*, their feet are many-cloven. In addition, they all have sensory organs including a tongue, except the Egyptian crocodile, which in this respect resembles certain fishes; for, generally speaking, the tongue of fishes is prickly and not properly separate, while some fishes exhibit in that position a quite smooth and unarticulated surface unless you pull their mouth well open. Again, all these animals lack ears; they have only the passage for hearing; and they have no breasts, or visible generative organ, and their testicles are not external but internal; and they have no hair, but are all covered with horny scales. Furthermore, they are all saw-toothed.

River crocodiles have the eyes of a pig. They have large teeth and tusks,[c] strong nails, and an impenetrable hide made of horny scales. They see poorly under water, but above water their sight is excellent. The day they spend for the most part on land, but

X INSTRUMENTAL PARTS OF BLOODED ANIMALS: (ii) Ovipara: External parts: Quadrupeds.

[b] *i.e.*, in the sense of "earthbound"; birds are not "terrestrial."

[c] These are, however, not tusks in the usual sense of the word.

ARISTOTLE

διατρίβει, τὴν δὲ νύκτα ἐν τῷ ὕδατι· ἀλεεινότερον γάρ ἐστι τῆς αἰθρίας.

XI Ὁ δὲ χαμαιλέων ὅλον μὲν τοῦ σώματος ἔχει τὸ σχῆμα σαυροειδές, τὰ δὲ πλευρὰ κάτω καθήκει συνάπτοντα πρὸς τὸ ὑπογάστριον, καθάπερ τοῖς ἰχθύσι, καὶ ἡ ῥάχις ἐπανέστηκεν ὁμοίως τῇ τῶν ἰχθύων. τὸ δὲ πρόσωπον ὁμοιότατον τῷ τοῦ χοιροπιθήκου. κέρκον δ' ἔχει μακρὰν σφόδρα, εἰς λεπτὸν καθήκουσαν καὶ συνελιττομένην ἐπὶ πολύ, καθάπερ ἱμάντα. μετεωρότερος δ' ἐστὶ τῇ ἀπὸ τῆς γῆς ἀποστάσει τῶν σαύρων, τὰς δὲ καμπὰς τῶν σκελῶν καθάπερ οἱ σαῦροι ἔχει. τῶν δὲ ποδῶν ἕκαστος αὐτοῦ διχῇ διῄρηται εἰς μέρη θέσιν ὁμοίαν πρὸς αὐτὰ ἔχοντα οἷάνπερ ὁ μέγας ἡμῶν δάκτυλος πρὸς τὸ λοιπὸν τῆς χειρὸς ἀντίθεσιν ἔχει. ἐπὶ βραχὺ δὲ καὶ τούτων τῶν μερῶν ἕκαστον διῄρηται εἴς τινας δακτύλους, τῶν μὲν ἔμπροσθεν ποδῶν τὰ μὲν πρὸς αὑτὸν τρίχα, τὰ δ' ἐκτὸς δίχα, τῶν δ' ὀπισθίων τὰ μὲν πρὸς αὑτὸν δίχα, τὰ δ' ἐκτὸς τρίχα. ἔχει δὲ καὶ ὀνύχια ἐπὶ τούτων ὅμοια τοῖς τῶν γαμψωνύχων. τραχὺ δ' ἔχει ὅλον τὸ σῶμα, καθάπερ ὁ κροκόδειλος. ὀφθαλμοὺς δ' ἔχει ἐν κοίλῳ τε κειμένους καὶ μεγάλους σφόδρα καὶ στρογγύλους καὶ δέρματι ὁμοίῳ τῷ τοῦ λοιποῦ σώματος περιεχομένους. κατὰ μέσους δ' αὐτοὺς διαλέλειπται μικρὰ τῇ ὄψει χώρα, δι' ἧς ὁρᾷ· οὐδέποτε δὲ τῷ δέρματι ἐπικαλύπτει τοῦτο. στρέφει δὲ τὸν ὀφθαλμὸν κύκλῳ καὶ τὴν ὄψιν ἐπὶ πάντας τοὺς τόπους μεταβάλλει, καὶ οὕτως ὁρᾷ ὃ βούλεται. τῆς δὲ χροιᾶς ἡ μεταβολὴ ἐμφυσωμένῳ αὐτῷ γίγνεται· ἔχει δὲ καὶ μέλαιναν ταύτην, οὐ πόρρω τῆς τῶν κροκοδείλων, καὶ ὠχρὰν καθάπερ οἱ σαῦροι, μέλανι ὥσπερ τὰ παρδάλια δια-

the night they spend in the water, because it is warmer than the open air.

The chamaeleon in the general formation of its body is like a lizard, but the ribs extend downwards and meet in the region of the belly, as occurs in fishes, and (also as in fishes) the spine projects upwards. Its face [a] is very similar to the pig-faced baboon's. It has a very long tail, which tapers off and is coiled up to a great extent, like a strap. It stands further off from the ground than the lizards, but the flexions of the legs are as in lizards. Each of its feet is divided into two parts, which have the same relative position as the human thumb has in opposition to the rest of the hand. Each of these parts in turn is for a short way divided into toes of a sort : on the front feet the inner part is divided into three, the outer part into two, while on the hind feet the inner is divided into two and the outer into three. On them it has claws, similar to those of the crook-taloned birds. Its body is rough all over, like the crocodile. Its eyes are placed in a hollow : they are very large and round and surrounded by skin similar to that of the rest of the body. In the middle there is a small space left for vision, through which the animal sees ; it never covers this opening with the skin. It turns its eye all round and keeps changing its line of vision in all directions, and so manages to see what it wishes. Its change of colour occurs when it becomes inflated : its colour then is dark, a shade not far removed from the crocodile's, or pale like the lizard, dark-spotted

---

[a] But see 491 b 10.

πεποικιλμένην. γίγνεται δὲ καθ' ἅπαν τὸ σῶμα αὐτοῦ ἡ τοιαύτη μεταβολή· καὶ γὰρ οἱ ὀφθαλμοὶ συμμεταβάλλουσιν ὁμοίως τῷ λοιπῷ σώματι καὶ ἡ κέρκος. ἡ δὲ κίνησις αὐτοῦ νωθὴς ἰσχυρῶς ἐστι, καθάπερ ἡ τῶν χελωνῶν. καὶ ἀποθνήσκων τε 10 ὠχρὸς γίγνεται, καὶ τελευτήσαντος αὐτοῦ ἡ χροιὰ τοιαύτη ἐστίν. τὰ δὲ περὶ τὸν στόμαχον καὶ τὴν ἀρτηρίαν ὁμοίως ἔχει τοῖς σαύροις κείμενα. σάρκα δ' οὐδαμοῦ ἔχει πλὴν πρὸς τῇ κεφαλῇ καὶ ταῖς σιαγόσιν ὀλίγα σαρκία, καὶ περὶ ἄκραν τὴν τῆς κέρκου πρόσφυσιν. καὶ αἷμα δ' ἔχει περί τε τὴν καρδίαν 15 μόνον καὶ τὰ ὄμματα καὶ τὸν ἄνω τῆς καρδίας τόπον, καὶ ὅσα ἀπὸ τούτων φλέβια ἀποτείνει· ἔστι δὲ καὶ ἐν τούτοις βραχὺ παντελῶς. κεῖται δὲ καὶ ὁ ἐγκέφαλος ἀνώτερον μὲν ὀλίγῳ τῶν ὀφθαλμῶν, συνεχὴς δὲ τούτοις. περιαιρεθέντος δὲ τοῦ ἔξωθεν 20 δέρματος τῶν ὀφθαλμῶν περιέχει τι διαλάμπον διὰ τούτων, οἷον κρίκος χαλκοῦς λεπτός. καθ' ἅπαν δ' αὐτοῦ τὸ σῶμα σχεδὸν διατείνουσιν ὑμένες πολλοὶ καὶ ἰσχυροὶ καὶ πολὺ ὑπερβάλλοντες τῶν περὶ τὰ λοιπὰ ὑπαρχόντων. ἐνεργεῖ δὲ καὶ τῷ πνεύματι ἀνατετμημένος ὅλος ἐπὶ πολὺν χρόνον, βραχείας 25 ἰσχυρῶς[1] ἔτι κινήσεως ἐν αὐτῷ περὶ τὴν καρδίαν οὔσης, καὶ συνάγει διαφερόντως μὲν τὰ περὶ τὰ πλευρά, οὐ μὴν ἀλλὰ καὶ τὰ λοιπὰ μέρη τοῦ σώ-

---

[1] ἰσχυρῶς damnant A.-W., sed cf. v. 8 supra.

like the leopard. This change affects the whole surface of the body, including the eyes and the tail. The animal's movements are very sluggish indeed, like the tortoise. When dying it becomes pale, and it retains this colour when dead. With regard to the position of the oesophagus and the windpipe it resembles the lizards. It has no flesh anywhere except some tiny portions on the head and jaws and the root of the tail. It has blood only round the heart, the eyes, and the region above the heart, and in the small blood-vessels ramifying from them, though even in these parts there is a very small quantity. The brain is situated slightly above the eyes, but is continuous with them. If the outer skin is removed from the eyes, some feature is disclosed which surrounds them and shines through them, like a thin copper ring. Numerous strong membranes extend over practically the whole of the animal's body, far exceeding those present in other animals. It actually goes on functioning with its breath [a] for a long time after being cut open completely, there being still a very slight movement in the breath around the heart, and while contraction takes place particularly round the ribs, it affects the remaining parts of the body as

[a] There is probably an implied reference here to the connate *pneuma* (see *G.A.*, Loeb ed., App. B). At *G.A.* 787 b 15 ff., Aristotle, speaking of the voices of bulls and cows, mentions that the generally sinewy character of bulls applies also to their hearts; hence, he says, this part with which they set the *pneuma* in movement is taut, like a sinewy string stretched tight. At 787 b 27 he speaks of the blood-vessel which has its starting-point at the heart near to the part which sets the voice in movement. *Cf.* also *G.A.* 776 b 12 ff., 781 a 32 ff., and App. B above cited, especially § 31. Thus it is the heart or the region of the heart which supplies movement to *pneuma* generally in the body.

μάτος. σπλῆνα δ' οὐδαμοῦ ἔχει φανερόν. φωλεύει δὲ καθάπερ οἱ σαῦροι.

XII. Ὁμοίως δ' ἔνια μόρια καὶ οἱ ὄρνιθες τοῖς εἰρημένοις ἔχουσι ζῴοις· καὶ γὰρ κεφαλὴν καὶ αὐχένα πάντ' ἔχει καὶ νῶτον καὶ τὰ ὕπτια τοῦ σώματος καὶ τὸ ἀνάλογον τῷ στήθει· σκέλη δὲ δύο καθάπερ ἄνθρωπος μάλιστα τῶν ζῴων· πλὴν κάμπτει εἰς τοὔπισθεν ὁμοίως τοῖς τετράποσιν, ὥσπερ εἴρηται πρότερον. χεῖρας δ' οὐδὲ πόδας προσθίους ἔχει, ἀλλὰ πτέρυγας ἴδιον πρὸς τὰ ἄλλα ζῷα. ἔτι δὲ τὸ ἰσχίον ὅμοιον μηρῷ μακρὸν καὶ προσπεφυκὸς μέχρι ὑπὸ μέσην τὴν γαστέρα, ὥστε δοκεῖν διαιρούμενον μηρὸν εἶναι, τὸν δὲ μηρὸν μέχρι¹ τῆς κνήμης, ἕτερόν τι μέρος. μεγίστους δὲ τοὺς μηροὺς ἔχει τὰ γαμψώνυχα τῶν ὀρνίθων, καὶ τὸ στῆθος ἰσχυρότερον τῶν ἄλλων. πολυώνυχοι δ' εἰσὶ πάντες οἱ ὄρνιθες, ἔτι δὲ πολυσχιδεῖς τρόπον τινὰ πάντες· τῶν μὲν γὰρ πλείστων διῄρηνται οἱ δάκτυλοι, τὰ δὲ πλωτὰ στεγανόποδά ἐστι, διηρθρωμένους δ' ἔχει καὶ χωριστοὺς ⟨τοὺς⟩² δακτύλους. εἰσὶ δ' ὅσοι αὐτῶν μετεωρίζονται πάντες τετραδάκτυλοι· τρεῖς μὲν γὰρ εἰς τὸ ἔμπροσθεν ἕνα δ' εἰς τὸ ὄπισθεν κείμενον ἔχουσιν οἱ πλεῖστοι ἀντὶ πτέρνης· ὀλίγοι δέ τινες δύο μὲν ἔμπροσθεν δύο δ' ὄπισθεν, οἷον ἡ καλουμένη ἴυγξ. αὕτη δ' ἐστὶ μικρῷ μὲν μείζων σπίζης, τὸ δ' εἶδος ποικίλον, ἴδια δ' ἔχει τά τε περὶ τοὺς δακτύλους καὶ τὴν γλῶτταν ὁμοίαν τοῖς ὄφεσιν· ἔχει γὰρ ἐπὶ μῆκος ἔκτασιν καὶ ἐπὶ τέτταρας δακτύλους, καὶ πάλιν συστέλλεται εἰς ἑαυτήν. ἔτι δὲ περιστρέφει τὸν τράχηλον εἰς τοὐπίσω τοῦ λοιποῦ σώματος ἠρεμοῦντος, καθάπερ οἱ ὄφεις. ὄνυχας δ'

---

¹ μέχρι coni. A.-W.: μεταξὺ vulg. ² ⟨τοὺς⟩ Dt.

well. It has no spleen which is visible. It lives in holes like the lizards.[a]

In some of their parts birds resemble the animals we have just been discussing: they all have a head, neck, back, underside, and a counterpart of the chest. They have two legs, like man—in this respect resembling him more than any other animal does—except that birds bend their legs backwards, like quadrupeds, as we remarked earlier. Birds have no forefeet or hands, but wings—a peculiarity distinguishing them from other animals. Their haunch-bone is long, like a thigh, and attached to the body as far as the middle of the belly: the result is that when taken separately it looks as though it is a thigh, although the thigh is actually a different part, extending from it to the shin. The largest thighs and the strongest breasts are found in the crook-taloned birds. All birds have numerous claws, and all in one way or another have numerous toes: in most birds the toes are quite distinct, and the swimmers, although web-footed, have separately articulated toes. Those birds which rise well into the air all have four toes: most of them have three toes in front and one behind instead of a heel: a few have two in front and two behind, such as the bird called the wryneck. This bird is slightly larger than the chaffinch, and mottled in appearance. The arrangement of its toes is peculiar, and its tongue is like the serpent's: it can extend its tongue for a distance of four fingerbreadths and then draw it in again. Further, it can turn its neck round right back while keeping the rest of its body unmoved, like the serpent. It

XII
Birds

---

[a] O. Regenbogen, "Bemerkungen zur *H.A.* des Aristoteles," in *Studi Ital. di Filos. class.*, XXVII-XXVIII, 1956, pp. 444-449, argues that ch. 11 about the chamaeleon is a later insertion. See add. note, p. 238.

ARISTOTLE

ἔχει μεγάλους μὲν ὁμοίως μέντοι πεφυκότας τοῖς τῶν κολιῶν[1]· τῇ δὲ φωνῇ τρίζει.

20 Στόμα δ' οἱ ὄρνιθες ἔχουσι μὲν ἴδιον δέ· οὔτε γὰρ χείλη οὔτ' ὀδόντας ἔχουσιν, ἀλλὰ ῥύγχος, οὔτ' ὦτα οὔτε μυκτῆρας, ἀλλὰ τοὺς πόρους τούτων τῶν αἰσθήσεων, τῶν μὲν μυκτήρων ἐν τῷ ῥύγχει, τῆς δ' ἀκοῆς ἐν τῇ κεφαλῇ. ὀφθαλμοὺς δὲ πάντες καθάπερ καὶ τἆλλα ζῷα δύο, ἄνευ βλεφαρίδων. μύ-25 ουσι δ' οἱ βαρεῖς τῷ κάτω βλεφάρῳ, σκαρδαμύττουσι δ' ἐκ τοῦ κανθοῦ δέρματι ἐπιόντι πάντες, οἱ δὲ γλαυκώδεις τῶν ὀρνίθων καὶ τῷ ἄνω βλεφάρῳ. τὸ δ' αὐτὸ τοῦτο ποιοῦσι καὶ τὰ φολιδωτά, οἷον οἱ σαῦροι καὶ τἆλλα τὰ ὁμοιογενῆ τούτοις τῶν ζῴων· μύουσι γὰρ τῇ κάτω βλεφαρίδι πάντες, οὐ μέντοι 30 σκαρδαμύττουσί γ' ὥσπερ οἱ ὄρνιθες.

Ἔτι δ' οὔτε φολίδας οὔτε τρίχας ἔχουσιν, ἀλλὰ πτερά· τὰ δὲ πτερὰ ἔχει καυλὸν ἅπαντα. καὶ οὐρὰν μὲν οὐκ ἔχουσιν, ὀρροπύγιον δέ, οἱ μὲν μακροσκελεῖς καὶ στεγανόποδες βραχύ, οἱ δ' ἐναντίοι μέγα. καὶ οὗτοι μὲν πρὸς τῇ γαστρὶ τοὺς πόδας ἔχοντες πέτονται, οἱ δὲ μικρορροπύγιοι ἐκτεταμένους. καὶ 35 γλῶτταν ἅπαντες, ταύτην δ' ἀνομοίαν· οἱ μὲν γὰρ μακρὰν οἱ δὲ πλατεῖαν. μάλιστα δὲ τῶν ζῴων μετὰ τὸν ἄνθρωπον γράμματα φθέγγεται ἔνια τῶν ὀρνίθων γένη· τοιαῦτα δ' ἐστὶ τὰ πλατύγλωττα αὐτῶν μάλιστα. τὴν δ' ἐπιγλωττίδα ἐπὶ τῆς ἀρτηρίας οὐδὲν τῶν ᾠοτοκούντων ἔχει, ἀλλὰ συνάγει καὶ 5 διοίγει τὸν πόρον ὥστε μηδὲν κατιέναι[2] τῶν ἐχόν-

---

[1] κολιῶν Rhen., Sn. : κελεῶν (quod idem valet) Dt. : κοιλιῶν A : κολοιῶν vulg. [2] AC : καθεῖναι vulg.

HISTORIA ANIMALIUM, II. xii

has large claws, though in nature they are similar to those of the green woodpecker.[a] Its note is a loud cry.

Birds have a mouth, but a peculiar one : they have neither lips nor teeth, but a beak. They have no ears or noses either, but passages serving these sensory faculties : the passages for the nostrils are in the beak, those for hearing in the head. They have, like all other animals, two eyes, but no eyelashes. The heavy birds[b] close their eyes with the lower eyelid, and all birds blink with a skin that comes up over the eye from the inner corner, whereas the owl-type of birds use also the upper eyelid. The same occurs in the animals covered with horny scales, *e.g.*, the lizards and other related animals : all of these close their eyes with the lower eyelid, but they do not blink like birds.

Furthermore, birds have neither horny scales nor hair, but feathers, and the feathers always have quills. Birds have no actual tail, but a rump[c] : the long legged and web-footed birds have a short one, the others a long one. The latter fly with their feet close up under the belly, whereas the small-rumped birds fly with their legs stretched out. All birds have a tongue, but it is not the same in all : some have a long tongue, some a broad one. More than any other animals, and second only to man, certain kinds of bird can utter articulate sounds : this faculty occurs chiefly in the broad-tongued birds. No oviparous animal has an epiglottis over the windpipe ; such creatures close and open the passage in such a way as to prevent any heavy matter passing down into the

[a] " Green woodpecker " (κολιῶν) is Schneider's restoration instead of the old reading " jackdaws " (κολοιῶν).

[b] See Table of Birds, *G.A.* (Loeb ed.), p. 368.

[c] In which the tail-feathers are set.

ARISTOTLE

504 b

των βάρος ἐπὶ τὸν πνεύμονα. γένη δ' ἔνια τῶν ὀρνίθων ἔχει καὶ πλῆκτρα· γαμψώνυχον δ' ἅμα καὶ πλῆκτρον ἔχον οὐδέν. ἔστι δὲ τὰ μὲν γαμψώνυχα τῶν πτητικῶν, τὰ δὲ πληκτροφόρα τῶν βαρέων.
10 ἔτι δ' ἔνια τῶν ὀρνέων λόφον ἔχουσι, τὰ μὲν αὐτῶν τῶν πτερῶν ἐπανεστηκότα, ὁ δ' ἀλεκτρυὼν μόνος ἴδιον· οὔτε γὰρ σάρξ ἐστιν οὔτε πόρρω σαρκὸς τὴν φύσιν.

XIII Τῶν δ' ἐνύδρων ζῴων τὸ τῶν ἰχθύων γένος ἓν ἀπὸ τῶν ἄλλων ἀφώρισται, πολλὰς περιέχον ἰδέας.
15 Κεφαλὴν μὲν γὰρ ἔχει καὶ τὰ πρανῆ καὶ τὰ ὕπτια, ἐν ᾧ τόπῳ ἡ γαστὴρ καὶ τὰ σπλάγχνα· καὶ ὀπίσθιον οὐραῖον συνεχὲς ἔχει καὶ ἄσχιστον· τοῦτο δ' οὐ πᾶσιν ὅμοιον. αὐχένα δ' οὐδεὶς ἔχει ἰχθύς, οὐδὲ κῶλον οὐδέν, οὐδ' ὄρχεις ὅλως, οὔτ' ἐντὸς οὔτ' ἐκτός, οὐδὲ μαστούς. τοῦτο μὲν οὖν ὅλως οὐδ' ἄλλο
20 οὐδὲν τῶν μὴ ζῳοτοκούντων, οὐδὲ τὰ ζῳοτοκοῦντα πάντα, ἀλλ' ὅσα εὐθὺς ἐν αὑτοῖς ζῳοτοκεῖ καὶ μὴ ᾠοτοκεῖ πρῶτον. καὶ γὰρ ὁ δελφὶς ζῳοτοκεῖ, διὸ ἔχει μαστοὺς δύο, οὐκ ἄνω δ' ἀλλὰ πλησίον τῶν ἄρθρων. ἔχει δ' οὐχ ὥσπερ τὰ τετράποδα ἐπιφανεῖς θηλάς, ἀλλ' οἷον ῥύακας δύο, ἑκατέρωθεν ἐκ τῶν
25 πλαγίων ἕνα, ἐξ ὧν τὸ γάλα ῥεῖ· καὶ θηλάζεται ὑπὸ τῶν τέκνων παρακολουθούντων· καὶ τοῦτο ὦπται ἤδη ὑπό τινων φανερῶς.

Οἱ δ' ἰχθύες, ὥσπερ εἴρηται, οὔτε μαστοὺς ἔχουσιν οὔτ' αἰδοίων πόρον ἐκτὸς οὐδένα φανερόν. ἴδιον δ' ἔχουσι τό τε τῶν βραγχίων, ᾗ τὸ ὕδωρ ἀφιᾶσι
30 δεξάμενοι κατὰ τὸ στόμα, καὶ τὰ πτερύγια, οἱ μὲν

lung. Some kinds of birds also have spurs; but no crook-taloned bird has them. The crook-taloned birds come into the class of good fliers; the spurred ones belong to the class of heavy birds. Further, some birds have a crest which normally is erect and consists of feathers; the domestic cock's, however, is exceptional: the substance of it is not flesh, yet it is a close approximation to it.

Water-animals. The tribe of fishes is a unified group, distinct from other water-animals, and includes many sorts of various appearance.

XIII Fishes.

To begin with, fishes have a head, dorsal and ventral sides; in the last-named situation are placed the belly and the viscera; and at the rear fishes have a tail which is continuous and undivided, but not identical in all fishes. No fish has a neck, or any limb, or testicles anywhere internally or externally, and no breasts. Breasts are absent also in all the non-viviparous animals, nor are they present in all the Vivipara, but only in those which are directly internally viviparous and not previously oviparous. Thus the dolphin is viviparous, and therefore has two breasts, not in the upper part of the body but near the generative organs. It has, however, no visible teats, like quadrupeds, but has as it were two flow-holes, one on each flank, from which the milk flows, and the young follow alongside to get suckled. There are some who have actually observed this quite clearly.

Fishes, then, as I have said, have no breasts, and no externally visible generative passage. But they have two peculiarities, gills and fins: they first take in water through the mouth and then expel it through the gills [a]: most fishes have four fins, though the

[a] See note on 487 a 17, and Notes, §§ 50 f.

ARISTOTLE

πλεῖστοι τέτταρα, οἱ δὲ προμήκεις δύο, οἷον ἐγχελυς, ὄντα πρὸς τὰ βράγχια. ὁμοίως δὲ καὶ κεστρεῖς, οἷον ἐν Σιφαῖς οἱ ἐν τῇ λίμνῃ, δύο, καὶ ἡ καλουμένη ταινία ὡσαύτως. ἔνια δὲ τῶν προμήκων οὐδὲ πτερύγια ἔχει, οἷον σμύραινα, οὐδὲ τὰ βράγχια διηρθρωμένα ὁμοίως τοῖς ἄλλοις ἰχθύσιν.

Αὐτῶν δὲ τῶν ἐχόντων βράγχια τὰ μὲν ἔχει ἐπικαλύμματα τοῖς βραγχίοις, τὰ δὲ σελάχη πάντα ἀκάλυπτα. καὶ τὰ μὲν ἔχοντα καλύμματα πάντα ἐκ πλαγίου ἔχει τὰ βράγχια, τῶν δὲ σελαχῶν τὰ μὲν πλατέα κάτω ἐν τοῖς ὑπτίοις, οἷον νάρκη καὶ βάτος, τὰ δὲ προμήκη ἐν τοῖς πλαγίοις, οἷον πάντα τὰ γαλεώδη. ὁ δὲ βάτραχος ἐκ πλαγίου μὲν ἔχει, καλυπτόμενα δ' οὐκ ἀκανθώδει καλύμματι ὥσπερ οἱ μὴ σελαχώδεις, ἀλλὰ δερματώδει.

Ἔτι δὲ τῶν ἐχόντων βράγχια τῶν μὲν ἁπλᾶ ἐστι τὰ βράγχια, τῶν δὲ διπλᾶ· τὸ δ' ἔσχατον πρὸς τὸ σῶμα πάντων ἁπλοῦν. καὶ πάλιν τὰ μὲν ὀλίγα βράγχια ἔχει, τὰ δὲ πλῆθος βραγχίων· ἴσα δ' ἐφ' ἑκάτερα πάντες. ἔχει δ' ὁ ἐλάχιστα ἔχων ἓν ἐφ' ἑκάτερα βράγχιον, διπλοῦν δὲ τοῦτο, οἷον κάπρος· οἱ δὲ δύο ἐφ' ἑκάτερα, τὸ μὲν ἁπλοῦν τὸ δὲ διπλοῦν, οἷον γόγγρος καὶ σκάρος· οἱ δὲ τέτταρα ἐφ' ἑκάτερα ἁπλᾶ, οἷον ἔλλοψ, συναγρίς, σμύραινα, ἔγχελυς· οἱ δὲ τέτταρα μὲν δίστοιχα δὲ πλὴν τοῦ ἐσχάτου, οἷον κίχλη καὶ πέρκη καὶ γλάνις καὶ κυπρῖνος. ἔχουσι

---

[a] In Boeotia, near Thespis, now Tipha. Aristotle refers to this mullet (*kestreus*) again, as being found in this lake, at *P.A.* 696 a 5 and *I.A.* 708 a 5.

[b] The identification is uncertain.

elongated ones, such as the eel, have only two, and these are situated near the gills. Similarly the various species of grey mullet, *e.g.*, those in the lake at Siphae,[a] have two ; so has the so-called tape-fish.[b] Some of the elongated fishes have no fins at all, *e.g.*, the muraena, and their gills are not well articulated like those of other fishes.

Of those which have gills, some have coverings on them, though none of the Selachia has any such covering. All which have this covering have their gills placed at the side ; whereas among Selachia the broad ones have them down below on the ventral surface (examples are the torpedo-fish and the ray) ; the elongated ones have them on the sides (examples are all the dogfish). The fishing-frog has them placed at the side, and covered, not with a spiny covering as in the non-selachians, but with a covering consisting of skin.

Of fishes possessing gills, some have single ones, others double : but in all of them the last gill towards the body is single. Again, some have few gills, but others a considerable number, though all have the same number on both sides. Those which have the least have one, a double one, on each side, *e.g.*, the boar-fish ; others have two on either side, one single and one double, *e.g.*, the conger and the parrot-wrasse ; others have four single ones on either side, *e.g.*, the *ellops*, the *synagris*,[c] the muraena, and the eel ; others have four arranged in two rows (excepting the extreme one), *e.g.*, the rainbow-wrasse, the perch, the *glanis*[d] and the carp. The dogfish all have

[c] The identification of *ellops* and *synagris* is uncertain ; *ellops* is perhaps *Acipenser sturio*, a sturgeon ; *synagris* is perhaps a species of *Dentex*.

[d] See note on 568 b 13.

δὲ καὶ οἱ γαλεώδεις διπλᾶ πάντες, καὶ πέντ' ἐφ' ἑκάτερα· ὁ δὲ ξιφίας ὀκτὼ διπλᾶ.

Περὶ μὲν οὖν πλήθους βραγχίων ἐν τοῖς ἰχθύσι τοῦτον ἔχει τὸν τρόπον.

Ἔτι δὲ πρὸς τἆλλα ζῷα οἱ ἰχθύες διαφέρουσι πρὸς τῇ διαφορᾷ τῇ περὶ τὰ βράγχια· οὔτε γὰρ ὥσπερ τῶν πεζῶν ὅσα ζῳοτόκα ἔχει τρίχας, οὔθ' ὥσπερ ἔνια τῶν ᾠοτοκούντων τετραπόδων φολίδας, οὔθ' ὡς τὸ τῶν ὀρνέων γένος πτερωτόν, ἀλλ' οἱ μὲν πλεῖστοι αὐτῶν λεπιδωτοί εἰσιν, ὀλίγοι δέ τινες τραχεῖς, ἐλάχιστον δ' ἐστὶ πλῆθος αὐτῶν τὸ λεῖον. τῶν μὲν οὖν σελαχῶν τὰ μὲν τραχέα ἐστὶ τὰ δὲ λεῖα, γόγγροι δὲ καὶ ἐγχέλυες καὶ θύννοι τῶν λείων.

Καρχαρόδοντες δὲ πάντες οἱ ἰχθύες ἔξω τοῦ σκάρου· καὶ πάντες ἔχουσιν ὀξεῖς τοὺς ὀδόντας καὶ πολυστοίχους, καὶ ἔνιοι ἐν τῇ γλώττῃ. καὶ γλῶτταν σκληρὰν καὶ ἀκανθώδη ἔχουσι, καὶ προσπεφυκυῖαν οὕτως ὥστ' ἐνίοτε μὴ δοκεῖν ἔχειν. τὸ δὲ στόμα οἱ μὲν ἀνερρωγός, ὥσπερ ἔνια τῶν ζῳοτοκούντων καὶ τετραπόδων, ⟨οἱ δὲ μύουρον⟩.[1]

Τῶν δ' αἰσθητηρίων τῶν μὲν ἄλλων οὐδὲν ἔχουσι φανερὸν οὔτ' αὐτὸ οὔτε τοὺς πόρους, οὔτ' ἀκοῆς οὔτ' ὀσφρήσεως· ὀφθαλμοὺς δὲ πάντες ἔχουσιν ἄνευ βλεφάρων, οὐ σκληρόφθαλμοι ὄντες. ἔναιμον μὲν οὖν ἐστιν ἅπαν τὸ τῶν ἰχθύων γένος, εἰσὶ δ' αὐτῶν οἱ μὲν ᾠοτόκοι οἱ δὲ ζῳοτόκοι, οἱ μὲν λεπιδωτοὶ πάντες ᾠοτόκοι, τὰ δὲ σελάχη πάντα ζῳοτόκα πλὴν βατράχου.

[1] coni. Dt. (cf. *P.A.* 662 a 31, 696 b 34), qui lacunam hic

double ones, five on each side; the swordfish has eight double ones.

So much then for the number of gills as found in fishes.

Furthermore, fishes differ from other animals otherwise than in possessing gills. Unlike viviparous land-animals they have no hair, and unlike some of oviparous quadrupeds they have no horny scales, and unlike birds they have no feathers. Generally, they are covered with ordinary scales, though a few are rough-skinned, and a very small number are smooth-skinned. Some of the Selachia are rough-skinned, some smooth-skinned; to the smooth-skinned also belong the conger, the eel, and the tunny.

All fishes are saw-toothed except the parrot-wrasse, and all have many rows of sharp teeth, some fishes having them on the tongue. Their tongue is hard and spiny, and it is so closely attached that they appear sometimes not to have one. In some the mouth is wide and gaping, as in some viviparous quadrupeds; ⟨in others, tapering⟩.[a]

Organs of sensation. Fishes have none of these, at least none observable, except eyes—neither the organ itself nor corresponding passages either for hearing or smell. But they all have eyes, without eyelashes, though their eyes are not hard. The whole tribe of fishes is blooded. Some are oviparous, some viviparous: the scaly ones are all oviparous; all the Selachia, except the fishing-frog, are viviparous.

---

[a] There is obviously a lacuna here in the Greek text, as editors point out: Dt. suggests the insertion which I have made, comparing *P.A.* 662 a 31; *cf.* also *P.A.* 696 b 34.

statuit; lacunam post οἱ μὲν stat. A.-W.; nullam exhibet Σ: οἱ μὲν ⟨σιμόν, οἱ δὲ⟩ Pi.

ARISTOTLE

XIV Λοιπὸν δὲ τῶν ἐναίμων ζῴων τὸ τῶν ὄφεων γένος. ἔστι δὲ κοινὸν ἀμφοῖν· τὸ μὲν γὰρ πλεῖστον αὐτῶν χερσαῖόν ἐστιν, ὀλίγον δὲ τὸ τῶν ἐνύδρων ἐν τοῖς ποτίμοις ὕδασι διατελεῖ. εἰσὶ δὲ καὶ θαλάττιοι ὄφεις, παραπλήσιοι τὴν μορφὴν τοῖς χερσαίοις τἆλλα· πλὴν τὴν κεφαλὴν ἔχουσι γογγροειδεστέραν. γένη δὲ πολλὰ τῶν θαλαττίων ὄφεών ἐστι, καὶ χρόαν ἔχουσι παντοδαπήν· οὐ γίγνονται δ' οὗτοι ἐν τοῖς σφόδρα βαθέσιν. ἄποδες δ' εἰσὶν οἱ ὄφεις ὥσπερ τὸ τῶν ἰχθύων γένος.

Εἰσὶ δὲ καὶ σκολόπενδραι θαλάττιαι, παραπλήσιαι τὸ εἶδος ταῖς χερσαίαις, τὸ δὲ μέγεθος μικρῷ ἐλάττους· γίγνονται δὲ περὶ τοὺς πετρώδεις τόπους. τὴν δὲ χροιάν εἰσιν ἐρυθρότεραι καὶ πολύποδες μᾶλλον καὶ λεπτοσκελέστεραι τῶν χερσαίων. οὐ γίγνονται δ' οὐδ' αὗται, ὥσπερ οὐδ' οἱ ὄφεις, ἐν τοῖς βαθέσι σφόδρα.

Ἔστι δ' ἰχθύδιόν τι τῶν πετραίων, ὃ καλοῦσί τινες ἐχενηΐδα, καὶ χρῶνταί τινες αὐτῷ πρὸς δίκας καὶ φίλτρα· ἔστι δὲ ἄβρωτον· τοῦτο δ' ἔνιοί φασιν

---

[a] Probably a species of *Nereïs*. See also 621 a 7.

[b] Probably a blenny or goby. Pliny (*N.H.* ix. 79) also mentions the *echeneïs* as frequenting rocks, its use as a love-charm and to stop proceedings at law, and Aristotle's denial that it has feet; he confuses it, however, with the sucking-fish, also called Echeneïs; it is, he says, believed to slow down ships by sticking to their hulls (whence its name); *cf. N.H.* xxxii. 2, in which book he tells how Mark Antony's ship was held up by one at the battle of Actium, and Caligula's was similarly delayed on the way to Antium. It is this other echeneïs or *remora* which has become familiar in literature: see, *e.g.*, R. R. Steele, *Medieval Lore*, London, 1905, p. 134, from Bartholomaeus Anglicus (*c.* 1260) " he cleaveth to the ship, and holdeth it still stedfastly in the sea, as though the ship were on ground therein "; and references in P. Ansell

HISTORIA ANIMALIUM, II. xiv

Of blooded animals there now remains the tribe of serpents. This shares the features of both classes, land- and water-animals. Most serpents are land-creatures, but a few, the aquatic ones, spend their time in fresh water. There are also sea-serpents, very similar in shape to the land-serpents in all respects, except that their heads are more like the conger's. There are many kinds of sea-serpents, and of all sorts of colours; these do not occur in really deep water. Like the tribe of fish, serpents have no feet.

XIV Serpents, etc.

There are also sea-millipedes,[a] similar in appearance to the terrestrial ones, but slightly smaller in size. They occur in rocky places. Compared with the land-millipedes they are redder, have more feet, and more delicate legs. Like the sea-serpents, they do not occur in really deep water.

Among fish which frequent rocks, there is a very small one, which some people call the "ship-brake"[b]: some use it as a charm for law-suits and love-affairs. It is not edible. Some say it has feet; but it has not:

Robin, *Animal Lore in English Literature*, London, 1932, p. 128, *e.g.*, Rabelais v. 30 "I saw a Remora, a little Fish call'd Echineïs by the Greeks, and near it a tall Ship, that did not get o'head an Inch, tho' she was in the Offin with Top and Top-gallants spread before the Wind. In the Days of Yore, two sorts of Fishes us'd to abound in our Courts of Judicature . . . the others [were] your beneficial Remorae's, that is, the Eternity of Law-Suits, the needless Letts that keep 'em undecided." Spenser, *Visions of the worlds vanitie*, 121 ff. "All sodainely there cloue vnto her keele | A little fish, that men call *Remora*, | Which stopt her course, and held her by the heele." Ben Jonson, *The Magnetick Lady*, II. ii. 29 f. "I say a *remora*; | For it will stay a Ship, that's under Saile," and *id.*, *Poetaster* III. ii. 3 f. "'Death, I am seaz'd on here | By a Land-*Remora*, I cannot stirre." See also Thompson, *G.G.F.*, *s.v.*; and *cf.* 557 a 31.

ἔχειν πόδας οὐκ ἔχον, ἀλλὰ φαίνεται διὰ τὸ τὰς πτέρυγας ὁμοίας ἔχειν ποσίν.[1]

Τὰ μὲν οὖν ἔξω μόρια, καὶ πόσα καὶ ποῖα τῶν ἐναίμων ζῴων, καὶ τίνας ἔχει πρὸς ἄλληλα διαφοράς, εἴρηται.

XV Τὰ δ' ἐντὸς πῶς ἔχει, λεκτέον ἐν τοῖς ἐναίμοις ζῴοις πρῶτον· τούτῳ γὰρ διαφέρει τὰ μέγιστα γένη πρὸς τὰ λοιπὰ τῶν ἄλλων ζῴων, τῷ τὰ μὲν ἔναιμα τὰ δ' ἄναιμα εἶναι. ἔστι δὲ ταῦτα τά τε ᾠοτόκα[a][2] καὶ τὰ ζῳοτόκα τῶν τετραπόδων, [ἔτι δὲ καὶ τὰ ᾠοτόκα τῶν τετραπόδων][3] καὶ ὄρνις καὶ ἰχθὺς καὶ κῆτος, καί εἴ τι ἄλλο ἀνώνυμόν ἐστι[4] διὰ τὸ μὴ εἶναι γένος ἀλλ' ἁπλοῦν τὸ εἶδος ἐπὶ τῶν καθ' ἕκαστον, οἷον ἄνθρωπος.[5][b]

Ὅσα μὲν οὖν ἐστι τετράποδα καὶ ζῳοτόκα, στόμαχον μὲν καὶ ἀρτηρίαν πάντ' ἔχει, καὶ κείμενα τὸν αὐτὸν τρόπον ὥσπερ ἐν τοῖς ἀνθρώποις· ὁμοίως δὲ καὶ ὅσα ᾠοτοκεῖ τῶν τετραπόδων, καὶ ἐν τοῖς ὄρνισιν· ἀλλὰ τοῖς εἴδεσι τῶν μορίων τούτων διαφέρουσιν. ὅλως δὲ πάντα ὅσα τὸν ἀέρα δεχόμενα ἀναπνεῖ

[1] ab initio cap. hucusque secl. A.-W.
[2] τά τε ᾠοτόκα Balme : ἄνθρωπός τε vulg.
[3] om. D, Karsch, Balme.
[4] scriberes fortasse ἀνώνυμόν ἐστιν· ⟨ἄλλα δ' ἀνώνυμά ἐστι⟩ διὰ τὸ κ.τ.λ.
[5] ἄνθρωπος Balme, cf. 490 b 16 sqq. : οἷον ὄφις καὶ κροκόδειλος vulg., eiciunt Scal., Sn. : ἔχις καὶ κορδύλος coni. Dt. locus hic vexatus ; fortasse 28 ἔστι δὲ ταῦτα hucusque secludenda.

[a] The whole of the chapter up to this point is considered an interpolation by A.-W. and Dittmeyer.
[b] There is no main group under which man falls ; hence there is no name for it. See above, note on 491 a 6.—This passage, however, even with some necessary corrections, does not square with what is stated above at 490 b 7 ff.,

HISTORIA ANIMALIUM, II. xiv–xv

though it appears to have them owing to the fact that its fins resemble feet.[a]

I have now described the external parts of blooded animals—their number, their character, and the various differences they exhibit among the various animals.

We must now go on to describe the arrangement of the internal parts, and first of all those of the blooded animals, because this is the feature in which the main groups differ from the rest of the animals : they are blooded, whereas the others are bloodless. These are the oviparous and the viviparous quadrupeds, bird, fish, cetacean, and any other there may be which is unnamed because there is no group but merely the simple type in the individual cases, for example, man.[b]

XV INSTRUMENTAL PARTS OF BLOODED ANIMALS: (b) Internal parts (resumed).

Very well then : all viviparous quadrupeds have an oesophagus and a windpipe, situated just as in human beings : and the same holds good of the oviparous quadrupeds and of birds, except that they differ in the conformation of these parts. Speaking generally, we may say that all animals which take up

b 31 ff. There, not only the 3 blooded groups (Birds, Fishes, Cetacea) appear to be reckoned as " main groups," but also the 4 bloodless groups, which are here contrasted with the " main groups." Furthermore, in the present passage the viviparous and the oviparous quadrupeds, as being blooded, are reckoned among the " main groups," whereas in the earlier passage they appear to be included among the groups which are " not μεγάλα," i.e., groups other than the 7 μέγιστα γένη. The reason for these discrepancies may be that in the earlier passage Ar. is concerned to indicate the main groupings and sub-groupings of animals, and at the same time to show how impossible it is to arrive at any tidy and exhaustive classification ; whereas here he is simply giving a practical list of the blooded groups (*plus* man), which he intends to deal with first in what follows. See further, Notes, §§ 9 ff.

ARISTOTLE

καὶ ἐκπνεῖ, πάντ' ἔχει πνεύμονα καὶ ἀρτηρίαν καὶ στόμαχον, καὶ τὴν θέσιν τοῦ στομάχου καὶ τῆς ἀρτηρίας ὁμοίως, ἀλλ' οὐχ ὅμοια, τὸν δὲ πνεύμονα οὔθ' ὅμοιον οὔτε τῇ θέσει ὁμοίως ἔχοντα. ἔτι δὲ καρδίαν ἅπαντ' ἔχει ὅσα αἷμα ἔχει, καὶ τὸ διάζωμα, ὃ καλοῦνται φρένες· ἀλλ' ἐν τοῖς μικροῖς διὰ λεπτότητα καὶ σμικρότητα οὐ φαίνεται ὁμοίως. (πλὴν ἐν τῇ καρδίᾳ ἴδιόν τι[1] ἐστὶν ἐπὶ τῶν βοῶν· ἔστι γάρ τι γένος βοῶν, ἀλλ' οὐ πάντες, ὃ ἔχει ἐν τῇ καρδίᾳ ὀστοῦν. ἔχει δὲ καὶ ἡ τῶν ἵππων καρδία ὀστοῦν.)[2]

Πνεύμονα δ' οὐ πάντα, οἷον ἰχθῦς οὐκ ἔχει, οὐδ' εἴ τι ἄλλο τῶν ζῴων ἔχει βράγχια. καὶ ἧπαρ ἅπαντ' ἔχει ὅσαπερ αἷμα. σπλῆνα δὲ τὰ πλεῖστα ἔχει ὅσαπερ καὶ αἷμα. τὰ δὲ πολλὰ τῶν μὴ ζῳοτόκων ἀλλ' ᾠοτόκων μικρὸν ἔχει τὸν σπλῆνα οὕτως ὥστε λανθάνειν ὀλίγου τὴν αἴσθησιν, ἔν τε τοῖς ὄρνισι τοῖς πλείστοις, οἷον ἐν περιστερᾷ καὶ ἰκτίνῳ καὶ ἱέρακι καὶ γλαυκί·[b] ὁ δ' αἰγοκέφαλος ὅλως οὐκ ἔχει. καὶ ἐπὶ τῶν ᾠοτόκων δὲ καὶ τετραπόδων τὸν αὐτὸν τρόπον ἔχει· μικρὸν γὰρ πάμπαν ἔχουσι καὶ ταῦτα, οἷον χελώνη, ἐμύς, φρύνη, σαῦρος, κροκόδειλος, βάτραχος.

Χολὴν δὲ τῶν ζῴων τὰ μὲν ἔχει τὰ δ' οὐκ ἔχει ἐπὶ τῷ ἥπατι. τῶν μὲν ζῳοτόκων καὶ τετραπόδων ἔλαφος οὐκ ἔχει οὐδὲ πρόξ, ἔτι δ' ἵππος, ὀρεύς, ὄνος, φώκη καὶ τῶν ὑῶν ἔνιοι. τῶν δ' ἐλάφων αἱ Ἀχαΐναι καλούμεναι δοκοῦσιν ἔχειν ἐν τῇ κέρκῳ χολήν·

---

[1] καρδίᾳ ἴδιόν τι Sn.: ὁμοίως, πλὴν ἐν τῇ καρδίᾳ. ἴδιον δ' vulg.
[2] πλὴν ... ὀστοῦν secl. A.-W.

---

[a] The same statement is found at *P.A.* 666 b 18 (of horses and a certain kind of ox) and at *G.A.* 787 b 18 (of some bulls).
[b] Perhaps a horned owl.

HISTORIA ANIMALIUM, II. xv

air, *i.e.*, which breathe in and out, have a lung, a windpipe and an oesophagus : the position of the oesophagus and windpipe is always the same, though the organs themselves are not ; the lung may both occupy a different position and be different in character. Furthermore, all blooded animals have a heart, and a diaphragm (which is called the midriff); in small animals, however, owing to its slightness and smallness it is not equally obvious. (Except that a peculiarity is found in the heart in the case of oxen : there is a sort of ox which has a bone in its heart— but not all have it. There is also a bone in the horse's heart.[a])

Not all these animals have a lung : fishes have none, for example, nor has any other animal which possesses gills. All blooded animals have a liver, and most of them have a spleen, but in most of the non-viviparous, oviparous, animals it is so small that it almost escapes observation : this is so in most birds (*e.g.*, the pigeon, the kite, the hawk, the owl), and the goat-poll [b] has none at all. The situation is the same with the oviparous quadrupeds : they too have quite a small spleen, *e.g.*, the tortoise, the freshwater tortoise, the toad, the lizard, the crocodile, and the frog.

Some animals have, and some have not, a gall-bladder up against the liver. The deer is an example of a viviparous quadruped which has none : other examples are the roe, the horse, the mule, the ass, the seal, and some kinds of pig. A certain sort of deer called Achaïnae [c] appear to have gall in their tail ; (Gall-bladder.)

[c] The origin and meaning of this word are doubtful ; it occurs again at 611 b 18.

25 ἔστι δ' ὃ λέγουσι τὸ μὲν χρῶμα ὅμοιον χολῇ, οὐ μέντοι ὅλον ὑγρὸν οὕτως, ἀλλ' ὅμοιον τῷ τοῦ σπληνὸς τὰ ἐντός.[1] σκώληκας μέντοι πάντες ἔχουσιν ἐν τῇ κεφαλῇ ζῶντας· ἐγγίγνονται δὲ ὑποκάτω τοῦ ὑπογλωττίου ἐν τῷ κοίλῳ καὶ περὶ τὸν σφόνδυλον, ᾗ ἡ κεφαλὴ προσπέφυκε, τὸ μέγεθος οὐκ ἐλάττους 30 ὄντες τῶν μεγίστων εὐλῶν· ἐγγίγνονται δ' ἀθρόοι καὶ συνεχεῖς, τὸν ἀριθμὸν δ' εἰσὶ μάλιστα περὶ εἴκοσι.[2]

Χολὴν μὲν οὖν οὐκ ἔχουσιν οἱ ἔλαφοι, ὥσπερ εἴρηται· τὸ δ' ἔντερον αὐτῶν ἐστι πικρὸν οὕτως ὥστε μηδὲ τοὺς κύνας ἐθέλειν ἐσθίειν, ἂν μὴ σφόδρα πίων ᾖ ὁ ἔλαφος. ἔχει δὲ καὶ ὁ ἐλέφας τὸ ἧπαρ ἄχολον μέν, τεμνομένου μέντοι περὶ τὸν τόπον οὗ τοῖς ἔχουσιν ἐπιφύεται ἡ χολή, ῥεῖ ὑγρότης χολώδης ἢ πλείων ἢ ἐλάττων. τῶν δὲ δεχομένων τὴν θάλατταν καὶ ἐχόντων πνεύμονα δελφὶς οὐκ ἔχει 5 χολήν. οἱ δ' ὄρνιθες καὶ οἱ ἰχθύες πάντες ἔχουσι, καὶ τὰ ᾠοτόκα καὶ τετράποδα, καὶ ὡς ἐπίπαν εἰπεῖν ἢ πλείω ἢ ἐλάττω· ἀλλ' οἱ μὲν πρὸς τῷ ἥπατι τῶν ἰχθύων, οἷον οἵ τε γαλεώδεις καὶ γλάνις καὶ ῥίνη καὶ λειόβατος καὶ νάρκη καὶ τῶν μακρῶν ἔγχελυς 10 καὶ βελόνη καὶ ζύγαινα. ἔχει δὲ καὶ ὁ καλλιώνυμος ἐπὶ τῷ ἥπατι, ὅσπερ ἔχει μεγίστην τῶν ἰχθύων ὡς κατὰ μέγεθος. οἱ δὲ πρὸς τοῖς ἐντέροις ἔχουσιν, ἀποτεταμένην ἀπὸ τοῦ ἥπατος πόροις ἐνίοις πάνυ λεπτοῖς. ἡ μὲν οὖν ἀμία παρὰ τὸ ἔντερον παρα-

---

[1] ἐκτός Bekker typoth. errore.
[2] 27 σκώληκας . . . εἴκοσι secl. Saint-Hilaire, Dt.

---

[a] Larvae of a gadfly, *Oestrus rufibarbis*.
[b] This sentence about maggots is considered an interpolation by Dittmeyer, following Saint-Hilaire.

the substance referred to is certainly gall-like in colour, although it is not so completely fluid, and internally the part is like the inside of the spleen. All deer, however, have living maggots[a] inside their heads: they infest the hollow region under the tongue and near the vertebra to which the head is attached: these maggots are as big as the biggest grubs; they grow in a bunch huddled up together, about twenty of them.[b]

Deer then, as I have said, have no gall-bladder: but their gut is so bitter that even dogs refuse to eat it unless the animal is exceptionally fat. In the elephant, also, the liver has no gall-bladder; but when the animal is cut in the region where other animals' gall-bladder is situated, there comes out a gall-like fluid in greater or less quantity. An example of an animal which takes in sea-water and has a lung, but lacks a gall-bladder, is the dolphin. Birds and fishes all have one, so have the oviparous quadrupeds, generally speaking, larger or smaller in size. Some fishes, however, have it close to the liver, *e.g.*, the dogfishes, the *glanis*, the angel-fish, the skate, the torpedo-fish, and (these are elongated fishes) the eel, the pipefish, and the hammer-headed shark. The *kallionymos*,[c] too, has it close to the liver, and this fish has a larger one for its size than any other fish. Others have it by the gut, attached to the liver by very narrow passages. The bonito[d] has its gall-bladder extending alongside the gut and equal to it

[c] The identification is uncertain; perhaps *Uranoscopus scaber*.

[d] *Sarda* [formerly *Pelamys*] *sarda*, a small tunny, the mainstay of the Black Sea fishery, whereas the true tunny is commoner in the Mediterranean (D'Arcy Thompson). See also 488 a 6.

ARISTOTLE

τεταμένην ἰσομήκη ἔχει, πολλάκις δὲ καὶ ἐπανα-
δίπλωμα· οἱ δ' ἄλλοι πρὸς τοῖς ἐντέροις οἱ μὲν πορ-
ρώτερον οἱ δ' ἐγγύτερον, οἷον βάτραχος, ἔλλοψ, συναγρίς, σμύραινα, ξιφίας. πολλάκις δὲ καὶ τὸ αὐτὸ
γένος ἐπ' ἀμφότερα φαίνεται ἔχον, οἷον γόγγροι οἱ
μὲν πρὸς τῷ ἥπατι, οἱ δὲ κάτω ἀπηρτημένην.
ὁμοίως δ' ἔχει τοῦτο καὶ ἐπὶ τῶν ὀρνίθων· ἔνιοι γὰρ
πρὸς τῇ κοιλίᾳ ἔχουσιν, οἱ δὲ πρὸς τοῖς ἐντέροις
τὴν χολήν, οἷον περιστερά, κόραξ, ὄρτυξ, χελιδών,
στρουθός. ἔνιοι δ' ἅμα πρὸς τῷ ἥπατι ἔχουσι καὶ
πρὸς τῇ κοιλίᾳ, οἷον αἰγοκέφαλος, οἱ δ' ἅμα πρὸς
τῷ ἥπατι καὶ τοῖς ἐντέροις, οἷον ἱέραξ καὶ ἰκτῖνος.

XVI Νεφροὺς δὲ καὶ κύστιν τὰ μὲν ζῳοτόκα τῶν τετραπόδων πάντ' ἔχει· ὅσα δ' ᾠοτοκεῖ, τῶν μὲν
ἄλλων οὐδὲν ἔχει, οἷον οὔτ' ὄρνις οὔτ' ἰχθύς, τῶν δὲ
τετραπόδων μόνη χελώνη ἡ θαλαττία μέγεθος κατὰ
λόγον τῶν ἄλλων μορίων. ὁμοίους δ' ἔχει τοὺς
νεφροὺς ἡ θαλαττία χελώνη τοῖς βοείοις· ἔστι δ' ὁ
τοῦ βοὸς οἷον ἐκ πολλῶν μικρῶν εἰς συγκείμενος.
(ἔχει δὲ καὶ ὁ βόνασος τὰ ἐντὸς ἅπαντα ὅμοια βοΐ.)[1]

XVII Τῇ δὲ θέσει, ὅσα ἔχει ταῦτα τὰ μόρια, ὁμοίως
κείμενα ἔχει, τήν τε καρδίαν περὶ τὸ μέσον, πλὴν
ἐν ἀνθρώπῳ· οὗτος δ' ἐν τῷ ἀριστερῷ μᾶλλον μέρει,
καθάπερ ἐλέχθη πρότερον. ἔχει δὲ καὶ τὸ ὀξὺ ἡ
καρδία πάντων εἰς τὸ πρόσθεν· πλὴν ἐπὶ τῶν ἰχθύων
οὐκ ἂν δόξειεν· οὐ γὰρ πρὸς τὸ στῆθος ἔχει τὸ ὀξύ,
ἀλλὰ πρὸς τὴν κεφαλὴν καὶ τὸ στόμα. ἀνήρτηται
δ' αὐλῷ[2] τὸ ἄκρον ᾗ συνάπτει τὰ βράγχια ἀλλήλοις

---

[1] ἔχει ... βοΐ secl. A.-W.
[2] αὐλῷ A.-W.: αὐτῷ PD: αὐτῶν vulg.

---

[a] See note on 505 a 15.

in length, and often there is a double fold to it. Others have the gall-bladder near the gut, some further off, some closer to it; examples are the fishing-frog, the *ellops*, the *synagris*,[a] the muraena, the sword-fish. Often we find two different positions in the same group of animals: *e.g.*, some congers have it near the liver, others have it below and detached from the liver. The same is found in birds: some have the gall-bladder near the stomach, others near the gut, *e.g.*, the pigeon, the raven, the quail, the swallow, and the sparrow; some have it at one and the same time close to the liver and close to the stomach, *e.g.*, the goat-poll; some have it at one and the same time close to the liver and close to the gut, *e.g.*, the hawk and the kite.

Kidneys and Bladder. All viviparous quadrupeds XVI have these parts. None of the non-quadrupedal Ovipara, whether birds or fishes, has them; and of the quadrupedal Ovipara the turtle is the only one which has kidneys and bladder proportionate in size to its other parts. The kidneys of the turtle are similar to those of the ox; the ox's kidney is as it were one organ consisting of many small ones. (All the internal organs of the bison, too, are similar to those of the ox.)

With regard to situation, these parts, in the animals that have them, are similarly placed, and the heart is XVII situated about the middle, except in man: in man it is somewhat over towards the left, as we have said before. In all animals the pointed end of the heart lies forward, though in fishes it might appear not to do so: the reason is that the pointed end is not towards the breast but towards the head and the mouth; and the apex is attached to a tube where the right and

τὰ δεξιὰ καὶ τὰ ἀριστερά. εἰσὶ δὲ καὶ ἄλλοι πόροι τεταμένοι ἐξ αὐτῆς εἰς ἕκαστον τῶν βραγχίων, μείζους μὲν τοῖς μείζοσιν, ἐλάττους δὲ τοῖς ἐλάττοσιν· ὁ δ' ἐπ' ἄκρας τῆς καρδίας τοῖς μεγάλοις αὐτῶν σφόδρα παχὺς αὐλός ἐστι καὶ λευκός.

Στόμαχον δ' ὀλίγοι ἔχουσι τῶν ἰχθύων, οἷον γόγγρος καὶ ἔγχελυς, καὶ οὗτοι μικρόν.

Καὶ τὸ ἧπαρ τοῖς ἔχουσι τοῖς μὲν ἀσχιδὲς ἔχουσίν ἐστιν[1] ἐν τοῖς δεξιοῖς ὅλον, τοῖς δ' ἐσχισμένον ἀπ' ἀρχῆς τὸ μεῖζον ἐν τοῖς δεξιοῖς. ἐνίοις γὰρ ἑκάτερον τὸ μόριον ἀπήρτηται καὶ οὐ συμπέφυκεν ἡ ἀρχή, οἷον τῶν τ' ἰχθύων τοῖς γαλεώδεσι, καὶ δασυπόδων τι γένος ἐστὶ καὶ ἄλλοθι καὶ περὶ τὴν λίμνην τὴν Βόλβην ἐν τῇ καλουμένῃ Συκίνῃ, οὓς ἄν τις δόξειε δύο ἥπατα ἔχειν διὰ τὸ πόρρω τοὺς πόρους[2] συνάπτειν, ὥσπερ καὶ ἐπὶ τοῦ τῶν ὀρνίθων πνεύμονος.

Καὶ ὁ σπλὴν δ' ἐστὶ πᾶσιν ἐν τοῖς ἀριστεροῖς κατὰ φύσιν. καὶ οἱ νεφροὶ τοῖς ἔχουσι κείμενοι τὸν αὐτὸν τρόπον. ἤδη δὲ διανοιχθέν τι τῶν τετραπόδων ὤφθη ἔχον τὸν σπλῆνα μὲν ἐν τοῖς δεξιοῖς, τὸ δ' ἧπαρ ἐν τοῖς ἀριστεροῖς· ἀλλὰ τὰ τοιαῦτα ὡς τέρατα κρίνεται.

Τείνει δ' ἡ μὲν ἀρτηρία πᾶσιν εἰς τὸν πνεύμονα (ὃν δὲ τρόπον, ὕστερον ἐροῦμεν), ὁ δὲ στόμαχος εἰς τὴν κοιλίαν διὰ τοῦ διαζώματος ὅσα ἔχει στόμαχον· οἱ γὰρ ἰχθύες, ὥσπερ εἴρηται πρότερον, οἱ πλεῖστοι οὐκ ἔχουσιν, ἀλλ' εὐθὺς πρὸς τὸ στόμα συνάπτει ἡ

---

[1] ἐστιν hic AC, ante ὅλον PD, vulg.
[2] λοβοὺς Sn.: μόρια Scal.: partes Σ.

[a] In Macedonia, to the north of Chalcidicê, not far from Stagirus. Mentioned by Thucydides i. 58.
[b] This phenomenon is mentioned again at *P.A.* 669 b 35,

left gills meet. There are other passages as well extending from the heart to each of the gills, larger in the larger fishes, smaller in the smaller ones; but in the large ones the passage at the apex of the heart is a very stout tube, and white in colour.

A few fishes have an oesophagus, *e.g.*, the conger and the eel; but even they have only a small one.

The liver, in those that have an undivided one, is entirely on the right side; in those which have one which is divided from the start, the larger portion is on the right side; for of course there are some fishes in which the two parts are completely detached and are not joined at the start, *e.g.*, the dogfishes, and there is also a kind of hare in what is known as the Fig country near Lake Bolbê,[a] as well as in other places, which one might well think had two livers[b] owing to the length of the connecting passages, similar to the arrangement with the lungs of birds.

The normal and natural place for the spleen in all animals is on the left side. And the kidneys, in those that have them, are situated identically.[c] Cases have been observed in dissected quadrupeds of the spleen situated on the right side and the liver on the left; but these are reckoned as monstrosities.

In all animals the windpipe extends to the lung (we shall explain the manner of this later), and the oesophagus (if present) extends to the stomach through the midriff. If present—for, as has been said earlier, most fishes have none, and the stomach connects immediately with the mouth, so that often, when

though the location is more vaguely stated: " in some districts."

[c] *i.e.*, in the same position in all animals that have them.

ARISTOTLE

κοιλία, διὸ πολλάκις ἐνίοις τῶν μεγάλων διώκουσι τοὺς ἐλάττους προπίπτει ἡ κοιλία εἰς τὸ στόμα.

Ἔχει δὲ κοιλίαν πάντα μὲν τὰ εἰρημένα, καὶ κειμένην ὁμοίως (κεῖται γὰρ ὑπὸ τὸ διάζωμα εὐθύς), καὶ τὸ ἔντερον ἐχόμενον καὶ τελευτῶν πρὸς τὴν ἔξοδον τῆς τροφῆς καὶ τὸν καλούμενον ἀρχόν. ἀνομοίας δ' ἔχουσι τὰς κοιλίας. πρῶτον μὲν γὰρ τῶν τετραπόδων καὶ ζῳοτόκων ὅσα μή ἐστιν ἀμφώδοντα τῶν κερατοφόρων, τέτταρας ἔχει τοὺς τοιούτους τόπους[1]· ἃ δὴ καὶ λέγεται μηρυκάζειν. διήκει γὰρ ὁ μὲν στόμαχος ἀπὸ τοῦ στόματος ἀρξάμενος ἐπὶ τὰ κάτω παρὰ τὸν πνεύμονα, ἀπὸ τοῦ διαζώματος ἐπὶ τὴν κοιλίαν τὴν μεγάλην· αὕτη δ' ἐστὶ τὰ ἔσω τραχεῖα καὶ διειλημμένη· συνήρτηται δ' αὐτῇ πλησίον τῆς τοῦ στομάχου προσβολῆς ὁ καλούμενος κεκρύφαλος ἀπὸ τῆς ὄψεως· ἔστι γὰρ τὰ μὲν ἔξωθεν ὅμοιος τῇ κοιλίᾳ, τὰ δ' ἐντὸς ὅμοιος τοῖς πλεκτοῖς κεκρυφάλοις· μεγέθει δὲ πολὺ ἐλάττων ἐστὶν ὁ κεκρύφαλος τῆς κοιλίας. τούτου δ' ἔχεται ὁ ἐχῖνος, τὰ ἐντὸς ὢν τραχὺς καὶ πλακώδης, τὸ δὲ μέγεθος παραπλήσιος τῷ κεκρυφάλῳ. μετὰ δὲ τοῦτον τὸ καλούμενον ἤνυστρόν ἐστι, τῷ μὲν μεγέθει τοῦ ἐχίνου μεῖζον, τὸ δὲ σχῆμα προμηκέστερον· ἔχει δ' ἐντὸς πλάκας πολλὰς καὶ μεγάλας καὶ λείας. ἀπὸ δὲ τούτου τὸ ἔντερον ἤδη.

Τὰ μὲν οὖν κερατοφόρα καὶ μὴ ἀμφώδοντα τοιαύτην ἔχει τὴν κοιλίαν, διαφέρει δὲ πρὸς ἄλληλα τοῖς σχήμασι καὶ τοῖς μεγέθεσι τούτων τε καὶ τῷ

---

[1] τόπους Sn. : ventres Σ : πόρους vulg.

some of the large fishes are chasing smaller ones, their stomach [a] falls forward into the mouth.

All the animals we have mentioned have a stomach, and it is similarly situated (*i.e.*, it is situated immediately under the midriff), and they have a gut connected to it and terminating at the outlet for the nourishment—what is called the rectum. But there are dissimilarities about the stomach. To begin with, those viviparous quadrupeds which are horned and do not have teeth in both jaws, have four chambers of this kind : these of course are the animals which are said to " chew the cud." In these animals, the oesophagus begins from the mouth and passes downwards beside the lung, from the midriff as far as the great stomach ; internally this stomach is rough and partitioned ; and connected to it near the entrance of the oesophagus is the " hair-net " [b] as it is called owing to its appearance : outside it is like the stomach, but inside it resembles a plaited hair-net ; it is much smaller in size than the stomach. Connected to it is the *echinus*,[c] which is rough internally and laminated, comparable in size to the " hair-net." After that is the *enystron*, as it is called,[d] larger than the *echinus*, and more elongated in shape : internally it has many large, smooth folds. Immediately after this is the gut.

Such is the formation of the stomach in the horned animals which have not teeth in both jaws, though they differ from one another in respect of the shape and size of these parts, and also because in some the

[a] Actually the air-bladder (Th.) ; *cf.* 591 b 6.
[b] The *reticulum*, or honeycomb(-bag).
[c] The many-plies. *Echinus* is a word applied to several other objects, including the hedgehog, the sea-urchin, a large wide-mouthed jar, the hard core of beechnuts, chestnuts, etc.
[d] The *abomasum*.

15 τὸν στόμαχον εἰς μέσην ἢ πλαγίαν τείνειν τὴν κοιλίαν.

Τὰ δ' ἀμφώδοντα μίαν ἔχει κοιλίαν, οἷον ἄνθρωπος, ὗς, κύων, ἄρκτος, λέων, λύκος. ἔχει δὲ καὶ ὁ θὼς πάντα τὰ ἐντὸς ὅμοια λύκῳ. πάντα μὲν οὖν ἔχει μίαν κοιλίαν, καὶ μετὰ ταῦτα τὸ ἔντερον· ἀλλὰ τὰ μὲν ἔχει μείζω τὴν κοιλίαν, ὥσπερ ὗς καὶ ἄρκτος
20 (καὶ ἥ γε τῆς ὑὸς ὀλίγας ἔχει λείας πλάκας), τὰ δὲ πολὺ ἐλάττω καὶ οὐ πολλῷ μείζω τοῦ ἐντέρου, καθάπερ κύων καὶ λέων καὶ ἄνθρωπος. καὶ τῶν ἄλλων δὲ τὰ εἴδη διέστηκε πρὸς τὰς τούτων κοιλίας· τὰ μὲν γὰρ ὑὶ ὁμοίαν ἔχει τὰ δὲ κυνί, καὶ τὰ μείζω
25 καὶ τὰ ἐλάττω τῶν ζῴων ὡσαύτως. διαφορὰ δὲ καὶ ἐν τούτοις κατὰ τὰ μεγέθη καὶ τὰ σχήματα καὶ πάχη καὶ λεπτότητας ὑπάρχει τὰς τῆς κοιλίας, καὶ κατὰ τοῦ στομάχου τὴν θέσιν καὶ σύντρησιν.[1]

Διαφέρει δὲ καὶ ἡ τῶν ἐντέρων φύσις ἑκατέροις τῶν εἰρημένων ζῴων, τοῖς τε μὴ ἀμφώδουσι καὶ
30 τοῖς ἀμφώδουσι, τῷ μεγέθει καὶ πάχει καὶ ταῖς ἐπαναδιπλώσεσιν.

Πάντα δὲ μείζω τὰ τῶν μὴ ἀμφωδόντων ἐστίν· καὶ γὰρ αὐτὰ πάντα μείζω· μικρὰ μὲν γὰρ ὀλίγα, πάμπαν δὲ μικρὸν οὐδέν ἐστι κερατοφόρον. ἔχουσι δ' ἔνια καὶ ἀποφυάδας τῶν ἐντέρων, εὐθυέντερον δ' οὐδέν ἐστι μὴ ἀμφώδουν.

35 Ὁ δ' ἐλέφας ἔντερον ἔχει συμφύσεις ἔχον, ὥστε φαίνεσθαι τέτταρας κοιλίας ἔχειν. ἐν τούτῳ καὶ ἡ τροφὴ ἐγγίγνεται, χωρὶς δ' οὐκ ἔχει ἀγγεῖον. καὶ τὰ σπλάγχνα ἔχει παραπλήσια τοῖς ὑείοις, πλὴν τὸ

---

[1] sic A.-W.: κατὰ τοῦ στομάχου τῇ θέσει τὴν σύντρησιν vulg.

[a] *Mustela erminea.* See Ingo Krumbiegel, " Die Thos-

oesophagus joins up to the stomach centrally, and in others sideways.

Animals with teeth in both jaws have one stomach only, *e.g.*, man, the pig, the dog, the bear, the lion, and the wolf. The stoat,[a] too, has all its internal parts similar to the wolf. All these, then, have one stomach, and following after that the gut. But there are differences. Some have a comparatively large stomach (*e.g.*, the pig and the bear ; the pig's stomach has a few smooth folds in it) ; some have a much smaller one, not much larger than the gut, e.g., the dog, the lion, and man. As for the others, the shape of their stomach tends towards one or other of those just mentioned : in some it resembles the pig's, in others the dog's, both in larger and smaller animals. In these animals also differences occur with regard to the size, shape, and thickness or thinness of the stomach, and also to the position where the orifice of the oesophagus comes into it.

The structure of the gut also differs in each of the two groups of animals mentioned, *viz.*, those with and without front teeth in the upper jaw, in size, in thickness, and in the character of its folds.

In the non-ambidentate animals, the intestines are all larger ; indeed, all the animals themselves are larger : few of them are small, and no horned animal is minute. Some animals have appendages to the gut ; but no non-ambidentate animal has a straight gut.

The elephant's gut has constrictions, which make it appear to have four stomachs. This gut is where the food has its place : the elephant has no separate receptacle. Its viscera resemble the pig's, except

Tiere des Aristoteles," *Archeion*, XVI (1934), pp. 24-37. Others identify with the jackal or civet.

μὲν ἧπαρ τετραπλάσιον τοῦ βοείου καὶ τἆλλα, τὸν δὲ σπλῆνα ἐλάττω ἢ κατὰ λόγον.

Τὸν αὐτὸν δὲ τρόπον ἔχει τὰ περὶ τὴν κοιλίαν καὶ τὴν τῶν ἐντέρων φύσιν καὶ τοῖς τετράποσι μὲν τῶν ζῴων ᾠοτόκοις δέ, οἷον χελώνῃ χερσαίᾳ καὶ χελώνῃ θαλαττίᾳ καὶ σαύρᾳ καὶ τοῖς κροκοδείλοις ἀμφοῖν καὶ πᾶσιν ὅλως[1] τοῖς τοιούτοις· ἁπλῆν τε γὰρ ἔχουσι καὶ μίαν τὴν κοιλίαν, καὶ τὰ μὲν ὁμοίαν τῇ ὑείᾳ, τὰ δὲ τῇ τοῦ κυνός.

Τὸ δὲ τῶν ὄφεων γένος ὅμοιόν ἐστι καὶ ἔχει παραπλήσια σχεδὸν πάντα τῶν πεζῶν καὶ ᾠοτόκων τοῖς σαύροις, εἴ τις μῆκος αὐτοῖς ἀποδοὺς ἀφέλοι τοὺς πόδας. φολιδωτόν τε γάρ ἐστι, καὶ τὰ πρανῆ καὶ τὰ ὕπτια παραπλήσια τούτοις ἔχει· πλὴν ὄρχεις οὐκ ἔχει, ἀλλ᾽ ὥσπερ ἰχθὺς δύο πόρους εἰς ἓν συνάπτοντας καὶ τὴν ὑστέραν μακρὰν καὶ δικρόαν. τὰ δ᾽ ἄλλα τὰ ἐντὸς τὰ αὐτὰ τοῖς σαύροις, πλὴν ἅπαντα διὰ τὴν στενότητα καὶ τὸ μῆκος στενὰ καὶ μακρὰ τὰ σπλάγχνα, ὥστε καὶ λανθάνειν διὰ τὴν ὁμοιότητα τῶν σχημάτων· τήν τε γὰρ ἀρτηρίαν ἔχει σφόδρα μακράν, ἔτι δὲ μακρότερον τὸν στόμαχον. ἀρχὴ δὲ τῆς ἀρτηρίας πρὸς αὐτῷ ἐστι τῷ στόματι, ὥστε δοκεῖν ὑπὸ ταύτην εἶναι τὴν γλῶτταν. προέχειν δὲ δοκεῖ τῆς γλώττης ἡ ἀρτηρία διὰ τὸ συσπᾶσθαι τὴν γλῶτταν καὶ μὴ μένειν ὥσπερ τοῖς ἄλλοις. ἔστι δ᾽ ἡ γλῶττα λεπτὴ καὶ μακρὰ καὶ μέλαινα, καὶ ἐξέρχεται μέχρι πόρρω· ἴδιον δὲ παρὰ τὰς τῶν ἄλλων γλώττας ἔχουσι καὶ οἱ ὄφεις καὶ οἱ σαῦροι τὸ δικρόαν αὐτῶν εἶναι τὴν γλῶτταν

---

[1] ὅλως AC : ὁμοίως PD, vulg.

that its liver is four times as large as the ox's, as are its other internal parts, though its spleen is disproportionately small.

The arrangements with regard to the stomach and the gut are the same in the oviparous quadrupeds as well, *e.g.*, the land-tortoise, the turtle, the lizard, both types of crocodile, and generally all animals of this sort: they have a single stomach, simple in formation, resembling in some cases the pig's, in others the dog's.

Serpents as a class are similar and their parts are (Serpents.) practically all comparable with a class which is included among the oviparous land-animals, *viz.*, the lizards—if one were to increase their length and take away their feet. In other words, the serpent has horny scales, and its dorsal and ventral parts resemble those of the lizard; it has, however, no testicles, but, like fishes, two passages which converge into one, and a long bifurcated ovary. The rest of the internal organs are identical with the lizard's, except that owing to its length and narrowness its viscera are long and narrow, so that it is difficult to recognize them on account of the similarity of their shapes. Thus, for instance, its windpipe is very long, and its oesophagus even longer. The beginning of the windpipe is right up against the mouth, so that the tongue appears to be underneath the windpipe; and the windpipe seems to stick out in front of the tongue because the tongue gets drawn back and does not stay where it normally does in other animals. Furthermore, the tongue is thin and long, and black, and will come out a long way. Compared with the tongues of other animals, those of serpents and lizards have the peculiarity of being forked at the tip; and this fea-

ARISTOTLE

ἄκραν, πολὺ δὲ μάλιστα οἱ ὄφεις· τὰ γὰρ ἄκρα αὐτῶν ἐστι λεπτὰ ὥσπερ τρίχες. ἔχει δὲ καὶ ἡ φώκη ἐσχισμένην τὴν γλῶτταν.

Τὴν δὲ κοιλίαν ὁ ὄφις ἔχει οἷον ἔντερον εὐρυχωρέστερον, ὁμοίαν τῇ τοῦ κυνός· εἶτα τὸ ἔντερον μακρὸν καὶ λεπτὸν καὶ μέχρι τοῦ τέλους ἕν. ἐπὶ δὲ τοῦ φάρυγγος ἡ καρδία μικρὰ[1] καὶ νεφροειδής· διὸ δόξειεν ἂν ἐνίοτε οὐ πρὸς τὸ στῆθος ἔχειν τὸ ὀξύ. εἶθ᾽ ὁ πνεύμων ἁπλοῦς, ἰνώδει πόρῳ διηρθρωμένος καὶ μακρὸς σφόδρα καὶ πολὺ ἀπηρτημένος τῆς καρδίας. καὶ τὸ ἧπαρ μακρὸν καὶ ἁπλοῦν, σπλῆνα δὲ μικρὸν καὶ στρογγύλον, ὥσπερ καὶ οἱ σαῦροι. χολὴν δ᾽ ἔχει ὁμοίως τοῖς ἰχθύσιν· οἱ μὲν γὰρ ὕδροι πρὸς[2] τῷ ἥπατι ἔχουσιν, οἱ δ᾽ ἄλλοι πρὸς τοῖς ἐντέροις ὡς ἐπὶ τὸ πολύ. καρχαρόδοντες δὲ πάντες εἰσίν. πλευρὰς δ᾽ ἔχουσιν ἴσας ταῖς ἐν τῷ μηνὶ ἡμέραις· τριάκοντα γὰρ ἔχουσιν.

Λέγουσι δέ τινες συμβαίνειν περὶ τοὺς ὄφεις τὸ αὐτὸ ὅπερ καὶ περὶ τοὺς νεοττοὺς τοὺς τῶν χελιδόνων· ἐὰν γάρ τις ἐκκεντήσῃ τὰ ὄμματα τῶν ὄφεων, φασὶ φύεσθαι πάλιν. καὶ αἱ κέρκοι δ᾽ ἀποτεμνόμεναι τῶν τε σαύρων καὶ τῶν ὄφεων φύονται.[3]

Ὡσαύτως δὲ καὶ τοῖς ἰχθύσιν ἔχει τὰ περὶ τὰ ἔντερα καὶ τὴν κοιλίαν. μίαν γὰρ καὶ ἁπλῆν ἔχουσι, διαφέρουσαν τοῖς σχήμασιν. ἔνιοι γὰρ πάμ-

---

[1] καρδία, μικρὰ δὲ vulg.: κ. μικρὰ καὶ μακρὰ (omisso δὲ) AC: κ. μικρὰ μακρὰ PD.
[2] πρὸς AC: ἐπὶ PD, vulg.
[3] λέγουσι hucusque secl. A.-W.

---

[a] *Cf. G.A.* 774 b 31, where Aristotle again mentions this phenomenon: the eyes of young swallows, if pricked out, will recover: the birds are born blind, hence the injury occurs

ture is specially marked in the serpents, the tips of whose tongues are as fine as hairs. The seal, too, has a divided tongue.

The serpent's stomach is rather like a more capacious gut, resembling the dog's; beyond it is the gut itself, which is long and narrow and single right to its end. The heart is near the pharynx: it is small and long and looks like a kidney; hence it may sometimes appear as though its pointed end were not facing towards the chest. Next is the lung, simple, and articulated with a fibrous passage; it is very long and well separated from the heart. The liver is long and simple; the spleen short and round, as in lizards. Its gall-bladder resembles that of fishes: in the water-snakes it is close to the liver, in others generally near the gut. All serpents are saw-toothed. The number of their ribs is the same as the days in the month—thirty.

Some say that serpents have the same faculty as swallow-chicks: that if anyone pricks their eyes out they grow again. The tails, too, of lizards and serpents grow again if they are cut off.[a]

A similar arrangement of the gut and stomach occurs also in fishes. They have one simple stomach, though its shape varies in different fishes. Some of

during the process of formation, not after its completion. Aristotle thus recognizes the connexion between regeneration and embryonic growth. A similar phenomenon is the well-known "Wolffian regeneration" in amphibia, where after removal of the lens of the eye a new lens regenerates from the margin of the iris, *i.e.*, from a place other than that of its normal origin, *viz.*, the embryonic skin; in newts and salamanders this may be repeated as often as twenty times. These animals retain the faculty of regeneration throughout adult life. See J. Needham, *Biochemistry and Morphogenesis*, 1942, pp. 295 ff. *Cf.* 563 a 14 ff.

παν ἐντεροειδῆ[1] ἔχουσιν, οἷον ὃν καλοῦσι σκάρον, ὃς δὴ καὶ δοκεῖ μόνος ἰχθὺς μηρυκάζειν. καὶ τὸ τοῦ ἐντέρου δὲ μέγεθος ἁπλοῦν, κἂν ἀναδίπλωσιν ἔχῃ,[2] ἀναλύεται εἰς ἕν.

Ἴδιον δὲ τῶν ἰχθύων ἐστὶ καὶ τῶν ὀρνίθων τῶν πλείστων τὸ ἔχειν ἀποφυάδας· ἀλλ' οἱ μὲν ὄρνιθες κάτωθεν καὶ ὀλίγας, οἱ δ' ἰχθύες ἄνωθεν περὶ τὴν κοιλίαν, καὶ ἔνιοι πολλάς, οἷον κωβιός, γαλῆ,[3] πέρκη, σκορπίος, κίθαρος, τρίγλη, σπάρος· ὁ δὲ κεστρεὺς ἐπὶ μὲν θάτερα τῆς κοιλίας πολλάς, ἐπὶ δὲ θάτερα μίαν. ἔνιοι δ' ἔχουσι μὲν ὀλίγας δέ, οἷον ἥπατος, γλαῦκος· ἔχει δὲ καὶ ὁ χρύσοφρυς ὀλίγας. διαφέρουσι δὲ καὶ αὐτοὶ αὑτῶν, οἷον χρύσοφρυς ὁ μὲν πλείους ἔχει ὁ δ' ἐλάττους. εἰσὶ δὲ καὶ οἳ ὅλως οὐκ ἔχουσιν, οἷον οἱ πλεῖστοι τῶν σελαχωδῶν· τῶν δ' ἄλλων οἱ μὲν ὀλίγας, οἱ δὲ καὶ πάνυ πολλάς. πάντες δὲ παρ' αὐτὴν ἔχουσι τὴν κοιλίαν τὰς ἀποφυάδας οἱ ἰχθύες.

Οἱ δ' ὄρνιθες ἔχουσι καὶ πρὸς ἀλλήλους καὶ πρὸς τἆλλα ζῷα περὶ τὰ ἐντὸς μέρη διαφοράν. οἱ μὲν γὰρ ἔχουσι πρὸ τῆς κοιλίας πρόλοβον, οἷον ἀλεκτρυών, φάττα, περιστερά, πέρδιξ· ἔστι δ' ὁ πρόλοβος δέρμα κοῖλον καὶ μέγα, ἐν ᾧ ἡ τροφὴ πρώτη εἰσιοῦσα ἄπεπτός ἐστιν. ἔστι δ' αὐτόθεν[4] μὲν ἀπὸ τοῦ

---

[1] sic PΣ, vulg.: ἑτεροειδῆ ACD.
[2] correxit Pi.: καὶ ἀναδ. ἔχει ὃ vulg., codd. varia.
[3] γαλῆ Sn.: γαλεός vulg.
[4] Sn. ex Rhen., m, Ambros.: αὐτόθι codd., vulg.

---

[a] The pyloric caeca.
[b] The mss. read γαλεός, the dogfish, but like all Selachia

HISTORIA ANIMALIUM, II. xvii

them have a completely gut-like stomach, *e.g.*, the fish known as the parrot-wrasse, which, we may note, appears to be the only fish that chews the cud. Further, the whole length of the gut is simple, and if it should have a convolution in it, it straightens out into a single passage.

The possession of appendages [a] to the gut is a peculiarity of fishes and of most birds. In birds they are low down and few in number; in fishes high up near the stomach, and sometimes they are quite numerous, as, *e.g.*, in the goby, the burbot,[b] the perch, the *skorpios*,[c] the *kitharos*,[d] the red mullet, and the sea-bream; the grey mullet has a number of them on one side of the stomach, and one on the other side. Some have them, but have only a few, *e.g.*, the *hepatos*[d] and the *glaukos*.[d] Another which has a few only is the gilthead. These fishes also differ from one another: thus one gilthead will have more, another fewer. There are some fishes which have none at all, *e.g.*, the majority of the Selachia. Of the remainder, some have few, some very many. Always, in fishes, these appendages are found near the stomach.

Birds differ both from each other and from the other animals in respect of their internal parts. Some for instance have a crop in front of the stomach, as the domestic fowl, the ringdove, the pigeon, and the partridge. This crop is a large hollow skin, into which the food comes at its first stage and is still unconcocted.[e] The crop from this point, where it leaves the oeso-

Birds.

---

this has no caeca. Schneider suggests γαλῇ, the burbot, a member of the cod family.

[c] *Skorpios*, a species of *Scorpaena* (bullhead or sculpin).

[d] *Kitharos, hepatos, glaukos*, unidentified fishes. Kitharos may be the lantern-flounder.

[e] See Notes, §§ 19 ff.

στομάχου στενότερος, ἔπειτα εὐρύτερος, ᾗ δὲ καθήκει πάλιν πρὸς τὴν κοιλίαν, λεπτότερος. τὴν δὲ κοιλίαν σαρκώδη καὶ στιφρὰν οἱ πλεῖστοι ἔχουσι, καὶ ἔσωθεν δέρμα ἰσχυρὸν καὶ[1] ἀφαιρούμενον ἀπὸ τοῦ σαρκώδους. οἱ δὲ πρόλοβον μὲν οὐκ ἔχουσιν, ἀλλ' ἀντὶ τούτου τὸν στόμαχον εὐρὺν καὶ πλατύν, ἢ δι' ὅλου ἢ τὸ πρὸς τὴν κοιλίαν τείνον, οἷον κολοιὸς καὶ κόραξ καὶ κορώνη. ἔχει δὲ καὶ ὁ ὄρτυξ τοῦ στομάχου τὸ πλατὺ κάτω, καὶ ὁ αἰγοκέφαλος μικρὸν εὐρύτερον τὸ κάτω καὶ ἡ γλαύξ. νῆττα δὲ καὶ χὴν καὶ λάρος καὶ καταρράκτης καὶ ὠτὶς τὸν στόμαχον εὐρὺν καὶ πλατὺν ὅλον, καὶ ἄλλοι δὲ πολλοὶ τῶν ὀρνίθων ὁμοίως. ἔνιοι δὲ τῆς κοιλίας αὐτῆς τι ἔχουσιν ὅμοιον προλόβῳ, οἷον ἡ κεγχρηΐς. ἔστι δ' ἃ οὐκ ἔχει οὔτε τὸν στόμαχον οὔτε τὸν πρόλοβον εὐρύν, ἀλλὰ τὴν κοιλίαν μακράν, ὅσα μικρὰ τῶν ὀρνίθων, οἷον χελιδὼν καὶ στρουθός. ὀλίγοι δ' οὔτε τὸν πρόλοβον ἔχουσιν οὔτε τὸν στόμαχον εὐρύν, ἀλλὰ σφόδρα μακρόν, ὅσοι τὸν αὐχένα μακρὸν ἔχουσιν, οἷον πορφυρίων· σχεδὸν δ' οὗτοι καὶ τὸ περίττωμα ὑγρότερον τῶν ἄλλων προΐενται πάντες. ὁ δ' ὄρτυξ ἰδίως ἔχει ταῦτα πρὸς τοὺς ἄλλους· ἔχει γὰρ καὶ πρόλοβον καὶ πρὸ τῆς γαστρὸς τὸν στόμαχον εὐρὺν καὶ πλάτος ἔχοντα· διέχει δ' ὁ πρόλοβος τοῦ πρὸ τῆς γαστρὸς στομάχου συχνὸν ὡς κατὰ μέγεθος.

Ἔχουσι δὲ καὶ λεπτὸν τὸ ἔντερον οἱ πλεῖστοι καὶ ἁπλοῦν ἀναλυόμενον. τὰς δ' ἀποφυάδας ἔχουσιν οἱ ὄρνιθες, καθάπερ εἴρηται, ὀλίγας, καὶ οὐκ ἄνωθεν ὥσπερ οἱ ἰχθύες, ἀλλὰ κάτωθεν κατὰ τὴν τοῦ

[1] καὶ AC: om. vulg.

phagus, is rather narrow, further on it is broader, and where it joins on to the stomach it is slighter. Most birds have a firm, fleshy stomach (gizzard); inside it there is a strong skin which can be pulled away from the fleshy part. Other birds have no crop, but instead of it their oesophagus is wide and spacious, either the whole way, or in the part which extends to the stomach, *e.g.*, the jackdaw, the raven, and the (carrion) crow. The quail's oesophagus, too, has this wide part at the lower end, and in the goat-poll [a] the lower part is slightly wider; so it is in the little owl. In the duck, the goose, the gull, the swooper, and the great bustard the oesophagus is wide and spacious throughout; and similarly in a great many other birds. In some, a portion of the stomach itself resembles a crop, as, *e.g.*, in the kestrel. There are some which have neither a wide oesophagus nor a wide crop but a long stomach—these are the small birds, *e.g.*, the swallow and the sparrow. A few have neither a crop nor a wide oesophagus, but a very long oesophagus, *viz.*, birds with a long neck, *e.g.*, the purple coot [b]; and practically all of these birds produce a more liquid residue [c] than other birds. In respect of these parts, the quail is exceptional: it has not only a crop, but also its oesophagus is wide and spacious in front of the stomach; and relatively to its size, the crop is quite a long way from the oesophagus where this is in front of the stomach.

Most birds' gut is narrow and simple when straightened out. Birds have few gut-appendages, as has been mentioned, and these are not high up, as in fishes, but low down near the termination of the gut.

[a] See note on 506 a 17.
[b] Mentioned again at 595 a 12.   [c] See Notes, §§ 22 ff.

ARISTOTLE

20 ἐντέρου τελευτήν. ἔχουσι δ' οὐ πάντες ἀλλ' οἱ πλεῖστοι, οἷον ἀλεκτρυών, πέρδιξ, νῆττα, νυκτικόραξ, λόκαλος,[1] ἀσκάλαφος, χήν, κύκνος, ὠτίς, γλαύξ. ἔχουσι δὲ καὶ τῶν μικρῶν τινές, ἀλλὰ μικρὰ πάμπαν, οἷον στρουθός.

[1] λόκαλος om. AC, secl. Dt.

---

[a] Lit., the night-raven. It is described at 597 b 23 ff.

Not all birds have them, but the majority of them do, *e.g.*, the domestic fowl, the partridge, the duck, the long-eared owl,[a] the *lokalos*,[b] the *askalaphos*,[c] the goose, the swan, the great bustard, and the little owl. Some of the small birds, too, have them, but very small ones, *e.g.*, the sparrow.

[b] Otherwise unknown; but see Thompson, *G.G.B.*[2], p. 77.
[c] Perhaps a kind of owl.

Γ

I 27 Περὶ μὲν οὖν τῶν ἄλλων μορίων τῶν ἐντὸς εἴρηται, καὶ πόσα καὶ ποῖ᾽ ἄττα, καὶ τίνας ἔχει πρὸς ἄλληλα διαφοράς· λοιπὸν δὲ περὶ τῶν εἰς τὴν γένε-
30 σιν συντελούντων μορίων εἰπεῖν. ταῦτα γὰρ τοῖς μὲν θήλεσι πᾶσιν ἐντός ἐστι, τὰ δὲ τῶν ἀρρένων διαφορὰς ἔχει πλείους.

Τὰ μὲν γὰρ ὅλως τῶν ἐναίμων ζῴων οὐκ ἔχει ὄρχεις, τὰ δ᾽ ἔχει μὲν ἐντὸς δ᾽ ἔχει, καὶ τῶν ἐντὸς ἐχόντων τὰ μὲν πρὸς τῇ ὀσφύϊ ἔχει περὶ τὸν τῶν νεφρῶν τόπον, τὰ δὲ πρὸς τῇ γαστρί, τὰ δ᾽ ἐκτός.
35 καὶ τὸ αἰδοῖον τούτων τοῖς μὲν συνήρτηται πρὸς τὴν γαστέρα, τοῖς δ᾽ ἀφεῖται καθάπερ καὶ οἱ ὄρχεις· πρὸς δὲ τὴν γαστέρα συνήρτηται ἄλλως τοῖς[1] ἐμπροσθουρητικοῖς καὶ τοῖς ὀπισθουρητικοῖς.

Τῶν μὲν οὖν ἰχθύων οὐδεὶς ὄρχεις ἔχει, οὐδ᾽ εἴ τι ἄλλο ἔχει βράγχια, οὐδὲ τὸ τῶν ὄφεων γένος ἅπαν,
5 οὐδ᾽ ὅλως ἄπουν οὐδέν, ὅσα μὴ ζωοτοκεῖ ἐν αὑτοῖς. οἱ δ᾽ ὄρνιθες ἔχουσι μὲν ὄρχεις, ἔχουσι δ᾽ ἐντὸς πρὸς τῇ ὀσφύϊ. καὶ τῶν τετραπόδων ὅσα ᾠοτοκεῖ, τὸν αὐτὸν ἔχει τρόπον, οἷον σαύρα καὶ χελώνη καὶ κροκόδειλος, καὶ τῶν ζῳοτόκων ἐχῖνος. τὰ δὲ τῶν ἐντὸς ἐχόντων πρὸς τῇ γαστρὶ ἔχει, οἷον τῶν ἀπό-

---

[1] τοῖς τ᾽ AC, vulg.: τ᾽ secl. A.-W.: τοῖς ἔμπροσθεν οὐρητικοῖς PD.

BOOK III

WITH one exception we have now discussed all the internal parts: we have stated their sizes, their character, and their mutual differences; and it now remains to speak of the parts which contribute to generation. These parts are internal in all female animals: in males they show numerous differences.

Generative parts:

To begin with, in the blooded animals some males have no testicles; others have them, but internally; and of those which have them internally some have them near the loin in the region of the kidneys, and others near the belly, while yet others have external ones: with some of these the penis is closely attached to the belly; with others it is more freely attached, as indeed are the testicles; and in the case of the closer attachment, this varies as between those animals which are retromingent and those which are the reverse.

in males;

Now no fish has testicles, nor has any creature which possesses gills, nor has any of the serpent tribe, nor generally any footless animal that is not internally viviparous. Birds have them, but internally, close to the loin. Similarly with oviparous quadrupeds, *e.g.*, the lizard, the tortoise and the crocodile, and—a viviparous example—the hedgehog. Of those that have them internally placed, some have them situated close to the belly: examples are the dolphin (a foot-

δων μὲν δελφίς, τῶν δὲ τετραπόδων καὶ ζῳοτόκων ἐλέφας· τὰ δ' ἄλλα φανεροὺς ἔχει.

Ἡ δ' ἐξάρτησις ἡ πρὸς τὴν κοιλίαν καὶ τὸν τόπον τὸν συνεχῆ τίνα διαφορὰν ἔχει, πρότερον εἴρηται· τοῖς μὲν γὰρ ἐκ τοῦ ὄπισθεν συνεχεῖς καὶ οὐκ ἀπηρτημένοι εἰσίν, οἷον τῷ γένει τῷ τῶν ὑῶν, τοῖς δ' ἀπηρτημένοι, καθάπερ τοῖς ἀνθρώποις.

Οἱ μὲν οὖν ἰχθύες ὄρχεις μὲν οὐκ ἔχουσιν, ὥσπερ εἴρηται πρότερον, οὐδ' οἱ ὄφεις· πόρους δὲ δύο ἔχουσιν ἀπὸ τοῦ ὑποζώματος ἠρτημένους ἐφ' ἑκάτερα τῆς ῥάχεως, συνάπτοντας εἰς ἕνα πόρον ἄνωθεν τῆς τοῦ περιττώματος ἐξόδου· τὸ δ' ἄνωθεν λέγομεν τὸ πρὸς τὴν ἄκανθαν. οὗτοι δὲ γίγνονται περὶ τὴν ὥραν τῆς ὀχείας θοροῦ πλήρεις, καὶ θλιβομένων ἐξέρχεται τὸ σπέρμα λευκόν. αὐτοὶ δὲ πρὸς αὑτοὺς ἣν ἔχουσι διαφοράν, ἔκ τε τῶν ἀνατομῶν δεῖ θεωρεῖν καὶ ὕστερον λεχθήσεται ἐν τοῖς περὶ ἕκαστον αὐτῶν ἰδίοις ἀκριβέστερον.

Ὅσα δ' ᾠοτοκεῖ ἢ δίποδα ὄντα ἢ τετράποδα, πάντ' ἔχει ὄρχεις πρὸς τῇ ὀσφύϊ κάτωθεν τοῦ διαζώματος, τὰ μὲν λευκοτέρους τὰ δ' ὠχροτέρους, λεπτοῖς πάμπαν φλεβίοις περιεχομένους. καὶ ἀφ' ἑκατέρου τείνει πόρος συνάπτων εἰς ἕν,[1] καθάπερ καὶ τοῖς ἰχθύσιν, ὑπὲρ τῆς τοῦ περιττώματος ἐξόδου. τοῦτο δ' ἐστὶν αἰδοῖον, ὃ τοῖς μὲν μικροῖς ἄδηλον, ἐν δὲ τοῖς μείζοσιν, οἷον ἐν χηνὶ καὶ τοῖς τηλικούτοις, φανερώτερον γίγνεται, ὅταν ἡ ὀχεία πρόσφατος ᾖ.

Οἱ δὲ πόροι καὶ τοῖς ἰχθύσι καὶ τούτοις προσπε-

---

[1] ἕν AC : ἕνα PD, vulg.

[a] Passages such as 566 a 15 and *G.A.* 746 a 15 suggest that

less example), and the elephant (a viviparous quadruped). Others have them visible externally.

We have already described the different ways in which these parts are attached to the belly and the adjoining region : *viz.*, that in some animals they are connected closely from the rear and do not hang free, as in the pig tribe ; in others they hang free, as in man.

Fishes, then, do not have testicles, as stated above, nor do serpents. But they have two passages connected to the midriff running to either side of the backbone, and converging into one passage above the outlet for the residue : by " above " I mean towards the spine. Towards the breeding season these passages get filled with the genital fluid, and when they are squeezed the semen (which is white) comes out. Fishes differ from one another in this respect ; these differences should be studied from the *Dissections*,[a] and will be described in greater detail when we discuss each species' peculiarities.

The males of oviparous animals, whether bipeds or quadrupeds, all have testicles, which are close to the loin below the midriff ; some are whitish, some rather pale in colour, and they are entirely covered with fine small blood-vessels. From each a passage extends, and, as in fishes, the two passages converge into one above the outlet for the residue. This is the penis : difficult to detect in the small animals, but in the larger ones (*e.g.*, the goose and creatures of similar size) it becomes more conspicuous when copulation has recently occurred.

The passages in fishes as well as in these oviparous

references to " the dissections " are to an actual treatise so called, now lost. See note on 497 a 32.

ARISTOTLE

509 b

φύκασι πρὸς τῇ ὀσφύϊ ὑποκάτω τῆς κοιλίας, μεταξὺ
35 τῶν ἐντέρων καὶ τῆς[1] μεγάλης φλεβός, ἀφ' ἧς τείνουσι πόροι εἰς ἑκάτερον τῶν ὄρχεων. ὥσπερ δὲ

510 a τοῖς ἰχθύσι περὶ μὲν τὴν ὥραν τῆς ὀχείας θορός τε
φαίνεται ἐνὼν καὶ οἱ πόροι σφόδρα δῆλοι, ὅταν δὲ
παρέλθῃ ἡ ὥρα, ἄδηλοι καὶ οἱ πόροι ἐνίοτε, οὕτω
καὶ τῶν ὀρνίθων οἱ ὄρχεις· πρὶν μὲν ὀχεύειν, οἱ μὲν
5 μικροὺς οἱ δὲ πάμπαν ἀδήλους ἔχουσιν, ὅταν δ'
ὀχεύωσι, σφόδρα μεγάλους ἴσχουσιν. ἐπιδηλότατα
δὲ τοῦτο συμβαίνει ταῖς φάτταις καὶ τοῖς πέρδιξιν,
ὥστ' ἔνιοι οἴονται οὐδ' ἔχειν τοῦ χειμῶνος ὄρχεις
αὐτά.

Τῶν δ' ἐν τῷ ἔμπροσθεν ἐχόντων τοὺς ὄρχεις οἱ
μὲν ἐντὸς ἔχουσι πρὸς τῇ γαστρί, καθάπερ δελφίς,
οἱ δ' ἐκτὸς ἐν τῷ φανερῷ πρὸς τῷ τέλει τῆς γα-
10 στρός. τούτοις δὲ τὰ μὲν ἄλλα ἔχει τὸν αὐτὸν τρόπον, διαφέρουσι δ' ὅτι οἱ μὲν αὐτῶν ἔχουσι καθ'
αὑτοὺς τοὺς ὄρχεις, οἱ δ' ἐν τῇ καλουμένῃ ὀσχέᾳ,
ὅσοι ἔξωθεν.[2]

Αὐτοὶ δ' οἱ ὄρχεις ἐν πᾶσι τοῖς πεζοῖς καὶ ζωο-
τόκοις τόνδ' ἔχουσι τὸν τρόπον. τείνουσιν ἐκ τῆς
15 ἀορτῆς πόροι φλεβικοὶ μέχρι τῆς κεφαλῆς ἑκατέρου
τοῦ ὄρχεως, καὶ ἄλλοι ἀπὸ τῶν νεφρῶν δύο· εἰσὶ δ'
οὗτοι μὲν αἱματώδεις, οἱ δ' ἐκ τῆς ἀορτῆς ἄναιμοι.
ἀπὸ δὲ τῆς κεφαλῆς πρὸς αὐτῷ τῷ ὄρχει πόρος ἐστὶ
πυκνότερος ἐκείνων[3] καὶ νευρωδέστερος, ὃς ἀνα-
κάμπτει πάλιν ἐν ἐσχάτῳ[4] τῷ ὄρχει πρὸς τὴν κε-
20 φαλὴν τοῦ ὄρχεως· ἀπὸ δὲ τῆς κεφαλῆς ἑκατέρας[5]

---

[1] sic Dt., A.-W. secutus: κοιλίας καὶ τῶν ἐντέρων, μεταξὺ τῆς codd., vulg.   [2] 7 τῶν δ' hucusque secl. Dt.
[3] ἐκείνων Dt. : ἐκείνου vulg.
[4] ἐσχάτῳ A.-W., Pi. : ἑκατέρῳ vulg.
[5] ἑκατέρας AC : ἑκάτεροι PD, vulg.

quadrupeds are organically attached to the loin under the stomach, between the guts and the Great Bloodvessel, from which passages run to each of the testicles. And as in fishes, the seminal fluid is seen to be present in them at the breeding season and the passages themselves become very conspicuous, but when the season is over the passages actually become invisible in some cases. And similarly with the testicles of birds. Before copulation, in some birds they are small, in others quite indistinguishable; but during the period of copulation they are very large indeed. A conspicuous example of this is afforded by the ring-dove and the partridge, with the result that some people suppose these birds have no testicles at all during the winter.

Of those whose testicles are in the front part of the body, in some they are internal near the belly, *e.g.*, the dolphin; in others they are external and visible, near the lower end of the belly. In most respects these animals exhibit precisely the same arrangement but they differ in this: some have their testicles placed separately by themselves, while others (where they are external) have them enclosed in what is called the scrotum.[a]

The following description may be given of the testicles themselves in all footed viviparous animals. From the Aorta passages of blood-vessel character run to the head of each testicle, and two more from the kidneys: the latter contain blood, the former two contain none. From the head of the testicle alongside it there is a passage stouter and more sinewy than the others, which bends back again at the end of the testicle towards the head of it; and again from the

[a] This paragraph is considered by Dittmeyer to be an interpolation.

ARISTOTLE

πάλιν εἰς ταὐτὸ συνάπτουσιν εἰς τὸ πρόσθεν ἐπὶ τὸ αἰδοῖον. οἱ δ᾽ ἐπανακάμπτοντες πόροι καὶ οἱ προσκαθήμενοι τοῖς ὄρχεσιν ὑμένι περιειλημμένοι εἰσὶ τῷ αὐτῷ, ὥστε δοκεῖν ἕνα εἶναι πόρον, ἂν μὴ διέλῃ τὸν ὑμένα τις. ὁ μὲν οὖν προσκαθήμενος πόρος ἔτι
25 αἱματῶδες ἔχει τὸ ὑγρόν, ἧττον μέντοι τῶν ἄνω τῶν ἐκ τῆς ἀορτῆς[1]· ἐν δὲ τοῖς ἐπανακάμπτουσιν εἰς τὸν καυλὸν τὸν ἐν τῷ αἰδοίῳ λευκή ἐστιν ἡ ὑγρότης. φέρει δὲ καὶ ἀπὸ τῆς κύστεως πόρος, καὶ συνάπτει ἄνωθεν εἰς τὸν καυλόν· περὶ τοῦτον δ᾽ οἷον κέλυφός ἐστι τὸ καλούμενον αἰδοῖον.

Θεωρείσθω δὲ τὰ εἰρημένα ταῦτα ἐκ τῆς ὑπο-
30 γραφῆς τῆσδε.

τῶν πόρων ἀρχὴ τῶν ἀπὸ τῆς ἀορτῆς, ἐφ᾽ οἷς Α·
κεφαλαὶ τῶν ὄρχεων καὶ οἱ καθήκοντες πόροι, ἐφ᾽ οἷς Κ[2]·

οἱ ἀπὸ τούτων πρὸς τῷ ὄρχει προσκαθήμενοι, ἐφ᾽ οἷς τὰ ΩΩ·

οἱ δ᾽ ἀνακάμπτοντες, ἐν οἷς ἡ ὑγρότης ἡ λευκή, ἐφ᾽ οἷς τὰ ΒΒ·

35 αἰδοῖον Δ, κύστις Ε, ὄρχεις ἐν[3] οἷς τὰ ΨΨ.

Ἀποτεμνομένων δ᾽ ἢ ἀφαιρουμένων τῶν ὄρχεων αὐτῶν ἀνασπῶνται οἱ πόροι ἄνω. διαφθείρουσι δ᾽ οἱ μὲν ἔτι νέων ὄντων τρίψει, οἱ δὲ καὶ ὕστερον ἐκτέμνοντες· συνέβη δ᾽ ἤδη ταῦρον ἐκτμηθέντα καὶ εὐθὺς ἐπιβάντα ὀχεῦσαι καὶ γεννῆσαι.

[1] τῶν ἐκ τῆς ἀορτῆς secl. Dt.
[2] ΚΚ Gaza, A.-W., Dt.
[3] ἐφ᾽ Sn., vulg.

head of each testicle the two passages continue until they converge in front at the penis. The passages that bend back and those which lie alongside the testicle are enveloped in one and the same membrane, so that they appear to be just one passage, unless the membrane is cut open. Further, the liquid contained in the passage which lies alongside the testicle is still bloodlike, though less so than in the passages higher up which are connected with the Aorta; whereas in the passages which bend back towards the canal of the penis the liquid is white in colour. Also from the bladder there runs a passage, which leads into the upper part of the canal; and round this, like a sheath, is what is called the penis.

These details which we have mentioned should be studied in the accompanying diagram.[a]

- A indicates the starting-point of the passages which run from the Aorta.
- K indicates the heads of the testicles and the passages which come down into them.
- $\Omega\Omega$ indicates the passages running from the preceding alongside the testicle.
- BB indicates the passages which turn back, containing the white-coloured fluid.
- Δ indicates the penis, E the bladder, $\Psi\Psi$ the testicles.

When the testicles themselves are excised or removed, the passages are drawn upwards. People sometimes destroy the testicles in young animals by rubbing, or cut them out later on. A case has been known of a bull immediately after castration serving a cow, copulating, and producing offspring.

[a] See diagram on p. 236.

ARISTOTLE

Τὰ μὲν οὖν περὶ τοὺς ὄρχεις τοῖς ζῴοις τοῦτον ἔχει τὸν τρόπον.

Αἱ δ' ὑστέραι τῶν ἐχόντων ὑστέρας ζῴων οὔτε τὸν αὐτὸν τρόπον ἔχουσιν οὔθ' ὅμοιαι πάντων εἰσίν, ἀλλὰ διαφέρουσι καὶ τῶν ζῳοτοκούντων πρὸς ἄλληλα καὶ τῶν ᾠοτοκούντων. δίκροαι μὲν οὖν εἰσιν ἁπάντων τῶν πρὸς τοῖς ἄρθροις ἐχόντων τὰς ὑστέρας, καὶ τὸ μὲν αὐτῶν ἐν τοῖς δεξιοῖς μέρεσι, τὸ δ' ἕτερον ἐν τοῖς ἀριστεροῖς ἐστιν· ἡ δ' ἀρχὴ μία καὶ τὸ στόμα ἕν, οἷον καυλὸς σαρκώδης σφόδρα καὶ χονδρώδης τοῖς πλείστοις καὶ μεγίστοις. καλεῖται δὲ τούτων τὸ μὲν ὑστέρα καὶ δελφύς (ὅθεν καὶ ἀδελφοὺς προσαγορεύουσι), μήτρα δ' ὁ καυλὸς καὶ τὸ στόμα τῆς ὑστέρας. ὅσα μὲν οὖν ἐστι ζῳοτόκα, καὶ δίποδα καὶ[1] τετράποδα, τούτων μὲν ἡ ὑστέρα πάντων ἐστὶ κάτω τοῦ ὑποζώματος, οἷον ἀνθρώπῳ καὶ κυνὶ καὶ ὑῒ καὶ ἵππῳ καὶ βοΐ· καὶ τοῖς κερατοφόροις ὁμοίως ταῦτά γ' ἔχει πᾶσιν. ἐπ' ἄκρων δ' αἱ ὑστέραι τῶν καλουμένων κερατίων εἱλιγμὸν ἔχουσιν αἱ τῶν πλείστων.

Τῶν δ' ᾠοτοκούντων εἰς τοὐμφανὲς οὐχ ὁμοίως ἁπάντων ἔχουσιν, ἀλλ' αἱ μὲν τῶν ὀρνίθων πρὸς τῷ ὑποζώματι, αἱ δὲ τῶν ἰχθύων κάτω, καθάπερ αἱ τῶν ζῳοτοκούντων διπόδων καὶ τετραπόδων, πλὴν λεπταὶ καὶ ὑμενώδεις καὶ μακραί,[2] ὥστ' ἐν τοῖς σφόδρα μικροῖς τῶν ἰχθύων δοκεῖν ἑκατέραν ᾠὸν εἶναι ἕν, ὡς δύο ἐχόντων ᾠὰ τῶν ἰχθύων τούτων, ὅσων λέγεται τὸ ᾠὸν εἶναι ψαθυρόν· ἔστι γὰρ οὐχ ἓν ἀλλὰ πολλά, διόπερ διαχεῖται εἰς πολλά.

Ἡ δὲ τῶν ὀρνίθων ὑστέρα κάτωθεν μὲν ἔχει τὸν

[1] καὶ Α : ἢ vulg.
[2] μικραί Α¹, fortasse recte : cf. 567 a 23 (Dt.).

This completes our description of the testicles in animals.

The uterus is not of identical formation in those In females. animals which possess one, nor is it similar in all; differences are found both among the Vivipara and among the Ovipara. In all animals which have the uterus close to the generative organs the uterus is forked, one fork being towards the right side and the other towards the left; its starting-point, however, is single, and so is its opening, as it were a tube consisting of much flesh and cartilage in most animals and in the largest. One of these parts is called the *hystera* or *delphys* (whence the word for brother, *adelphos*, is derived); the other part, the tube or opening of the uterus, is called the *mētra*. In all two-footed and four-footed Vivipara, the uterus is below the midriff, *e.g.*, man, dog, pig, horse, ox; and similarly in all the horned animals. In most amimals, at the extremities of the so-called *keratia* (horns), the uterus has a convolution.

The uterus is not similarly situated in all those animals which are externally oviparous: thus in birds the uterus is near the diaphragm, in fishes down below, just as in two- and four-footed Vivipara, except that in fishes the uterus is of light structure, membranous and long, with the result that, in very small fishes, each of the two forks appears to be a single egg, and these fishes thus appear to have a pair of eggs: these are the fishes whose egg is described as crumbling.[a] In actuality, each portion is not one egg, but many eggs, and that is why the roe breaks up into numerous pieces.

In birds the uterus in its lower regions has its tubu-

---

[a] *i.e.*, the roe. *Cf.* 517 b 7.

ARISTOTLE

καυλὸν σαρκώδη καὶ στιφρόν, τὸ δὲ πρὸς τῷ ὑποζώματι ὑμενῶδες καὶ λεπτὸν πάμπαν, ὥστε δόξαι
ἂν ἔξω τῆς ὑστέρας εἶναι τὰ ᾠά. ἐν μὲν οὖν τοῖς
μείζοσι τῶν ὀρνίθων δῆλός ἐστιν ὁ ὑμὴν[1] μᾶλλον,
καὶ φυσώμενος διὰ τοῦ καυλοῦ αἴρεται καὶ κολ-
ποῦται· ἐν δὲ τοῖς μικροῖς ἀδηλότερα πάντα ταῦτα.

Τὸν αὐτὸν δὲ τρόπον ἔχει ἡ ὑστέρα καὶ ἐν τοῖς
τετράποσι μὲν τῶν ζώων ᾠοτόκοις δέ, οἷον χελώνῃ
καὶ σαύρᾳ καὶ βατράχοις καὶ τοῖς ἄλλοις τοῖς τοιού-
τοις· ὁ μὲν γὰρ καυλὸς κάτωθεν εἷς καὶ σαρκωδέ-
στερος, ἡ δὲ σχίσις καὶ τὰ ᾠὰ ἄνω πρὸς τῷ ὑποζώματι. ὅσα δὲ τῶν ἀπόδων εἰς τὸ φανερὸν μὲν
ζῳοτοκεῖ ἐν αὑτοῖς δ' ᾠοτοκεῖ, οἷον οἵ τε γαλεοὶ
καὶ τἆλλα τὰ καλούμενα σελάχη (καλεῖται δὲ σέλα-
χος ὃ ἂν ἄπουν ὂν καὶ βράγχια ἔχον ζῳοτόκον ᾖ),
τούτων δὴ δικρόα μὲν ἡ ὑστέρα, ὁμοίως δὲ καὶ πρὸς
τὸ ὑπόζωμα τείνει κάτωθεν ἀρξαμένη[2] καθάπερ καὶ
τῶν ὀρνίθων, ἔστι[3] δὲ διὰ μέσου τῶν δικρόων [κάτ-
ωθεν ἀρξαμένη] μέχρι πρὸς τὸ ὑπόζωμα στενή,[4] καὶ
τὰ ᾠὰ ἐνταῦθα γίγνεται καὶ ἄνω ἐπ' ἀρχῇ[5] τοῦ ὑπο-
ζώματος· εἶτα προελθόντα εἰς τὴν εὐρυχωρίαν ζῷα
γίγνεται ἐκ τῶν ᾠῶν. αὐτῶν δὲ τούτων πρὸς ἀλ-
λήλά τε καὶ πρὸς τοὺς ἄλλους ἰχθῦς ἡ διαφορὰ τῶν
ὑστερῶν ἀκριβέστερον ἂν θεωρηθείη τοῖς σχήμασιν
ἐκ τῶν ἀνατομῶν.

Ἔχει δὲ καὶ τὸ τῶν ὄφεων γένος πρός τε ταῦτα
καὶ πρὸς ἄλληλα διαφοράν. τὰ μὲν γὰρ ἄλλα γένη
τῶν ὄφεων ᾠοτοκεῖ πάντα, ὁ δ' ἔχις[6] ζῳοτοκεῖ μόνον, ᾠοτοκήσας ἐν αὑτῷ πρῶτον· διὸ παραπλησίως

---

[1] ὁ ὑμήν ἐστι PD, vulg.
[2] κάτωθεν ἀρξαμένη huc ex v. 8 transtul. Sn., Pi.
[3] Pi.: ἔτι vulg. [4] Pi.: τείνει vulg.

lar part fleshy and firm, and the part which is close to the diaphragm membranous and very fine, to such a degree that the eggs might well seem to be right outside the uterus. In the larger birds the membrane is more clearly visible, and when it is inflated through the tube it rises and swells out ; in the smaller birds all these parts are less clearly seen.

The disposition of the uterus is similar to this in the oviparous quadrupeds, *e.g.*, the tortoise, the lizard, the frog, and the like ; the tube down below is single and tends to be fleshy, and the divided part and the eggs are up at the top close to the diaphragm. In those footless animals which are externally viviparous though internally oviparous, *e.g.*, the dogfish and the other Selachia as they are called (a term which denotes any footless creature which has gills and is viviparous), the uterus is forked and similarly it begins down below and extends as far as the midriff, as with birds. In between the two portions of the bifurcation it is narrow as far as the diaphragm ; and the eggs are produced here and above at the starting-point of the diaphragm ; afterwards they advance into the wider part and the young hatch out of them. But the differences to be observed in the uterus as among these fishes and also when compared with the other fishes can be more precisely studied in their various formations from the *Dissections*.

The serpent tribe also shows differences, both compared with the animals just mentioned and within itself. All kinds of serpents are oviparous, except the viper, which is the only viviparous serpent, though it has a previous stage of being internally oviparous,

---

[5] ἀρχῇ Bk. per typoth. errorem.
[6] ἔχις AC : ἔχις δὲ PD, vulg., omisso πάντα, ὁ.

ARISTOTLE

ἔχει τὰ περὶ τὴν ὑστέραν τοῖς σελάχεσιν. ἡ δὲ τῶν ὄφεων ὑστέρα μακράν,[1] καθάπερ τὸ σῶμα, τείνει, κάτωθεν ἀρξαμένη ἀφ' ἑνὸς πόρου, συνεχὴς ἔνθεν καὶ ἔνθεν τῆς ἀκάνθης, οἷον πόρος ἑκατέρωθεν,[2] μέχρι πρὸς τὸ ὑπόζωμα, ἐν ᾗ τὰ ᾠὰ κατὰ στοῖχον ἐγγίγνεται, καὶ ἐκτίκτει οὐ καθ' ἓν ἀλλὰ συνεχές. [ἔχει δὲ τὴν ὑστέραν, ὅσα μὲν ζῳοτοκεῖ καὶ ἐν αὑτοῖς καὶ εἰς τοὐμφανές, ἄνωθεν τῆς κοιλίας, ὅσα δ' ᾠοτοκεῖ, πάντα κάτωθεν πρὸς τῇ ὀσφύϊ. ὅσα δ' εἰς τὸ φανερὸν μὲν ζῳοτοκεῖ ἐν αὑτοῖς δ' ᾠοτοκεῖ, ἐπαμφοτερίζει· τὸ μὲν γὰρ κάτωθεν πρὸς τὴν ὀσφὺν αὐτῆς μέρος ἐστίν, ἐν ᾧ τὰ ᾠά, τὸ δὲ περὶ τὴν ἔξοδον ἐπάνω τῶν ἐντέρων.][3]

Ἔτι δὲ διαφορὰ καὶ ἥδε πρὸς ἀλλήλας ἐστὶ τῶν ὑστερῶν. τὰ μὲν γὰρ κερατοφόρα καὶ μὴ ἀμφώδοντα ἔχει κοτυληδόνας ἐν τῇ ὑστέρᾳ, ὅταν ἔχῃ τὸ ἔμβρυον, καὶ τῶν ἀμφωδόντων οἷον δασύπους καὶ μῦς καὶ νυκτερίς· τὰ δ' ἄλλα τὰ ἀμφώδοντα καὶ ζῳοτόκα καὶ ὑπόποδα πάντα λείαν ἔχει τὴν ὑστέραν, καὶ ἡ τῶν ἐμβρύων ἐξάρτησις ἐξ αὐτῆς ἐστι τῆς ὑστέρας, ἀλλ' οὐκ ἐκ κοτυληδόνος.

Τὰ μὲν οὖν ἀνομοιομερῆ ἐν τοῖς ζῴοις μέρη τοῦτον ἔχει τὸν τρόπον, καὶ τὰ ἐκτὸς καὶ τὰ ἐντός.

II Τῶν δ' ὁμοιομερῶν κοινότατον μέν ἐστι τὸ αἷμα πᾶσι τοῖς ἐναίμοις ζῴοις καὶ τὸ μόριον ἐν ᾧ πέφυκεν ἐγγίγνεσθαι (τοῦτο δὲ καλεῖται φλέψ), ἔπειτα δὲ

---

[1] μακράν Warmington : μακρά vulg.
[2] ⟨χωρὶς⟩ ἑκατέρωθεν Pi. : ἑκάτερος ὤν vulg.
[3] secl. Sn., Dt.

and for this reason its uterus is similar to that of Selachia. The uterus of the serpents, like these creatures' whole body, reaches a long way, and beginning below from a single passage it extends continuously on either side of the spine, as though there were a passage on each side, until it reaches the diaphragm; and here the eggs are formed, in a row; and they are laid, not one by one, but all adhering together. [Those animals which are viviparous both internally and externally have the uterus situated above the stomach; those which are oviparous all have it below, near the loin. Those which are externally viviparous but internally oviparous are dualizers[a]: the lower part of the uterus (where the eggs are) is near the loin; the part round the outlet is above the gut.][b]

Here is a further point of difference when one uterus is compared with another. The horned non-ambidentate animals have cotyledons in the uterus when they are pregnant; so have such ambidentate animals as the hare, the mouse and the bat. The other animals, ambidentate, viviparous, and footed, all have a smooth uterus, and the attachment of the embryo is to the uterus itself, and not to any cotyledon.

Such then is the nature of the non-uniform parts, both the external and the internal ones, which are found in animals.

We go on now to the uniform parts.[c] The most universal of these in all blooded animals is the blood, and the part wherein the blood has its natural place, viz., the blood-vessel. Next come the counterparts of

II THE UNIFORM PARTS.

[a] See note on 488 a 2, and Notes, §§ 28 ff.
[b] This passage is considered by Schneider and Dittmeyer to be an interpolation.
[c] See Notes, § 1.

τὰ¹ ἀνάλογον τούτοις, ἰχὼρ καὶ ἶνες, καὶ ὃ μάλιστα
δή ἐστι τὸ σῶμα τῶν ζῴων, ἡ σὰρξ καὶ τὸ τούτῳ
ἀνάλογον ἐν ἑκάστῳ μόριον, ἔτι ὀστοῦν καὶ τὸ ἀνάλογον τούτῳ, οἷον ἄκανθα καὶ χόνδρος· ἔτι δὲ δέρμα,
ὑμήν, νεῦρα, τρίχες, ὄνυχες, καὶ τὰ ὁμολογούμενα
τούτοις· πρὸς δὲ τούτοις πιμελή, στέαρ καὶ τὰ
περιττώματα· ταῦτα δ' ἐστὶ κόπρος, φλέγμα, χολὴ
ξανθὴ καὶ μέλαινα.

Ἐπεὶ δ' ἀρχῇ ἔοικεν ἡ τοῦ αἵματος φύσις καὶ ἡ
τῶν φλεβῶν, πρῶτον περὶ τούτων λεκτέον, ἄλλως
τ' ἐπειδὴ καὶ τῶν πρότερον εἰρηκότων τινὲς οὐ
καλῶς λέγουσιν. αἴτιον δὲ τῆς ἀγνοίας τὸ δυσ-
θεώρητον αὐτῶν. ἐν μὲν γὰρ τοῖς τεθνεῶσι τῶν
ζῴων ἄδηλος ἡ φύσις τῶν κυριωτάτων φλεβῶν διὰ
τὸ συμπίπτειν εὐθὺς ἐξιόντος τοῦ αἵματος μάλιστα
ταύτας (ἐκ τούτων γὰρ ἐκχεῖται ἀθρόον ὥσπερ ἐξ
ἀγγείου· καθ' αὑτὸ γὰρ οὐδὲν ἔχει αἷμα, πλὴν ὀλίγον ἐν τῇ καρδίᾳ, ἀλλὰ πᾶν ἐστιν ἐν ταῖς φλεψίν), ἐν
δὲ τοῖς ζῶσιν ἀδύνατόν ἐστι θεάσασθαι πῶς ἔχου-
σιν· ἐντὸς γὰρ ἡ φύσις αὐτῶν. ὥσθ' οἱ μὲν ἐν τεθνεῶσι καὶ διῃρημένοις τοῖς ζῴοις θεωροῦντες τὰς
μεγίστας ἀρχὰς οὐκ ἐθεώρουν, οἱ δ' ἐν τοῖς λελεπτυσμένοις σφόδρα ἀνθρώποις ἐκ τῶν τότε ἔξωθεν
φαινομένων τὰς ἀρχὰς τῶν φλεβῶν διώρισαν. Συέννεσις μὲν ⟨γὰρ⟩² ὁ Κύπριος ἰατρὸς τόνδε τὸν τρόπον

¹ τὰ Peck : τὸ vulg.; cf. 487 a 5.
² ⟨γὰρ⟩ Pi.

---

[a] See note on 489 a 23.

HISTORIA ANIMALIUM, III. 11

these, ichor and fibres [a]; and that which *par excellence* constitutes the bodies of animals, *viz.*, flesh and the counterpart [b] of it in each animal; then bone and its counterparts, such as fish-spine and cartilage; also skin, membrane, sinew, hair, nails, and the substances which correspond to these; and in addition, fat, suet, and the residues [c] (dung, phlegm, yellow and black bile).

Now, since blood and blood-vessels, as natural substances, give the impression of being fundamental,[d] we must discuss them first, especially in view of the fact that some earlier writers have not given satisfactory accounts of them. Their mistakes were due to the difficulties of observation. In dead animals the nature of the most important blood-vessels cannot be discovered because they above all else collapse immediately as soon as the blood goes out of them: the blood pours out of them all at once as though out of a vessel (since no part of the body possesses blood loose, as it were, except that there is a little in the heart—all of it is in the blood-vessels). And in living animals it is impossible to inspect the blood-vessels to ascertain their nature, because they are internal. Consequently, those who have inspected dead and dissected animals have failed to observe the main sources ⟨of the blood-vessels⟩, while those who have examined extremely emaciated patients have drawn their conclusions about the sources of the blood-vessels from what could then be seen externally. For instance, Syennesis, the physician of Cyprus, writes

Blood and blood-vessels.

Syennesis.

[b] See Notes, §§ 3 f.
[c] See Notes, §§ 22 ff.
[d] Or, "are pretty clearly a first principle." See Notes, § 12, and *cf.* 520 b 10 below: "blood is an indispensable part."

ARISTOTLE

λέγει.[1]

Αἱ φλέβες αἱ παχεῖαι ὧδε πεφύκασιν· ἐκ τοῦ ὀφθαλμοῦ[2] παρὰ τὴν ὀφρὺν[3] διὰ τοῦ νώτου παρὰ τὸν πνεύμονα ὑπὸ τοὺς μαστούς, ἡ μὲν ἐκ τοῦ δεξιοῦ εἰς τὰ ἀριστερά, ἡ δ' ἐκ τοῦ ἀριστεροῦ εἰς τὸ δεξιόν. ἡ μὲν οὖν ἐκ τοῦ ἀριστεροῦ διὰ τοῦ ἥπατος εἰς τὸν νεφρὸν καὶ εἰς τὸν ὄρχιν, ἡ δ' ἐκ τοῦ δεξιοῦ εἰς τὸν σπλῆνα καὶ νεφρὸν καὶ ὄρχιν, ἐντεῦθεν δ' εἰς τὸ αἰδοῖον.

Διογένης δ' ὁ Ἀπολλωνιάτης τάδε λέγει·

Αἱ φλέβες ἐν τῷ ἀνθρώπῳ ὧδ' ἔχουσιν. εἰσὶ δύο μέγισται· αὗται τείνουσι διὰ τῆς κοιλίας παρὰ τὴν νωτιαίαν ἄκανθαν, ἡ μὲν ἐπὶ δεξιὰ ἡ δ' ἐπ' ἀριστερά, εἰς τὰ σκέλη, ἑκατέρα ⟨εἰς τὸ⟩[4] παρ' ἑαυτῇ, καὶ ἄνω εἰς τὴν κεφαλὴν παρὰ τὰς κλεῖδας διὰ τῶν σφαγῶν. ἀπὸ δὲ τούτων καθ' ἅπαν τὸ σῶμα φλέβες διατείνουσιν, ἀπὸ μὲν τῆς δεξιᾶς εἰς τὰ δεξιά, ἀπὸ δὲ τῆς ἀριστερᾶς εἰς τὰ ἀριστερά, μέγισται μὲν δύο εἰς τὴν καρδίαν περὶ αὐτὴν τὴν νωτιαίαν ἄκανθαν, ἕτεραι δ' ὀλίγον ἀνωτέρω διὰ τῶν στηθῶν ὑπὸ τὴν μασχάλην εἰς ἑκατέραν τὴν χεῖρα τὴν παρ' ἑαυτῇ· καὶ καλεῖται ἡ μὲν σπληνῖτις, ἡ δ' ἡπατῖτις. σχίζεται δ' αὐτῶν ἄκρα ἑκατέρα, ἡ μὲν ἐπὶ τὸν μέγαν δάκτυλον, ἡ δ' ἐπὶ

---

[1] λέγει AC : dicens Σ : om. PD, vulg.
[2] ὀφθαλμοῦ PD, text. Hipp., Pi., Dt. : ex parte oculorum et superciliorum Σ : ὀμφαλοῦ AC, vulg., Th.
[3] ὀφρὺν PCDA¹, text. Hipp., Pi., Dt., Σ : ὀσφὺν A², vulg., Th. exinde Σ et procedunt ad partes colli, deinde ad dorsum, etc.
[4] ⟨εἰς τὸ⟩ A.-W. : τὰ D, vulg., om. PCA².

---

[a] This passage, with an addition, occurs also in the Hippo-

as follows [a]:

The nature of the stout blood-vessels is this: from the eye across the eyebrow,[b] along the back, past the lung, below the breasts; one runs from right to left, the other from left to right: the latter runs through the liver to the kidney and the testicle, the former to the spleen and kidney and testicle, and thence to the penis.

Diogenes [c] of Apollonia gives the following account: Diogenes.

The blood-vessels in man are as follows: There are two which are outstanding in size, and these extend through the belly along the backbone, one to the right, the other to the left, each to the leg on its own side of the body, and upwards to the head past the collar bones through the throat. From these two, blood-vessels extend all over the body: from the right-hand blood-vessel to the right side, and from the left-hand blood-vessel to the left side; of which the most important are two which run to the heart near the backbone, and two others which run slightly higher up through the chest under the armpits, each to the hand on its own side of the body: one of these is called the splenetic, the other the hepatic. Each of them splits into two at its extremity, one branch going to the thumb,

cratic treatise *de natura ossium*, ch. 8 (ix. 174 L.); *cf.* also Littré i. 419 f.

[b] Another reading is "from the navel across the loin." The Hippocratic text agrees with that given opposite.

[c] Diogenes of Apollonia, "the last of the Presocratics," *floruit c.* 440–430 B.C. This passage is not known elsewhere, but a statement in Simplicius (*Phys.* 153. 15) seems to suggest that it came in his book on *Nature* (περὶ Φύσεως). It is, no doubt, the "precise anatomy of the blood-vessels" which Simplicius says Diogenes gave.

τὸν ταρσόν· ἀπὸ δὲ τούτων λεπταὶ καὶ πολύοζοι
ἐπὶ τὴν ἄλλην χεῖρα καὶ δακτύλους. ἕτεραι δὲ
λεπτότεραι ἀπὸ τῶν πρώτων φλεβῶν τείνουσιν,
ἀπὸ μὲν τῆς δεξιᾶς εἰς τὸ ἧπαρ ἀπὸ δὲ τῆς ἀριστερᾶς εἰς τὸν σπλῆνα καὶ τοὺς νεφρούς. αἱ δ' εἰς
τὰ σκέλη τείνουσαι σχίζονται κατὰ τὴν πρόσφυσιν,[1] καὶ διὰ παντὸς τοῦ μηροῦ τείνουσιν. ἡ δὲ
μεγίστη αὐτῶν ὄπισθεν τείνει τοῦ μηροῦ, καὶ ἐκφαίνεται παχεῖα· ἑτέρα δ' εἴσω τοῦ μηροῦ, μικρὸν
ἧττον παχεῖα ἐκείνης. ἔπειτα παρὰ τὸ γόνυ τείνουσιν εἰς τὴν κνήμην τε καὶ τὸν πόδα, καθάπερ
αἱ[2] εἰς τὰς χεῖρας· καὶ ἐπὶ τὸν ταρσὸν τοῦ ποδὸς
καθήκουσι, καὶ ἐντεῦθεν ἐπὶ τοὺς δακτύλους διατείνουσιν. σχίζονται δὲ καὶ ἐπὶ τὴν κοιλίαν καὶ
τὸ πλευρὸν πολλαὶ ἀπ' αὐτῶν καὶ λεπταὶ φλέβες.

Αἱ δ' εἰς τὴν κεφαλὴν τείνουσαι διὰ τῶν σφαγῶν φαίνονται ἐν τῷ αὐχένι μεγάλαι· ἀφ' ἑκατέρας δ' αὐτῶν, ᾗ τελευτᾷ, σχίζονται εἰς τὴν
κεφαλὴν πολλαί, αἱ μὲν ἐκ τῶν δεξιῶν εἰς τὰ
ἀριστερά, αἱ δ' ἐκ τῶν ἀριστερῶν εἰς τὰ δεξιά·
τελευτῶσι δὲ παρὰ τὸ οὖς ἑκάτεραι. ἔστι δ' ἑτέρα
φλὲψ ἐν τῷ τραχήλῳ παρὰ τὴν μεγάλην ἑκατέρωθεν, ἐλάττων ἐκείνης ὀλίγον, εἰς ἣν αἱ πλεῖσται ἐκ
τῆς κεφαλῆς συνέχουσιν αὐτῆς· καὶ αὗται τείνουσιν διὰ τῶν σφαγῶν εἴσω. καὶ ἀπ' αὐτῶν ἑκατέρας ὑπὸ τὴν ὠμοπλάτην τείνουσι καὶ εἰς τὰς
χεῖρας, καὶ φαίνονται παρά τε τὴν σπληνῖτιν καὶ

---

[1] ὀσφύν coni. A.-W.: *separantur a radicibus coxe* Σ.
[2] αἱ C: καὶ vulg.: om. A (καὶ καθάπερ A.-W.).

the other to the palm of the hand ; and a number of tiny blood-vessels, much branched, lead off from them to the fingers and the rest of the hand. Other blood-vessels, tinier still, lead off from the principal ones : from the blood-vessel on the right, to the liver, from that on the left, to the spleen and the kidney. Those blood-vessels which run to the legs fork into two at the junction of the legs and trunk, and extend the whole length of the thigh. The largest of them runs down the thigh at the rear, and can readily be seen as a stout one ; the other runs inside the thigh, and is slightly less stout than the first. After that they continue past the knee to the shin and the foot, just like those which run to the hands, and ultimately reach the sole of the foot, whence they ramify to the toes. There are also numerous small blood-vessels which split off from the great blood-vessels towards the stomach and the ribs.

The blood-vessels which run to the head through the throat can be seen as large ones in the neck ; and from each of the two, at the point where it terminates, a number of blood-vessels branch off to the head, some from the right towards the left, some from the left towards the right ; and each set finishes up beside the ear. There is another blood-vessel in the neck running beside the large one on either side : each of these is slightly smaller than the large one which it accompanies, and to it most of the blood-vessels from the head are connected. These two run through the throat inside. From each of them blood-vessels extend below the shoulder-blades and towards the hands and can be observed alongside the splenetic and hepatic as

ARISTOTLE

30 τὴν ἡπατῖτιν ἕτεραι ὀλίγον ἐλάττους, ἃς ἀποσχάζουσιν,[1] ὅταν τι ὑπὸ τὸ δέρμα λυπῇ· ἐὰν δέ τι περὶ τὴν κοιλίαν, τὴν ἡπατῖτιν καὶ τὴν σπληνῖτιν. τείνουσι δὲ καὶ ὑπὸ τοὺς μαστοὺς ἀπὸ τούτων ἕτεραι.

Ἕτεραι δ' εἰσὶν αἱ ἀπὸ ἑκατέρας τείνουσαι διὰ τοῦ νωτιαίου μυελοῦ εἰς τοὺς ὄρχεις, λεπταί. ἕτεραι δ' ὑπὸ τὸ δέρμα καὶ διὰ τῆς σαρκὸς τείνουσιν εἰς τοὺς νεφρούς, καὶ τελευτῶσιν εἰς τοὺς ὄρχεις τοῖς ἀνδράσι, ταῖς δὲ γυναιξὶν εἰς τὰς 5 ὑστέρας· αὗται δὲ σπερματίτιδες καλοῦνται.[2] αἱ δὲ φλέβες αἱ μὲν πρῶται ἐκ τῆς κοιλίας εὐρύτεραί εἰσιν, ἔπειτα λεπτότεραι γίγνονται, ἕως ἂν μεταβάλωσιν ἐκ τῶν δεξιῶν εἰς τὰ ἀριστερὰ καὶ ἐκ τούτων εἰς τὰ δεξιά.

Τὸ δ' αἷμα τὸ μὲν παχύτατον ὑπὸ τῶν σαρκῶν 10 ἐκπίνεται·[3] ὑπερβάλλον δ' εἰς τοὺς τόπους τούτους λεπτὸν καὶ θερμὸν καὶ ἀφρῶδες γίγνεται.

III   Συέννεσις μὲν οὖν καὶ Διογένης οὕτως εἰρήκασιν, Πόλυβος δ' ὧδε·

[Τὰ δὲ][4] τῶν φλεβῶν τέτταρα ζεύγη ἐστίν, ἓν μὲν ἀπὸ τοῦ ἐξόπισθεν τῆς κεφαλῆς διὰ τοῦ αὐχένος 15 ἔξωθεν παρὰ τὴν ῥάχιν ἔνθεν καὶ ἔνθεν μέχρι τῶν ἰσχίων εἰς τὰ σκέλη, ἔπειτα διὰ τῶν κνημῶν ἐπὶ τῶν σφυρῶν τὰ ἔξω[5] καὶ εἰς τοὺς πόδας· διὸ καὶ

---

[1] ἀποσχῶσιν D, vulg.: ἀπο(ὑπο- A)σχίζουσιν καὶ (καὶ om. A) αἱ ὑπερέχουσαι C.
[2] haec iv verba huc transtulit Th., suadente Dt.; post v. 8 δεξιά codd., vulg.
[3] ἐκπίνεται PD: *imbibitur et suggitur a carne* Σ: ἐγγί(γ)νεται AC, vulg.
[4] secl. Dt.

another pair slightly smaller in size. These are the vessels which surgeons lance when any pain is felt below the surface of the skin; but when any pain is felt in the region of the stomach it is the hepatic and the splenetic which are lanced. From these, other blood-vessels run below the breasts.

There is yet another pair, which run on each side through the spinal marrow to the testicles; these are thin ones. Another pair runs just below the skin through the flesh to the kidneys; and in men these terminate at the testicles, in women, at the uterus. These are known as the spermatic blood-vessels. The blood-vessels which run from the stomach are at first fairly broad, but become narrower as they proceed, until they change over from right to left and from left to right.

The thickest of the blood is drawn off by the fleshy parts of the body; the overflow which passes into the regions just mentioned becomes thin, hot, and frothy.

Such are the accounts given by Syennesis and Diogenes. Here is Polybus's [a] account:

III

Polybus.

There are four pairs of blood-vessels. One pair runs from the back of the head through the neck on the outside past the backbone on either side till it reaches the loins and so to the legs, and after that through the shins to the outer part of the ankles

---

[a] Polybus, *alias* Polybius, the pupil and son-in-law of Hippocrates: " Polybius, his [*i.e.* Hippocrates's] daughter's husband," Hipp. ix. 420 L. This passage, with a continuation not quoted by Aristotle, occurs also in the two Hippocratic treatises *de natura ossium*, ch. 9 (ix. 174-176 L.) and *de natura hominis*, ch. 11 (vi. 58-60 L.).

---

[5] sic text. Hipp., Dt.: codd. varia.

ARISTOTLE

τὰς φλεβοτομίας ποιοῦνται τῶν περὶ τὸν νῶτον ἀλγημάτων καὶ ἰσχίον ἀπὸ[1] τῶν ἰγνύων καὶ τῶν σφυρῶν ἔξωθεν.[2] ἕτεραι δὲ φλέβες ἐκ τῆς κεφαλῆς παρὰ τὰ ὦτα διὰ τοῦ αὐχένος, αἳ καλοῦνται σφαγίτιδες, ἔνδοθεν παρὰ τὴν ῥάχιν ἑκάτεραι φέρουσι[3] παρὰ τὰς ψύας[4] εἰς τοὺς ὄρχεις καὶ εἰς τοὺς μηρούς, καὶ διὰ τῶν ἰγνύων τοῦ ἔνδοθεν μορίου καὶ διὰ τῶν κνημῶν ἐπὶ τὰ σφυρὰ τὰ εἴσω καὶ τοὺς πόδας· διὸ καὶ τὰς φλεβοτομίας ποιοῦνται τῶν περὶ τὰς ψύας[5] καὶ τοὺς ὄρχεις ἀλγημάτων ἀπὸ τῶν ἰγνύων καὶ τῶν σφυρῶν ⟨εἴσωθεν⟩.[6] τὸ δὲ τρίτον ζεῦγος ἐκ τῶν κροτάφων διὰ τοῦ αὐχένος ὑπὸ τὰς ὠμοπλάτας εἰς τὸν πνεύμονα ἀφικοῦνται, ἡ[7] μὲν ἐκ τῶν δεξιῶν εἰς τὰ ἀριστερὰ ὑπὸ τὸν μαστὸν καὶ εἰς τὸν σπλῆνά τε καὶ εἰς τὸν νεφρόν, ἡ δ' ἀπὸ τῶν ἀριστερῶν εἰς τὰ δεξιὰ[8] ἐκ τοῦ πνεύμονος ὑπὸ τὸν μαστὸν καὶ ⟨εἰς τὸ⟩[9] ἧπαρ καὶ εἰς τὸν νεφρόν· ἄμφω δὲ τελευτῶσιν εἰς τὸν ἀρχόν.[10] αἱ δὲ τέταρται ἀπὸ τοῦ ἔμπροσθεν τῆς κεφαλῆς καὶ τῶν ὀφθαλμῶν ὑπὸ τὸν αὐχένα καὶ τὰς κλεῖδας[11]· ἐντεῦθεν δὲ τείνουσι διὰ τῶν βραχιόνων ἄνωθεν εἰς τὰς καμπάς, εἶτα διὰ τῶν πήχεων ἐπὶ τοὺς καρποὺς καὶ τὰς συγκαμπάς, καὶ διὰ τῶν βραχιόνων τοῦ κάτωθεν μορίου εἰς τὰς μασχάλας, καὶ ἐπὶ τῶν πλευρῶν[12] ἄνωθεν, ἕως ἡ μὲν ἐπὶ τὸν σπλῆνα ἡ δ' ἐπὶ τὸ ἧπαρ ἀφίκηται·

[1] μεταξὺ vel ἐντὸς loco ἀπὸ velit Dt.
[2] sic AC, text. Hipp.: τῶν ἐξ. σφ. P, vulg.
[3] sic PD, text. Hipp.: φέρουσαι AC, vulg.
[4] ψύας A, text. Hipp.: ψοιὰς PD, vulg.

and to the feet. That is why surgeons make incisions in the hams and outer parts of the ankles to relieve pains in the back and loin. Another pair of blood-vessels runs from the head past the ears, through the neck: these are called the jugular veins. They continue inside along the backbone, past the loin-muscles to the testicles and to the thighs, and through the inner part of the hams and through the shins to the inside of the ankles and to the feet; and that is why surgeons make incisions in the hams and ⟨inner side of⟩ the ankles for pains in the loin-muscles and testicles. The third pair runs from the temples through the neck under the shoulder-blades till they reach the lung: the one from the right going to the left under the breast and on to the spleen and kidney; the one from the left going to the right from the lung under the breast and to the liver and kidney: both terminate at the rectum. The fourth pair runs from the front of the head and the eyes underneath the neck and the collar-bones; from there they extend from above through the upper part of the arms to the elbows, then through the forearms to the wrists and the joints at the base of the fingers, and also through the lower part of the upper arms to the armpits, and keeping above the ribs, until one of them reaches the spleen and the other reaches the liver;

---

[5] ψυὰς AC, ψοιὰς PD, vulg.
[6] suppl. Dt. e text. Hipp.
[7] ἡ . . . ἡ text. Hipp.: αἱ . . . αἱ vulg.
[8] τὰ δεξιὰ text. Hipp., Pi.: τὸν δεξιὸν vulg.
[9] add. Pi. e text. Hipp.
[10] ἀρχόν AC, text. Hipp.: ὄρχιν PD, vulg.
[11] κλεῖδας AC: κληῖδας text. Hipp.: κλεῖς PD, vulg.
[12] ⟨τὸ⟩ add. Dt.

εἶθ' ὑπὲρ τῆς γαστρὸς εἰς τὸ αἰδοῖον ἄμφω τελευτῶσιν.

Τὰ μὲν οὖν ὑπὸ τῶν ἄλλων εἰρημένα σχεδὸν ταῦτ' ἐστίν· εἰσὶ δὲ καὶ τῶν περὶ φύσιν οἳ τοιαύτην μὲν οὐκ ἐπραγματεύθησαν ἀκριβολογίαν περὶ τὰς φλέβας, πάντες δ' ὁμοίως τὴν ἀρχὴν αὐτῶν ἐκ τῆς κεφαλῆς καὶ τοῦ ἐγκεφάλου ποιοῦσι, λέγοντες οὐ καλῶς. χαλεπῆς δ' οὔσης, ὥσπερ εἴρηται πρότερον,[1] τῆς θεωρίας ἐν μόνοις τοῖς ἀποπεπνιγμένοις τῶν ζῴων προλεπτυνθεῖσιν ἔστιν ἱκανῶς καταμαθεῖν, εἴ τινι περὶ τῶν τοιούτων ἐπιμελές.

Ἔχει δὲ τοῦτον τὸν τρόπον ἡ τῶν φλεβῶν φύσις. δύο φλέβες εἰσὶν ἐν τῷ θώρακι κατὰ τὴν ῥάχιν ἐντός, ἔστι δὲ κειμένη αὐτῶν ἡ μὲν μείζων ἐν τοῖς ἔμπροσθεν, ἡ δ' ἐλάττων ὄπισθεν ταύτης, καὶ ἡ μὲν μείζων ἐν τοῖς δεξιοῖς μᾶλλον, ἡ δ' ἐλάττων ἐν τοῖς ἀριστεροῖς, ἣν καλοῦσί τινες ἀορτὴν ἐκ τοῦ τεθεᾶσθαι καὶ ἐν τοῖς τεθνεῶσιν τὸ νευρῶδες[2] αὐτῆς μόριον. αὗται δ' ἔχουσι τὰς ἀρχὰς ἀπὸ τῆς καρδίας· διὰ μὲν γὰρ τῶν ἄλλων σπλάγχνων, ᾗ τυγχάνουσι τείνουσαι, ὅλαι δι' αὐτῶν διέρχονται σωζόμεναι καὶ οὖσαι φλέβες, ἡ δὲ καρδία ὥσπερ μόριον αὐτῶν ἐστι, καὶ μᾶλλον τῆς ἐμπροσθίας καὶ μείζονος, διὰ

[1] πρότερον om. PD, vulg.
[2] τὸ νευρῶδες vulg., edd.: *partem quae est in eo ex nervo* Σ: ἀερῶδες coni. Th.

[a] It is not clear that this view is expressed in Diogenes's account. Littré (i. 220) thought that all Hippocratic treatises which assign the origin of the blood-vessels to the heart were of later date than Aristotle. In Plato, *Timaeus* 70 B, however,

HISTORIA ANIMALIUM, III. III

then, passing over the stomach, they both terminate at the penis.

These passages give a pretty fair idea of what other writers have said. There are, too, some workers on the subject of Nature who have not gone into the subject of the blood-vessels with such precision. All alike, however, specify the source of them as being in the head or brain [a]; and this assertion is mistaken. Now since, as I said before, observation is difficult, it is only in strangled animals which have been previously emaciated that it is possible adequately to discover the facts, if one makes the subject one's business.[b] {Aristotle's comments,}

Actually the natural scheme of the blood-vessels is as follows: There are two blood-vessels in the thorax beside the backbone, on the inside, and the larger of the two is placed in front, and the smaller behind it. The larger is rather over to the right side of the body, and the smaller to the left, and this one some call the Aorta because even in dead bodies they have observed the sinewy part of it.[c] These blood-vessels have their starting-point in the heart, as is shown by the fact that whatever other viscera they happen to go through on their course, they pass through them intact, retaining their character of blood-vessels, whereas the heart is as it were a part of them, and especially of the front one (the large one), because these two {and his own account.}

the heart is described as "the tie of the blood-vessels and the fountain-spring of the blood which runs round through all the limbs."

[b] This is Aristotle's method in contrast with the methods described at 496 b 5 ff. and 511 b 15 ff; *cf.* 515 b 1 ff.

[c] For τὸ νευρῶδες Thompson conjectures ἀερῶδες, "they have seen part of it to be full of air"; but this seems inappropriate, because (*a*) "Aorta" means, literally, "hanger," "suspender," "strap," and (*b*) Aristotle insists on the sinewy nature of the Aorta; see, *e.g.*, 513 b 9, 514 b 23, 515 a 30.

ARISTOTLE

τὸ ἄνω μὲν καὶ κάτω τὰς φλέβας εἶναι ταύτας, ἐν μέσῳ δ' αὐτῶν τὴν καρδίαν.

Ἔχουσι δ' αἱ καρδίαι πᾶσαι μὲν κοιλίας ἐν αὐταῖς, ἀλλ' αἱ μὲν τῶν σφόδρα μικρῶν ζῴων μόλις φανερὰν τὴν μεγίστην ἔχουσι, τὰ δὲ μέσα τῷ μεγέθει τῶν ζῴων καὶ τὴν ἑτέραν, τὰ δὲ μέγιστα τὰς τρεῖς. ἔστι δὲ τῆς καρδίας τὸ ὀξὺ ἐχούσης εἰς τὸ πρόσθεν, καθάπερ εἴρηται πρότερον, ἡ μεγίστη μὲν κοιλία ἐν τοῖς δεξιοῖς καὶ ἀνωτάτω αὐτῆς, ἡ δ' ἐλαχίστη ἐν τοῖς ἀριστεροῖς, ἡ δὲ μέση μεγέθει τούτων ἐν τῷ μέσῳ ἀμφοῖν· ἀμφότεραι δὲ πολλῷ ἐλάττους εἰσὶ τῆς μεγίστης. συντέτρηνται μέντοι πᾶσαι αὗται πρὸς τὸν πνεύμονα, ἀλλ' ἄδηλοι ⟨αἱ τρήσεις⟩[1] διὰ σμικρότητα τῶν πόρων πλὴν μιᾶς.

Ἡ μὲν οὖν μεγάλη φλὲψ ἐκ τῆς μεγίστης ἤρτηται κοιλίας τῆς ἄνω καὶ ἐν τοῖς δεξιοῖς, εἶτα διὰ τοῦ κοίλου τοῦ μέσου τείνασα γίγνεται[2] πάλιν φλέψ, ὡς οὔσης τῆς κοιλίας μορίου τῆς φλεβὸς ἐν ᾧ λιμνάζει τὸ αἷμα. ἡ δ' ἀορτὴ ἀπὸ τῆς μέσης· πλὴν οὐχ οὕτως ἀλλὰ κατὰ στενοτέραν σύριγγα πολλῷ κοινωνεῖ.

Καὶ ἡ μὲν φλὲψ διὰ τῆς καρδίας, †εἰς δὲ τὴν ἀορτὴν[3] ἀπὸ τῆς καρδίας† τείνει. καὶ ἔστιν ἡ μὲν μεγάλη φλὲψ ὑμενώδης καὶ δερματώδης, ἡ δ' ἀορτὴ στενοτέρα μὲν ταύτης, σφόδρα δὲ νευρώδης· καὶ ἀποτεινομένη πόρρω πρός τε τὴν κεφαλὴν καὶ πρὸς τὰ κάτω μόρια στενή τε γίγνεται καὶ νευρώδης πάμπαν.

[1] add. Dt.: *non manifestatur illa perforatio* Σ.
[2] τείνασα γίγνεται Dt.: τείνεται AC, vulg.: γίνεται PD.
[3] add. ⟨πόρος⟩ Dt.: ἡ δ' ἀορτὴ Sylb., Sn., Karsch, alii: ἡ δ' ἀορτὴ εἰς τὴν ἀριστερὰν Th.

blood-vessels are above and below, while the heart lies in the middle between them.

In all animals the heart contains cavities. In those of the very small animals even the largest of the cavities is hardly distinguishable: in those of medium size the second also is distinguishable, whereas in the largest animals all three can be seen. In the heart, then, which as already stated has its pointed end forward, the largest of the cavities is on the right-hand side, and highest up; the smallest is on the left, and the medium-sized one is between the two. Both of the first-named are much smaller than the largest. All three, however, have connecting-passages with the lung, though these connexions except one are obscure owing to the smallness of all the passages.

Now the Great Blood-vessel is attached to the largest cavity, *i.e.*, the one which is uppermost and on the right side; then it runs through the middle cavity and reappears as a blood-vessel again, as though the cavity were part of it, like a lake formed by the blood. The Aorta is attached to the middle cavity, though the arrangement is dissimilar: the connexion here is by a much narrower pipe.

Thus the blood-vessel passes through the heart [and it extends into the Aorta from the heart].[a] The Great Blood-vessel is membranous and skinlike in appearance, whereas the Aorta is narrower, and very sinewy; and as it continues on farther towards the head and the lower parts of the body it becomes very narrow and sinewy indeed.

---

[a] This statement is incorrect, and inconsistent with Aristotle's account. Some proposals for alteration are noted opposite.

Τείνει δὲ πρῶτον μὲν ἄνω ἀπὸ τῆς καρδίας τῆς μεγάλης φλεβὸς μόριον πρὸς τὸν πνεύμονα καὶ τὴν σύναψιν τῆς ἀορτῆς, ἄσχιστος καὶ μεγάλη οὖσα φλέψ. σχίζεται δ' ἀπ' αὐτῆς μόρια δύο, τὸ μὲν ἐπὶ τὸν πνεύμονα, τὸ δ' ἐπὶ τὴν ῥάχιν καὶ τὸν ὕστατον τοῦ τραχήλου σφόνδυλον.

Ἡ μὲν οὖν ἐπὶ τὸν πνεύμονα τείνουσα φλὲψ εἰς διμερῆ ὄντ' αὐτὸν διχῇ σχίζεται πρῶτον, εἶτα παρ' ἑκάστην σύριγγα καὶ ἕκαστον τρῆμα τείνει, μείζων μὲν παρὰ τὰ μείζω, ἐλάττων δὲ παρὰ τὰ ἐλάττω, οὕτως ὥστε μηδὲν εἶναι μόριον λαβεῖν ἐν ᾧ οὐ τρῆμά τ' ἔνεστι καὶ φλέβιον· τὰ γὰρ τελευταῖα τῷ μεγέθει ἄδηλα διὰ τὴν μικρότητά ἐστιν, ἀλλὰ πᾶς ὁ πνεύμων φαίνεται μεστὸς ὢν αἵματος. ἐπάνω δ' οἱ ἀπὸ τῆς φλεβός εἰσι πόροι τῶν ἀπὸ τῆς ἀρτηρίας συρίγγων τεινουσῶν. ἡ δ' ἐπὶ τὸν σφόνδυλον τοῦ τραχήλου τείνουσα φλὲψ καὶ τὴν ῥάχιν πάλιν παρὰ τὴν ῥάχιν τείνει· ἣν καὶ Ὅμηρος ἐν τοῖς ἔπεσιν εἴρηκε ποιήσας·

ἀπὸ δὲ φλέβα πᾶσαν ἔκερσεν,
ἥ τ' ἀνὰ νῶτα θέουσα διαμπερὲς αὐχέν' ἱκάνει.

ἀπὸ δὲ ταύτης τείνουσι παρά τε τὴν πλευρὰν ἑκάστην φλέβια καὶ πρὸς ἕκαστον τὸν σφόνδυλον, κατὰ δὲ τὸν ὑπὲρ τῶν νεφρῶν σφόνδυλον σχίζεται διχῇ.

Ταῦτα μὲν οὖν τὰ μόρια ἀπὸ[1] τῆς μεγάλης φλεβὸς τοῦτον ἔσχισται τὸν τρόπον.

Ὑπεράνω δὲ τούτων ἀπὸ τῆς ἐκ τῆς καρδίας τεταμένης πάλιν ἡ ὅλη σχίζεται εἰς δύο τόπους. αἱ μὲν γὰρ φέρουσιν εἰς τὰ πλάγια καὶ τὰς κλεῖδας,

---

[1] add. ἀπὸ AC, vulg., om. PD.

First, then, there is a part of the Great Blood-vessel which runs up from the heart towards the lung and the attachment of the Aorta : this is a large and undivided blood-vessel. But two parts split off from it : one towards the lung and one towards the backbone and the last vertebra of the neck.

So then the blood-vessel which extends to the lung, because the lung itself is double, splits into two to begin with, and then runs alongside every pipe and every perforation, being larger by the larger ones and smaller by the smaller, in such a way that one can find no part at all where there is not both perforation and blood-vessel ; indeed, the extremities are invisible owing to their excessively small size, and in fact the whole lung looks full of blood. The passages from the blood-vessel are situated above the tubes which extend from the windpipe. And the blood-vessel which runs to the vertebra of the neck and the backbone extends back again alongside the backbone. This is the blood-vessel mentioned by Homer in the passage where he says [a] :

> All that vein he severed through
> Which to the neck runs right along the back.

And from this blood-vessel there run small blood-vessels past each rib and to each vertebra, and at the vertebra above the kidneys the blood-vessel divides into two.

I have now described how these parts branch off from the Great Blood-vessel.

Above these, from that part of the blood-vessel which extends from the heart, the whole blood-vessel divides in two directions. The one set of blood-vessels leads off to the sides and the collar-bones, after that

[a] *Iliad*, xiii. 546 f.

ARISTOTLE

κἄπειτα διὰ τῶν μασχαλῶν τοῖς μὲν ἀνθρώποις εἰς τοὺς βραχίονας, τοῖς δὲ τετράποσιν εἰς τὰ πρόσθια σκέλη τείνουσι, τοῖς δ' ὄρνισιν εἰς τὰς πτέρυγας, τοῖς δ' ἰχθύσιν εἰς τὰ πτερύγια τὰ πρανῆ. αἱ δ' ἀρχαὶ τούτων τῶν φλεβῶν, ᾗ σχίζονται τὸ πρῶτον, καλοῦνται σφαγίτιδες· ᾗ δὲ σχίζονται εἰς τὸν αὐχένα [ἀπὸ τῆς μεγάλης φλεβός],[1] παρὰ τὴν ἀρτηρίαν τείνουσιν [τὴν τοῦ πνεύμονος][2]· ὧν ἐπιλαμβανομένων ἐνίοτε ἔξωθεν ἄνευ πνιγμοῦ καταπίπτουσιν αἱ ἄνθρωποι μετ' ἀναισθησίας, τὰ βλέφαρα συμβεβληκότες. οὕτω δὲ τείνουσαι καὶ μεταξὺ λαμβάνουσαι τὴν ἀρτηρίαν φέρουσι μέχρι τῶν ὤτων, ᾗ συμβάλλουσιν αἱ γένυες τῇ κεφαλῇ.[3] πάλιν δ' ἐντεῦθεν εἰς τέτταρας σχίζονται φλέβας, ὧν μία μὲν ἐπανακάμψασα καταβαίνει διὰ τοῦ τραχήλου καὶ τοῦ ὤμου, καὶ συμβάλλει τῇ πρότερον ἀποσχίσει τῆς φλεβὸς κατὰ τὴν τοῦ βραχίονος καμπήν, τὸ δ' ἕτερον μόριον εἰς τὴν χεῖρα τελευτᾷ καὶ τοὺς δακτύλους.

Μία δ' ἑτέρα ἀφ' ἑκατέρου τοῦ τόπου τοῦ περὶ τὰ ὦτα ἐπὶ τὸν ἐγκέφαλον τείνει, καὶ σχίζεται εἰς πολλὰ καὶ λεπτὰ φλέβια εἰς τὴν καλουμένην μήνιγγα τὴν περὶ τὸν ἐγκέφαλον. αὐτὸς δ' ὁ ἐγκέφαλος ἄναιμος πάντων ἐστί, καὶ οὔτε μικρὸν οὔτε μέγα φλέβιον τελευτᾷ εἰς αὐτόν. τῶν δὲ λοιπῶν τῶν ἀπὸ τῆς φλεβὸς ταύτης σχισθεισῶν φλεβῶν αἱ μὲν τὴν κεφαλὴν κύκλῳ περιλαμβάνουσιν, αἱ δ' εἰς τὰ αἰσθητήρια ἀποτελευτῶσι καὶ τοὺς ὀδόντας λεπτοῖς πάμπαν φλεβίοις.

IV Τὸν δ' αὐτὸν τρόπον καὶ τὰ τῆς ἐλάττονος φλεβός, καλουμένης δ' ἀορτῆς, ἔσχισται μέρη, συμπαρακολουθοῦντα τοῖς τῆς μεγάλης· πλὴν ἐλάττους

they continue in men through the armpits to the arms, in quadrupeds to the forelegs, in birds to the wings, and in fishes to the pectoral fins. The primary portions of these blood-vessels, where they first divide off, are called the jugular veins; and where they divide off to the neck they extend alongside the windpipe; and sometimes if these blood-vessels are pressed from outside, persons though not choked shut their eyes, and fall down unconscious. Running on in this way and keeping the windpipe between them they continue as far as the ears, where the lower jaws connect with the skull. Then again from this point they divide off into four, ⟨two on each side⟩, one of which bends back and runs down through the neck and the shoulder, and connects up with the former branch of the blood-vessel at the bend of the arm, while the other part ends up in the hand and fingers.

Each one of the others runs on its own side from the region round the ear to the brain, and divides up into a number of small delicate blood-vessels into the meninx (as it is called) which surrounds the brain. The brain itself is bloodless in all animals; no blood-vessel, large or small, terminates there. Of the remaining blood-vessels which divide off from the one just mentioned, some encircle the head, others find their terminus in the sense-organs and the teeth in extremely fine blood-vessels.

In the same way too the parts of the lesser Blood- IV vessel called the Aorta divide off, following along together with those of the Great Blood-vessel; the only difference is that the passages and the blood-vessels

---

[1] om. Gaza, secl. Dt. [2] secl. Dt.
[3] τῇ κεφαλῇ A, Dt. : τῆς κεφαλῆς PCD, vulg.

ARISTOTLE

οἱ πόροι καὶ τὰ φλέβια πολλῷ ἐλάττω ταύτης[1] ἐστὶ τῶν τῆς μεγάλης φλεβός.

Τὰ μὲν οὖν ἄνωθεν τῆς καρδίας τοῦτον ἔχουσι τὸν τρόπον αἱ φλέβες.

Τὸ δ' ὑποκάτω τῆς καρδίας μέρος τῆς μεγάλης φλεβὸς τείνει μετέωρον διὰ τοῦ ὑποζώματος, συνέχεται δὲ καὶ πρὸς τὴν ἀορτὴν καὶ πρὸς τὴν ῥάχιν πόροις ὑμενώδεσι καὶ χαλαροῖς. τείνει δ' ἀπ' αὐτῆς μία μὲν διὰ τοῦ ἥπατος φλέψ, βραχεῖα μὲν πλατεῖα δέ, ἀφ' ἧς πολλαὶ καὶ λεπταὶ εἰς τὸ ἧπαρ ἀποτείνουσαι ἀφανίζονται. δύο δ' ἀπὸ τῆς διὰ τοῦ ἥπατος φλεβὸς ἀποσχίσεις εἰσίν, ὧν ἡ μὲν εἰς τὸ ὑπόζωμα τελευτᾷ καὶ τὰς καλουμένας φρένας, ἡ δὲ πάλιν ἐπανελθοῦσα διὰ τῆς μασχάλης εἰς τὸν βραχίονα τὸν δεξιὸν συμβάλλει ταῖς ἑτέραις φλεψὶ κατὰ τὴν ἐντὸς καμπήν· διὸ ἀποσχαζόντων τῶν ἰατρῶν ταύτην ἀπολύονταί τινων πόνων περὶ τὸ ἧπαρ. ἐκ δὲ τῶν ἀριστερῶν αὐτῆς μικρὰ μὲν παχεῖα δὲ φλὲψ τείνει εἰς τὸν σπλῆνα, καὶ ἀφανίζεται τὰ ἀπ' αὐτῆς φλέβια εἰς τοῦτον. ἕτερον δὲ μέρος ἀπὸ τῶν ἀριστερῶν τῆς μεγάλης φλεβὸς ἀποσχισθὲν τὸν αὐτὸν τρόπον ἀναβαίνει ἐπὶ τὸν ἀριστερὸν βραχίονα· πλὴν ἐκείνη μὲν ἡ διὰ τοῦ ἥπατός ἐστιν, αὕτη δ' ἑτέρα τῆς εἰς τὸν σπλῆνα τεινούσης.

Ἔτι δ' ἄλλαι ἀπὸ τῆς μεγάλης φλεβὸς ἀποσχίζονται, ἡ μὲν ἐπὶ τὸ ἐπίπλοον, ἡ δ' ἐπὶ τὸ καλούμενον πάγκρεας. ἀπὸ δὲ ταύτης πολλαὶ φλέβες διὰ τοῦ μεσεντερίου τείνουσιν. πᾶσαι δ' αὗται εἰς μίαν φλέβα μεγάλην τελευτῶσιν,[2] παρὰ πᾶν τὸ ἔντερον καὶ τὴν κοιλίαν μέχρι τοῦ στομάχου τεταμένην. καὶ

---

[1] ταύτης A, Dt. : ταῦτ' PCD. vulg.
[2] μ. τ. AC, Dt. ; τ. μ. vulg.

belonging to the Aorta are much smaller in size than those of the Great Blood-vessel.

With regard to the regions above the heart, such is the disposition of the blood-vessels.

That part of the Great Blood-vessel which lies below the heart runs unattached through the midriff, and is held to the Aorta and the back-bone by slack membranous passages. From it there extends one blood-vessel, which is short and wide, through the liver, and from this numerous fine blood-vessels lead off into the liver and become invisible. From the blood-vessel which passes through the liver two branches divide off: one of them ends up in the diaphragm or midriff as it is called; the other goes back again through the armpit into the right arm and joins up with the other blood-vessels at the inside of the bend; and that is why when doctors lance this blood-vessel people get relief from certain pains in the liver. From the left side of it a small but thick blood-vessel runs to the spleen, and the small blood-vessels which branch off from it enter the spleen and become invisible. Another part divides off from the left side of the Great Blood-vessel and goes up in the same way to the left arm—with this difference, that the former blood-vessel is the one that goes through the liver, whereas this one is distinct from that which goes into the spleen.

There are still further blood-vessels which divide off from the Great Blood-vessel, one to the Omentum, and one to the pancreas as it is called, and from the latter a number of blood-vessels extend through the mesentery. All these terminate into a single large blood-vessel, which extends along the whole length of the gut and the stomach as far as the oesophagus;

ARISTOTLE

περὶ ταῦτα τὰ μόρια πολλαὶ ἀπ' αὐτῶν σχίζονται φλέβες.

Μέχρι μὲν οὖν τῶν νεφρῶν μία οὖσα ἑκατέρα τείνει, καὶ ἡ ἀορτὴ καὶ ἡ μεγάλη φλέψ· ἐνταῦθα δὲ πρός τε τὴν ῥάχιν μᾶλλον προσπεφύκασι καὶ σχίζονται εἰς δύο ὡσπερεὶ λάμβδα[1] ἑκατέρα, καὶ γίγνεται εἰς τοὔπισθεν μᾶλλον ἡ μεγάλη φλὲψ τῆς ἀορτῆς. προσπέφυκε δ' ἡ ἀορτὴ μάλιστα τῇ ῥάχει περὶ τὴν καρδίαν· ἡ δὲ πρόσφυσίς ἐστι φλεβίοις νευρώδεσι καὶ μικροῖς. ἔστι δ' ἡ ἀορτὴ ἀπὸ μὲν τῆς καρδίας ἀγομένη εὖ μάλα κοίλη, προϊοῦσα δ' ἐστὶ στενοτέρα[2] καὶ νευρωδεστέρα. τείνουσι δὲ καὶ ἀπὸ τῆς ἀορτῆς εἰς τὸ μεσεντέριον φλέβες ὥσπερ αἱ ἀπὸ τῆς μεγάλης φλεβός, πλὴν πολλῷ λειπόμεναι τῷ μεγέθει· στεναὶ γάρ εἰσι καὶ ἰνώδεις, λεπτοῖς δὲ[3] καὶ ποικίλοις[4] καὶ ἰνώδεσι τελευτῶσι φλεβίοις. εἰς δὲ τὸ ἧπαρ καὶ τὸν σπλῆνα οὐδεμία τείνει ἀπὸ τῆς ἀορτῆς φλέψ.

Αἱ δὲ σχίσεις ἑκατέρας τῆς φλεβὸς τείνουσιν εἰς τὸ ἰσχίον ἑκάτερον, καὶ καθάπτουσιν εἰς τὸ ὀστοῦν ἀμφότεραι. φέρουσι δὲ καὶ εἰς τοὺς νεφροὺς ἀπό τε τῆς μεγάλης φλεβὸς καὶ τῆς ἀορτῆς φλέβες· πλὴν οὐκ εἰς τὸ κοῖλον ἀλλ' εἰς τὸ σῶμα καταναλίσκονται τῶν νεφρῶν. ἀπὸ μὲν οὖν τῆς ἀορτῆς ἄλλοι δύο πόροι φέρουσιν εἰς τὴν κύστιν, ἰσχυροὶ καὶ συνεχεῖς· καὶ ἄλλοι ἐκ τοῦ κοίλου τῶν νεφρῶν, οὐδὲν κοινωνοῦντες τῇ μεγάλῃ φλεβί. ἐκ μέσου δὲ τῶν νεφρῶν ἑκατέρου φλὲψ κοίλη καὶ νευρώδης ἐξήρτηται, τεί-

---

[1] λάμβδα ACD : λάβδα P, vulg.
[2] ἐστὶ στενωτέρα Camot.: ἔτι στενο(ω)τέρα AC: ἐπιστενότερα P, ἐπιστενωτέρα vulg.   [3] δὲ Dt. : γὰρ vulg.
[4] ποικίλοις AC, Dt.: κοίλοις PD, vulg.: κυκλικοῖς coni. Th.: *subtilis vacua* Σ.

HISTORIA ANIMALIUM, III. iv

and round these parts many blood-vessels branch off from them.

The Aorta and the Great Blood-vessel, each still single and undivided, extend as far as the kidneys : here they become more closely attached to the backbone, and each divides into two, like the shape of the letter lambda, and the Great Blood-vessel gets round somewhat to the back of the Aorta. The Aorta is most firmly attached to the backbone near the heart, and this attachment is by means of very small sinewy blood-vessels. The Aorta, when it strikes off from the heart, is a hollow tube of considerable size, but as it proceeds it is narrower and more sinewy. Also, from the Aorta there are blood-vessels extending to the mesentery, just like those which come from the Great Blood-vessel, except that they are much smaller in size : they are narrow and fibrous, and they end up in very fine and complicated fibrous blood-vessels. No blood-vessel runs from the Aorta to the liver or to the spleen.

Branches from each of the two great blood-vessels extend to each flank, and they both attach themselves to the bone. [a] There are also blood-vessels leading to the kidneys both from the Great Blood-vessel and from the Aorta, except that they do not go into the cavity of the kidney but peter out in the mass of the kidneys themselves. Now there are, leading from the Aorta, two other passages into the bladder, which are strong and continuous ; and there are others from the cavity of the kidneys, which have no communication with the Great Blood-vessel. From the middle of each kidney a hollow sinewy blood-vessel

---

[a] The following passage, down to 515 a 2, closely follows the passage in the first Book, 497 a 4 ff.

ARISTOTLE

514 b

515 a

νουσα παρ' αὐτὴν τὴν ῥάχιν διὰ τῶν στενῶν[1]· εἶτα εἰς ἑκάτερον τὸ ἰσχίον ἀφανίζεται ἑκατέρα πρῶτον, ἔπειτα δῆλαι γίγνονται πάλιν διατεταμέναι πρὸς τὸ ἰσχίον.[2] καθάπτουσι δὲ πρὸς [τὴν κύστιν καὶ][3] τὸ αἰδοῖον τὰ πέρατα αὐτῶν ἐν τοῖς ἄρρεσιν, ἐν δὲ τοῖς 5 θήλεσι πρὸς τὰς ὑστέρας. τείνει δ' ἀπὸ μὲν τῆς μεγάλης φλεβὸς οὐδεμία εἰς τὰς ὑστέρας, ἀπὸ δὲ τῆς ἀορτῆς πολλαὶ καὶ πυκναί.

Τείνουσι δ' ἀπό τε[4] τῆς ἀορτῆς καὶ τῆς μεγάλης φλεβὸς σχιζομένων[5] καὶ ἄλλαι, αἱ μὲν ἐπὶ τοὺς βουβῶνας πρῶτον μεγάλαι καὶ κοῖλαι, ἔπειτα διὰ τῶν 10 σκελῶν τελευτῶσιν εἰς τοὺς πόδας καὶ τοὺς δακτύλους· καὶ πάλιν ἕτεραι διὰ τῶν βουβώνων καὶ τῶν μηρῶν φέρουσιν ἐναλλάξ, ἡ μὲν ἐκ τῶν ἀριστερῶν εἰς τὰ δεξιά, ἡ δ' εἰς τὰ ἀριστερὰ ἐκ τῶν δεξιῶν· καὶ συνάπτουσι περὶ τὰς ἰγνύας ταῖς ἑτέραις φλεψίν.

Ὃν μὲν οὖν τρόπον ἔχουσιν αἱ φλέβες καὶ πόθεν 15 ἤρτηνται τὰς ἀρχάς, φανερὸν ἐκ τούτων.

Ἔχει δ' ἐν ἅπασι μὲν τοῖς ἐναίμοις ζῴοις οὕτω[6] τὰ περὶ τὰς ἀρχὰς καὶ τὰς μεγίστας φλέβας (τὸ γὰρ ἄλλο πλῆθος τῶν φλεβῶν οὐχ ὡσαύτως ἔχει πᾶσιν· οὐδὲ γὰρ τὰ μέρη τὸν αὐτὸν τρόπον ἔχουσιν, οὐδὲ ταὐτὰ πάντα[7] ἔχουσιν), οὐ μὴν οὐδ' ὁμοίως ἐν ἅπα- 20 σίν ἐστι φανερόν, ἀλλὰ μάλιστα ἐν τοῖς μάλιστα πολυαίμοις καὶ μεγίστοις. ἐν γὰρ τοῖς μικροῖς καὶ μὴ πολυαίμοις ἢ διὰ φύσιν ἢ διὰ πιότητα τοῦ σώ-

---

[1] στενῶν Dt. ex 497 a 15 : φλεβῶν vulg. : νευρῶν A : νεφρῶν Buss. : *et deinde ad hancas* (=ἰσχίον) Σ.

[2] v. 34 καὶ ἄλλοι ... ἰσχίον secl. A.-W., Dt. ut ex 497 a 11 sqq. decerpta.

[3] τὴν κ. καὶ secl. Dt. : habet Σ.

[4] τε PD, om. AC, vulg.

[5] ἀπὸ τῶν ante σχ. vulg., delent A.-W.

184

HISTORIA ANIMALIUM, III. iv

is attached, which runs alongside the backbone itself through the narrow regions; afterwards each of these blood-vessels disappears into its own flank, and farther on reappears extended towards the flank. The ends of these attach themselves [to the bladder and] to the penis in males, and in females to the uterus. From the Great Blood-vessel there is no blood-vessel running to the uterus, but there are many clustered together which do so from the Aorta.

Also, from the Aorta and the Great Blood-vessel where they branch off there are other blood-vessels, some of which—large hollow ones—run first to the groins, and then through the legs and finish up in the feet and toes; while others run through the groins and the thighs criss-cross, across from left to right and from right to left, and they join up with the other blood-vessels in the hams.

It will be clear from the foregoing account how the blood-vessels are disposed and where they have their starting-points and attachments.

The description just given of the starting-points and the chief blood-vessels applies to all blooded animals, though the remainder of the blood-vessels (which are numerous) are not arranged in that way in all of them. Indeed, the parts of the body are not arranged in the same way in all, nor do all animals possess exactly the same parts. And furthermore, the state of affairs is not equally discernible in all animals: it is clearest in those animals which are largest and contain the greatest abundance of blood; for in the small ones which lack an abundance of blood, either because that is their natural condition

---

[6] οὕτω(ς) hic A; cett., vulg. post μέν.
[7] τὰ αὐτὰ π. C: π. τὰ αὐτὰ PD: ταῦτα π. A, vulg.

ARISTOTLE

ματος οὐχ ὁμοίως ἔστι καταμαθεῖν· τῶν μὲν γὰρ
οἱ πόροι συγκεχυμένοι καθάπερ ὀχετοί τινες ὑπὸ
πολλῆς ἰλύος εἰσίν, οἱ δ' ὀλίγας καὶ ταύτας ἶνας
ἀντὶ φλεβῶν ἔχουσιν. ἡ δὲ μεγάλη φλὲψ ἐν πᾶσι
μάλιστα διάδηλος, καὶ τοῖς μικροῖς.

V Τὰ δὲ νεῦρα τοῖς ζῴοις ἔχει τόνδε τὸν τρόπον. ἡ
μὲν ἀρχὴ καὶ τούτων ἐστὶν ἀπὸ[1] τῆς καρδίας· καὶ
γὰρ ἐν αὐτῇ ἡ καρδία ἔχει νεῦρα ἐν τῇ μεγίστῃ
κοιλίᾳ, καὶ ἡ καλουμένη ἀορτὴ νευρώδης ἐστὶ φλέψ,
τὰ μέντοι[2] τελευταῖα καὶ παντελῶς αὐτῆς· ἄκοιλα
γάρ ἐστι, καὶ τάσιν ἔχει τοιαύτην οἵαν περ τὰ νεῦρα,
ᾗ τελευτᾷ πρὸς τὰς καμπὰς τῶν ὀστῶν. οὐ μὴν
ἀλλ' οὐκ ἔστι συνεχὴς ἡ τῶν νεύρων φύσις ἀπὸ μιᾶς
ἀρχῆς, ὥσπερ αἱ φλέβες. αἱ μὲν γὰρ φλέβες, ὥσπερ
ἐν τοῖς γραφομένοις κανάβοις, τὸ τοῦ σώματος ἔ-
χουσι σχῆμα παντὸς οὕτως ὥστ' ἐν τοῖς σφόδρα
λελεπτυσμένοις πάντα τὸν ὄγκον φαίνεσθαι πλήρη
φλεβίων (γίνεται γὰρ ὁ αὐτὸς τόπος λεπτῶν μὲν
ὄντων φλέβια, παχυνθέντων δὲ σάρκες), τὰ δὲ νεῦρα
διεσπασμένα περὶ τὰ ἄρθρα καὶ τὰς τῶν ὀστῶν ἐστι
κάμψεις. εἰ δ' ἦν συνεχὴς ἡ φύσις αὐτῶν, ἐν τοῖς
λελεπτυσμένοις ἂν καταφανὴς ἐγίγνετο ἡ συνέχεια
πάντων.

Μέγιστα δὲ μέρη τῶν νεύρων τό τε περὶ τὸ

---

[1] ἀπὸ AC : ἐκ PD, vulg.
[2] μέντοι Sn. : μὲν vulg.

---

[a] A similar comparison is found at *P.A.* 668 a 14 ff., a 27 ff.
[b] *Cf. G.A.* 743 a 2, where the same illustration is used. Hesychius's and Photius's definitions of κάναβοι describe them as the woodwork round which modellers, when they begin their modelling, mould the wax or plaster. There is a similar

or because their bodies are so fat, it is less easy to ascertain what the arrangement is. Thus, in some of them the passages get choked up, as it might be so many irrigation-channels silted up with mud,[a] and in others there are just a few perforated fibres instead of blood-vessels. Nevertheless, the Great Blood-vessel is clearly visible in all, even in the smallest.

The sinews of animals are arranged as follows. The starting-point of them, as of the blood-vessels, is the heart: the heart has sinews within itself, in the largest cavity; and the Aorta as it is called is a sinewy blood-vessel; indeed, its extremities are wholly sinewy; they are not hollow, and it can be stretched in the same way as the sinews where they terminate at the joints of the bones. Nevertheless, the sinews do not constitute a continuous system from one starting-point, as the blood-vessels do. No; the difference is this. The blood-vessels are like the lines of the wooden skeletons used in modelling [b]— they display the shape of the entire body, so that in very attenuated persons the whole bulk of the body appears to be full of small blood-vessels: the whole space which in well-fattened persons is occupied by flesh, in thin ones is occupied by small blood-vessels. The sinews, on the other hand, are scattered about round the joints and the flexions of the bones. If they were a continuous system, their continuity would be evident in attenuated persons.

A very important part of the sinews is that in the

V
Sinews.

passage in *P.A.* (654 b 29), though without mention of the term κάναβος; there Aristotle speaks of a "hard and solid core or foundation" round which the figure is modelled, though in that passage he is speaking of the bones. Clearly something more substantial than a mere outline or sketch seems to be intended. *Cf.* also *G.A.* 764 b 31.

ARISTOTLE

515 b

μόριον τὸ τῆς ἄλσεως κύριον (καλεῖται δὲ τοῦτο
ἰγνύα), καὶ ἕτερον νεῦρον διπτυχές, ὁ τένων, καὶ τὰ
10 πρὸς τὴν ἰσχὺν βοηθητικά, ἐπίτονός τε καὶ ὠμιαία.
τὰ δ' ἀνώνυμα περὶ τὴν τῶν ὀστῶν ἐστὶ κάμψιν·
πάντα γὰρ τὰ ὀστά, ὅσα ἁπτόμενα πρὸς ἄλληλα
σύγκειται, συνδέδενται νεύροις, καὶ περὶ πάντα
ἐστὶ τὰ ὀστᾶ πλῆθος νεύρων. πλὴν ἐν τῇ κεφαλῇ
οὐκ ἔστιν οὐδέν, ἀλλ' αἱ ῥαφαὶ αὐτῶν τῶν ὀστῶν
συνέχουσιν αὐτήν.

15 Ἔστι δ' ἡ τοῦ νεύρου φύσις σχιστὴ κατὰ μῆκος,
κατὰ δὲ πλάτος ἄσχιστος καὶ τάσιν ἔχουσα πολλήν.
ὑγρότης δὲ περὶ αὐτὰ[1] μυξώδης γίγνεται, λευκὴ καὶ
κολλώδης, ᾗ τρέφεται καὶ ἐξ ἧς γιγνόμενα φαίνεται.
ἡ μὲν οὖν φλέψ δύναται πυροῦσθαι, νεῦρον δὲ πᾶν
φθείρεται πυρωθέν· κἂν διακοπῇ, οὐ συμφύεται πά-
20 λιν. οὐ λαμβάνει δ' οὐδὲ νάρκη, ὅπου μὴ νεῦρόν ἐστι
τοῦ σώματος.

Πλεῖστα δ' ἐστὶ νεῦρα περὶ τοὺς πόδας καὶ τὰς
χεῖρας καὶ πλευρὰς καὶ ὠμοπλάτας καὶ περὶ τὸν
αὐχένα καὶ τοὺς βραχίονας.

Ἔχει δὲ νεῦρα πάντα ὅσα ἔχει αἷμα· ἀλλ' ἐν οἷς
μή εἰσι καμπαὶ ἀλλ' ἄποδα καὶ ἄχειρά ἐστι, λεπτὰ
25 καὶ ἄδηλα· διὸ τῶν ἰχθύων μάλιστά ἐστι δῆλα πρὸς
τοῖς πτερυγίοις.

VI Αἱ δ' ἶνές εἰσι μεταξὺ νεύρου καὶ φλεβός. ἔνιαι
δ' αὐτῶν ἔχουσιν ὑγρότητα τὴν τοῦ ἰχῶρος, καὶ δι-

---

[1] αὐτὰ AC : ταῦτα PD, vulg.

---

[a] At Plato, *Timaeus* 84 E, ἐπίτονος is used of the great sinews of the shoulder and arm. The term is also used of the backstay of a mast (Homer, *Od.* xii. 423).

[b] The phrase used here, " bound together by sinews," is

region which controls the act of jumping (the part known as the ham); another sinew, a double one, is the *tenon* (tendon), and those which are called upon for feats of strength, *viz.*, the *epitonos*[a] and the shoulder-sinews. The sinews which have no special name are at the flexions of the bones: all the bones which are attached to one another are bound together by sinews,[b] and there is a large number of sinews round all the bones. An exception is the head, which contains none, being held together by the sutures of the bones themselves.

The texture of sinew is such that it can be split lengthwise, but not crosswise; and it admits of a high degree of tension. Round the sinews a mucous liquid is formed, which is white and glutinous; the sinews are nourished by this, and we can see them being formed out of it. Now blood-vessel can be cauterized, but sinew when cauterized is completely destroyed; and if sinew is cut it does not grow together again. And numbness never takes effect in any part of the body where no sinew is present.

The greatest number of sinews is found in connexion with the feet, the hands, the ribs, the shoulder-blades, the neck, and the arms.

All blooded animals have sinews; but in those animals which have no flexions in their limbs, *i.e.*, which are without hands and feet, the sinews are thin and difficult to see; and that is why in fishes they are most clearly to be seen near the fins.

VI Fibres.
Fibres[c] are intermediate between sinew and blood-vessel. Some fibres have fluid, *viz.*, *ichor*, and they

probably the source of interpolations into the description of the bones in *P.A.* 654 b 15-25 (see Loeb ed.). Sinews are not separately treated of in *P.A.*

[c] See note on 489 a 23.

ἔχουσιν ἀπό τε τῶν νεύρων πρὸς τὰς φλέβας καὶ
ἀπ' ἐκείνων πρὸς τὰ νεῦρα. ἔστι δὲ καὶ ἄλλο γένος
ἰνῶν, ὃ γίγνεται μὲν ἐν αἵματι, οὐκ ἐν ἅπαντος δὲ
ζῴου αἵματι· ὧν ἐξαιρουμένων ἐκ τοῦ αἵματος οὐ
πήγνυται τὸ αἷμα, ἐὰν δὲ μὴ ἐξαιρεθῶσι, πήγνυται.
ἐν μὲν οὖν τῷ τῶν πλείστων αἵματι ζῴων ἔνεισιν,
ἐν δὲ τῷ τῆς ἐλάφου καὶ προκὸς καὶ βουβαλίδος
καὶ ἄλλων τινῶν οὐκ ἔνεισιν ἶνες· διὸ καὶ οὐ πήγνυται αὐτῶν τὸ αἷμα ὁμοίως τοῖς ἄλλοις, ἀλλὰ τὸ
μὲν τῶν ἐλάφων παραπλησίως τῷ τῶν δασυπόδων
(ἔστι δ' ἀμφοτέρων αὐτῶν ἡ πῆξις οὐ στιφρά, καθάπερ ἡ τῶν ἄλλων, ἀλλὰ πλαδῶσα, καθάπερ ἡ τοῦ
γάλακτος, ἄν τις εἰς αὐτὸ τὸ πῆγμα μὴ ἐμβάλῃ),
τὸ δὲ τῆς βουβαλίδος πήγνυται μᾶλλον· παραπλησίως γὰρ συνίσταται ἢ μικρῷ ἧττον τοῦ τῶν προβάτων.

Περὶ μὲν οὖν φλεβὸς καὶ νεύρου καὶ ἰνὸς τοῦτον
ἔχει τὸν τρόπον.

VII Τὰ δ' ὀστᾶ τοῖς ζῴοις ἀφ' ἑνὸς πάντα συνηρτη-
μένα ἐστὶ καὶ συνεχῆ ἀλλήλοις ὥσπερ αἱ φλέβες·
αὐτὸ δὲ καθ' αὑτὸ οὐδέν ἐστιν ὀστοῦν. ἀρχὴ δ' ἡ
ῥάχις ἐστὶν ἐν πᾶσι τοῖς ἔχουσιν ὀστᾶ. σύγκειται
δ' ἡ ῥάχις ἐκ σφονδύλων, τείνει δ' ἀπὸ τῆς κεφαλῆς
μέχρι πρὸς τὰ ἰσχία. οἱ μὲν οὖν σφόνδυλοι πάντες
τετρημένοι εἰσίν, ἄνω δὲ τὸ τῆς κεφαλῆς ὀστοῦν
συνεχές ἐστι τοῖς ἐσχάτοις σφονδύλοις, ὃ καλεῖται
κρανίον. τούτου δὲ τὸ πριονωτὸν μέρος ῥαφή. ἔστι
δ' οὐ πᾶσιν ὁμοίως ἔχον τοῦτο τοῖς ζῴοις· τὰ μὲν

---

[a] βουβαλίς, an African antelope. Also mentioned (as βού-
βαλος) at *P.A.* 663 a 11.

[b] *Cf.* Goethe's famous "vertebral theory," according to
which the structure of the skull can be explained as a fused

HISTORIA ANIMALIUM, III. vi–vii

extend from the sinews to the blood-vessels and *vice versa*. There is also another kind of fibres, which occurs in blood, but not in the blood of every animal. If the fibres are taken out of the blood it does not congeal, whereas if they are not taken out it does. Now fibres are present in the blood of most animals, but there are none in the blood of the deer, the roe, the antelope,[a] and some others, and that is why their blood does not coagulate as much as other animals' does. That of the deer coagulates about as much as the hare's: in both these animals the coagulation is not stiff, as in other cases, but flaccid, like the coagulation of milk if one does not put the rennet well into it. The blood of the antelope coagulates better: it coagulates similarly to that of sheep, or perhaps a little less.

Such then is the nature of blood-vessel, sinew, and fibre.

The bones in animals are all connected with one single bone and are a continuous system, like the blood-vessels: there is no such thing as a bone all on its own. In all animals which possess bones the backbone is their starting-point and origin. It consists of vertebrae, and extends from the head down to the loins. The vertebrae are all perforated; and at the top the bone of the head is continuous with the endmost vertebrae[b]: it is called the skull. The serrated part of this is a suture. The skull is not identical in

VII
Bones.

series of bone-groups, comparable to the series of vertebrae. See E. S. Russell, *Form and Function*, London, 1916, pp. 49, 96 ff., where he gives the reference in Goethe, *Zur Morphologie*, i. 2, p. 250 (1820), and ii. 2, pp. 122-124 (1824). Goethe made his discovery in 1790 while contemplating a dried sheep's skull in the Jewish cemetery at Venice, but he did not publish it until 30 years later.

191

γὰρ ἔχει μονόστεον τὸ κρανίον, ὥσπερ κύων, τὰ δὲ συγκείμενον, ὥσπερ ἄνθρωπος, καὶ τούτου τὸ μὲν θῆλυ κύκλῳ ἔχει τὴν ῥαφήν, τὸ δ' ἄρρεν τρεῖς ῥαφὰς ἄνωθεν συναπτούσας, τριγωνοειδεῖς· ἤδη δ' ὤφθη καὶ ἀνδρὸς κεφαλὴ οὐκ ἔχουσα ῥαφάς. σύγκειται δ' ἡ κεφαλὴ οὐκ ἐκ τεττάρων ὀστῶν, ἀλλ' ἐξ ἕξ· ἔστι δὲ δύο τούτων περὶ τὰ ὦτα, μικρὰ πρὸς τὰ λοιπά. ἀπὸ δὲ τῆς κεφαλῆς αἱ σιαγόνες τείνουσιν ὀστᾶ. [κινεῖται δὲ τοῖς μὲν ἄλλοις ζῴοις ἅπασιν ἡ κάτωθεν σιαγών· ὁ δὲ κροκόδειλος ὁ ποτάμιος μόνος τῶν ζῴων κινεῖ τὴν σιαγόνα τὴν ἄνωθεν.][1] ἐν δὲ ταῖς σιαγόσιν ἔνεστι τὸ τῶν ὀδόντων γένος, ὀστοῦν τῇ μὲν ἄτρητον τῇ δὲ τρητόν, καὶ ἀδύνατον γλύφεσθαι τῶν ὀστῶν μόνον.

Ἀπὸ δὲ τῆς ῥάχεως ἥ τε περωνίς[2] ἐστι καὶ αἱ κλεῖδες καὶ αἱ πλευραί. ἔστι δὲ καὶ τὸ στῆθος ἐπὶ πλευραῖς κείμενον· ἀλλ' αὗται μὲν συνάπτουσιν, αἱ δ' ἄλλαι ἀσύναπτοι· οὐδὲν γὰρ ἔχει ζῷον ὀστοῦν περὶ τὴν κοιλίαν. ἔτι δὲ τά τ' ἐν τοῖς ὤμοις ὀστᾶ, καὶ αἱ καλούμεναι ὠμοπλάται, καὶ τὰ τῶν βραχιόνων ἐχόμενα, καὶ[3] τούτων, τὰ ἐν ταῖς χερσίν. ὅσα δ' ἔχει σκέλη πρόσθια, καὶ ἐν τούτοις τὸν αὐτὸν ἔχει τρόπον.

Κάτω δ' ᾗ περαίνει, μετὰ τὸ ἰσχίον ἡ κοτυληδὼν ἐστι καὶ τὰ τῶν σκελῶν ἤδη ὀστᾶ, τά τ' ἐν τοῖς

[1] secl. A.-W., Dt., ut ex 492 b 23 decerpta.
[2] περόνη D, vulg.; mox κλεῖς PD, vulg.; ἥ τε ... κλεῖδες suspic. A.-W.; ἄνω δὲ τῆς ῥ. ᾗ περαίνει conі. Th.
[3] καὶ ⟨τὰ⟩ Dt.

all animals : in some it consists of a single bone (*e.g.*, the dog), in others it is composite (*e.g.*, man), and here the female has a circular suture, whereas in the male there are three sutures which unite at the top, and are shaped like a triangle.[a] A man's head has actually been observed with no sutures at all.[b] The head is composed not of four bones but of six : two of these are by the ears, and are small compared with the rest. From the head extend the jaws, which are bones. [All animals except one move the lower jaw —all except the river crocodile, which moves the upper one.][c] In the jaws are the teeth, which consist of bone, and are partly perforated, partly not : this is the only bone which resists the graving-tool.

From the backbone extend the *perōnis*,[d] the collar-bones, and the ribs. The chest lies upon ribs ; but these ribs meet together, whereas the others do not, since no animal has any bone round the stomach. Further, there are the bones of the shoulder—what are called the shoulder-blades, and the bones of the arms, which are connected with them, and in their turn, connected with these, the bones in the hands. The bones in the forelegs (in those animals which have forelegs) are of similar formation.

At the lower terminus of the backbone, after the haunch-bone, is the hip-socket, and then immediately the leg-bones, *i.e.*, the thigh- and shank-bones, which

[a] *Cf.* 491 b 2. For a suggested explanation of Aristotle's belief that male and female skulls differ see W. Ogle's note on *P.A.* 653 b 1 (page 168).

[b] As stated in Herodotus ix. 83.

[c] This sentence is considered by A.-W. to be an interpolation.

[d] περωνίς appears to be another form of περόνη, the basic meaning of which is pin, bolt, or rivet. Various suggestions have been made for altering the text.

516 b μηροῖς καὶ κνήμαις, οἳ καλοῦνται κωλῆνες, ὧν μέρος τὰ σφυρὰ καὶ τούτων τὰ καλούμενα πλῆκτρα ἐν τοῖς ἔχουσι σφυρόν[1]· καὶ τούτοις συνεχῆ τὰ ἐν τοῖς ποσίν.

Ὅσα μὲν οὖν τῶν ἐναίμων καὶ πεζῶν ζῳοτόκα ἐστίν, οὐ πολὺ διαφέρει τὰ ὀστᾶ, ἀλλὰ κατ' ἀνα-
5 λογίαν μόνον σκληρότητι καὶ μαλακότητι καὶ μεγέθει. ἔτι δὲ τὰ μὲν ἔχει μυελὸν τὰ δ' οὐκ ἔχει τῶν ἐν τῷ αὐτῷ ζῴῳ ὀστῶν. ἔνια δὲ ζῷα οὐδ' ἂν ἔχειν δόξειεν ὅλως μυελὸν ἐν τοῖς ὀστοῖς, οἷον λέων, διὰ τὸ πάμπαν ἔχειν μικρὸν καὶ λεπτὸν καὶ ἐν ὀλίγοις· ἔχει γὰρ ἐν τοῖς μηροῖς καὶ βραχίοσιν. στερεὰ δὲ
10 πάντων μάλιστα ὁ λέων ἔχει τὰ ὀστᾶ· οὕτω γάρ ἐστι σκληρὰ ὥστε συντριβομένων ὥσπερ ἐκ λίθων ἐκλάμπειν πῦρ. ἔχει δὲ[2] καὶ ὁ δελφὶς ὀστᾶ, ἀλλ' οὐκ ἄκανθαν.

Τὰ δὲ τῶν ἄλλων ζῴων τῶν ἐναίμων τὰ μὲν μικρὸν παραλλάττει, οἷον τὰ τῶν ὀρνίθων, τὰ δὲ τῷ ἀνάλογόν ἐστι ταὐτά, οἷον ἐν τοῖς ἰχθύσι. τούτων
15 γὰρ τὰ μὲν ζῳοτοκοῦντα χονδράκανθά ἐστιν, οἷον τὰ καλούμενα σελάχη, τὰ δ' ᾠοτοκοῦντα ἄκανθαν ἔχει, ἥ ἐστιν ὥσπερ ἐν τοῖς τετράποσιν ἡ ῥάχις. ἴδιον δ' ἐν τοῖς ἰχθύσιν, ὅτι ἐν ἐνίοις εἰσὶ κατὰ τὴν σάρκα κεχωρισμένα ἀκάνθια λεπτά. ὁμοίως δὲ καὶ
20 ὁ ὄφις ἔχει τοῖς ἰχθύσιν· ἀκανθώδης γὰρ ἡ ῥάχις αὐτοῦ ἐστιν. τὰ δὲ τῶν τετραπόδων μὲν ᾠοτο-

[1] καὶ τούτων ... σφυρόν secl. A.-W., Dt.: τοῖς ἔχουσι ⟨καὶ ἔχουσι⟩ σφ. Gohlke.   [2] δὲ A : om. PD, vulg.

[a] The term πλῆκτρον is otherwise regularly used of a bird's

are called *cōlēnes* (*i.e.*, limb-bones), of which the ankle is a part, and another part of them is the so-called *plectra* [a] which are found in those animals which have an ankle. Continuous with these are the bones in the feet.

In those blooded and footed animals which are viviparous, the bones do not differ much : they differ only " by analogy," [b] *i.e.*, in hardness and softness, and in size. A further difference is that, in the same animal, some bones will contain marrow and others will not. Some animals would appear to have no marrow at all in their bones, *e.g.*, the lion, on account of having very little marrow indeed, and that thin, and in a few bones only ; actually, there is marrow in the thigh and arm-bones. The bones of the lion are more solid than in any other animal ; they are so hard that when they are rubbed together sparks fly out as though from flint. The dolphin is included among those that have bone, not fish-spine.

Of other blooded animals, the bones of some show but little deviation from what has been described, *e.g.*, birds ; others are the same " by analogy," *e.g.*, the bones of fishes : thus, the viviparous fishes are cartilaginous-spined, *e.g.*, the Selachia as they are called, while the oviparous ones have a spine corresponding to the backbone in quadrupeds. A peculiar feature in fishes is that in some species there are small thin spines unconnected together throughout the flesh. The serpent is similar to fishes, having a spinous backbone. With the oviparous quadrupeds,

spur, *e.g.*, 504 b 7, and at 526 a 5 certain projections on the feet of crayfish are compared to spurs. The word means, literally, something used for striking ; hence its application to the instrument used for striking a lyre.

[b] See Notes, § 3.

ARISTOTLE

κούντων δὲ τῶν μὲν μειζόνων ὀστωδέστερά ἐστι, τῶν δ' ἐλαττόνων ἀκανθωδέστερα. πάντα δὲ τὰ ζῷα ὅσα ἔναιμά ἐστιν, ἔχει ῥάχιν ἢ ὀστώδη ἢ ἀκανθώδη· τὰ δ' ἄλλα μόρια τῶν ὀστῶν ἐν ἐνίοις μέν ἐστιν, ἐν[1] ἐνίοις δ' οὐκ ἔστιν, ἀλλ' ὡς ὑπάρχει τοῦ
25 ἔχειν τὰ μόρια, οὕτω καὶ τοῦ ἔχειν τὰ ἐν τούτοις ὀστᾶ. ὅσα γὰρ μὴ ἔχει σκέλη καὶ βραχίονας, οὐδὲ κωλῆνας ἔχει, οὐδ' ὅσα ταὐτὰ μὲν ἔχει μόρια, μὴ ὅμοια δέ· καὶ γὰρ ἐν τούτοις ἢ τῷ μᾶλλον καὶ ἧττον διαφέρει ἢ τῷ ἀνάλογον.

30 Τὰ μὲν οὖν περὶ τὴν τῶν ὀστῶν φύσιν τοῦτον ἔχει τὸν τρόπον τοῖς ζῴοις.

VIII Ἔστι δὲ καὶ ὁ χόνδρος τῆς αὐτῆς φύσεως τοῖς ὀστοῖς, ἀλλὰ τῷ μᾶλλον διαφέρει καὶ ἧττον. καὶ ὥσπερ οὐδ' ὀστοῦν οὐδ' ὁ χόνδρος αὐξάνεται, ἂν ἀποκοπῇ. εἰσὶ δ' ἐν μὲν τοῖς χερσαίοις καὶ ζῳοτό-
35 κοις τῶν ἐναίμων ἄτρητοι οἱ χόνδροι, καὶ οὐ γίγνεται ἐν αὐτοῖς ὥσπερ ἐν τοῖς ὀστοῖς μυελός· ἐν δὲ τοῖς σελάχεσιν (ταῦτα γάρ ἐστι χονδράκανθα) ἔνεστιν [αὐτῶν ἐν τοῖς πλατέσι][2] τὸ κατὰ τὴν ῥάχιν ἀνάλογον τοῖς ὀστοῖς χονδρῶδες, ἐν ᾧ[3] ὑπάρχει ὑγρότης μυελώδης. τῶν δὲ ζῳοτοκούντων καὶ [τῶν] πεζῶν[4] περί τε τὰ ὦτα χόνδροι εἰσὶ καὶ τοὺς μυ-
5 κτῆρας καὶ περὶ ἔνια ἀκρωτήρια τῶν ὀστῶν.

IX Ἔτι δ' ἐστὶν ἄλλα γένη μορίων, οὔτε τὴν αὐτὴν ἔχοντα φύσιν τούτοις οὔτε πόρρω τούτων, οἷον ὄνυχές τε καὶ ὁπλαὶ καὶ χηλαὶ καὶ κέρατα—καὶ ἔτι παρὰ ταῦτα ῥύγχος, οἷον ἔχουσιν οἱ ὄρνιθες—ἐν οἷς

[1] ἐν (bis) AC : om. PD, vulg.
[2] aut suspicantur aut eiciunt edd. : om. Σ.
[3] ᾧ A.-W. : οἷς codd., vulg.
[4] sic AC (ζῳοτόκων A) : τῶν delent A.-W. : τῶν ζ. δὲ π. PD, vulg.

HISTORIA ANIMALIUM, III. vii–ix

the larger of them have more bone-like skeletons, the smaller more spine-like ones. But all blooded animals have a backbone, whether bony or spinous. The other parts of the bone-system are present in some animals, absent in others; but whatever parts they have or do not have, there is a corresponding presence or absence of the bones proper to those parts. Animals which have no legs or arms have no limb-bones (*cōlēnes*) either; and the same is true of those which, though they have the same parts, have them in a modified form: in these animals the bones differ either by " the more and less " or else " by analogy."[a]

I have now described the nature of the bones in animals.

VIII Cartilage.

Cartilage is of the same nature as the bones, but differs from them by way of " the more and less,"[b] and, like bone,[c] cartilage does not grow again if it is cut. In those blooded animals which are terrestrial and viviparous, the cartilaginous parts are not perforated, and there is no marrow in them as there is in bones; but in the Selachia (because they are cartilaginous-spined) the cartilage which constitutes their backbone is for them analogous to bones, and in it there is a marrow-like fluid.[d] In footed Vivipara there are cartilaginous parts at the ears and nostrils and some of the extremities of the bones.

IX Nail, hoof, etc.

Further, there are other kinds of parts whose substance is neither the same as those already mentioned nor far removed from it, *e.g.*, nail, hoofs, claws, and horns, and in addition to these the beak, as in birds—

[a] See Notes, § 3.
[b] *Cf. P.A.* 655 a 32.
[c] Regeneration of bone is in fact quite common, and occurs rapidly, in amphibia such as newts and salamanders.
[d] *Cf. P.A.* 652 a 13, 655 a 23, 37.

ARISTOTLE

517 a

ὑπάρχει ταῦτα τὰ μόρια τῶν ζῴων. ταῦτα μὲν γὰρ καὶ καμπτὰ καὶ σχιστά, ὀστοῦν δ' οὐδὲν καμπτὸν οὐδὲ σχιστόν, ἀλλὰ θραυστόν.

Καὶ τὰ χρώματα τῶν κεράτων καὶ ὀνύχων καὶ χηλῆς καὶ ὁπλῆς κατὰ τὴν τοῦ δέρματος καὶ τὴν τῶν τριχῶν ἀκολουθεῖ χρόαν. τῶν τε γὰρ μελανοδερμάτων μέλανα τὰ κέρατα καὶ αἱ χηλαὶ καὶ αἱ ὁπλαί, ὅσα χηλὰς ἔχει, καὶ τῶν λευκῶν λευκά, μεταξὺ δὲ τὰ τῶν ἀνὰ μέσον. ἔχει δὲ καὶ περὶ τοὺς ὄνυχας τὸν αὐτὸν τρόπον. οἱ δ' ὀδόντες κατὰ τὴν τῶν ὀστῶν εἰσι φύσιν. διόπερ τῶν μελάνων ἀνθρώπων, ὥσπερ Αἰθιόπων καὶ τῶν τοιούτων, οἱ μὲν ὀδόντες λευκοὶ καὶ τὰ ὀστᾶ, οἱ δ' ὄνυχες μέλανες, ὥσπερ καὶ τὸ πᾶν δέρμα. τῶν δὲ κεράτων τὰ μὲν πλεῖστα κοῖλά ἐστιν ἀπὸ τῆς προσφύσεως περὶ τὸ ἐντὸς ἐκπεφυκὸς ἐκ τῆς κεφαλῆς ὀστοῦν, ἐπ' ἄκρου δ' ἔχει τὸ στερεόν, καί ἐστιν ἁπλᾶ· τὰ δὲ τῶν ἐλάφων μόνα δι' ὅλου στερεὰ καὶ πολυσχιδῆ. καὶ τῶν μὲν ἄλλων τῶν ἐχόντων κέρας οὐδὲν ἀποβάλλει τὰ κέρατα, ἔλαφος δὲ μόνος[1] καθ' ἕκαστον ἔτος, ἐὰν μὴ ἐκτμηθῇ· περὶ δὲ τῶν ἐκτετμημένων ἐν τοῖς ὕστερον λεχθήσεται. τὰ δὲ κέρατα προσπέφυκε μᾶλλον τῷ δέρματι ἢ τῷ ὀστῷ· διὸ καὶ ἐν Φρυγίᾳ εἰσὶ βόες καὶ ἄλλοθι οἳ κινοῦσι[2] τὰ κέρατα ὥσπερ τὰ ὦτα.

Τῶν δ' ἐχόντων ὄνυχας (ἔχει δ' ὄνυχας ἅπαντα ὅσαπερ δακτύλους, δακτύλους δ' ὅσα πόδας, πλὴν ἐλέφας· οὗτος δὲ καὶ δακτύλους ἀσχίστους καὶ ἠρέμα διηρθρωμένους καὶ ὄνυχας ὅλως οὐκ ἔχει) τῶν δ' ἐχόντων τὰ μέν ἐστιν εὐθυώνυχα, ὥσπερ

---

[1] μόνος AC : μόνον PD, vulg.
[2] οἱ κινοῦσι AC : κινοῦσαι PD, vulg.

in the several animals in which they occur. All these parts permit of bending and splitting ; bone permits of neither, but it can be broken.

Besides this, the colours of horns, nails, claws, and hoofs follow suit with the colour of the skin and hair : animals with black skin have black horns, claws, and hoofs (not all, of course, have claws), white-skinned animals have white ones, those with intermediate colours have ones to match. The same holds good for nails. The teeth, however, match the bones. Thus black men, like Ethiopians and such, have white teeth as well as white bones, but their nails, like the whole of their skin, are black. Most horns are hollow from the point where they are attached to the bone which projects inside them from the head, but at their tip there is a solid piece.[a] Their structure is simple. The stag is the only animal whose horns are solid throughout and branching. No other horned animal sheds its horns ; the stag is the only one which does this— every year, unless it has been castrated. (This subject will be dealt with later on.) Horns grow attached to the skin rather than to the bone. This explains why both in Phrygia and elsewhere cattle are found which can move their horns just as though they were ears.[b]

With reference to animals that possess nails—and all do possess them if they possess toes, and all have toes that have feet, except the elephant, which has toes that are not divided and faintly articulated, but no nails at all—of the animals that possess nails, then, some have straight ones (*e.g.*, man), others

[a] *Cf.* 500 a 6 ff., *P.A.* 663 b 15 ff., and add. n., p. 237.
[b] Thompson gives several references to ancient authors, and one dated 1661 relating to cattle in Madagascar.

ἄνθρωπος, τὰ δὲ γαμψώνυχα, ὥσπερ καὶ τῶν πεζῶν λέων καὶ τῶν πτηνῶν ἀετός.

X Περὶ δὲ τριχῶν καὶ τῶν ἀνάλογον καὶ δέρματος τόνδ' ἔχει τὸν τρόπον.

Τρίχας μὲν ἔχει τῶν ζῴων ὅσα πεζὰ καὶ ζωοτόκα, φολίδας δ' ὅσα πεζὰ καὶ ᾠοτόκα, λεπίδας δ' ἰχθύες μόνοι, ὅσοι ᾠοτοκοῦσι τὸ ψαθυρὸν ᾠόν· τῶν γὰρ μακρῶν γόγγρος μὲν οὐ τοιοῦτον ἔχει ᾠόν, οὐδ' ἡ μύραινα, ἔγχελυς δ' ὅλως οὐκ ἔχει.

Τὰ δὲ πάχη τῶν τριχῶν καὶ αἱ λεπτότητες καὶ τὰ μεγέθη διαφέρουσι κατὰ τοὺς τόπους, ἐν οἷς ἂν ὦσι τῶν μερῶν, καὶ ὁποῖον ἂν ᾖ τὸ δέρμα· ὡς γὰρ ἐπὶ τὸ πολὺ ἐν τοῖς παχυτέροις δέρμασι σκληρότεραι αἱ τρίχες καὶ παχύτεραι, πλείους δὲ καὶ μακρότεραι ἐν τοῖς κοιλοτέροις καὶ ὑγροτέροις, ἄνπερ ὁ τόπος ᾖ τοιοῦτος οἷος ἔχειν τρίχας. ὁμοίως δὲ καὶ περὶ τῶν λεπιδωτῶν ἔχει καὶ τῶν φολιδωτῶν. ὅσα μὲν οὖν μαλακὰς ἔχει τὰς τρίχας, εὐβοσίᾳ χρώμενα σκληροτέρας ἴσχει, ὅσα δὲ σκληράς, μαλακωτέρας καὶ ἐλάττους. διαφέρουσι δὲ καὶ κατὰ τοὺς τόπους τοὺς θερμοτέρους καὶ ψυχροτέρους, οἷον αἱ τῶν ἀνθρώπων τρίχες ἐν μὲν τοῖς θερμοῖς σκληραί, ἐν δὲ τοῖς ψυχροῖς μαλακαί. εἰσὶ δ' αἱ μὲν εὐθεῖαι μαλακαί, αἱ δὲ κεκαμμέναι σκληραί.

XI Ἡ δὲ φύσις τῆς τριχός ἐστι σχιστή. τῷ μᾶλλον δὲ καὶ ἧττον διαφέρουσι πρὸς ἀλλήλας. ἔνιαι δὲ τῇ σκληρότητι μεταβαίνουσαι κατὰ μικρὸν οὐκέτι

---

[a] Cf. 510 b 26 above.
[b] For an account of the breeding habits of the eel see *G.A.*, Loeb ed., p. 565. To the references given there should be added: Léon Bertin, *Eels, a biological study* (London, 1956). A modification of Schmidt's explanation is pro-

HISTORIA ANIMALIUM, III. ix–xi

crooked or curved ones (an example from walking animals is the lion, from flying ones, the eagle).

We now go on to describe hair and its counterparts, and skin.

X
Hair and skin.

All footed viviparous animals have hair; footed Ovipara have horny scales; fish alone have ordinary scales (*i.e.*, fishes which produce a crumbling egg [a]: there are, of course, long-bodied fishes such as the conger and the muraena which do not produce that type of egg; and the eel [b] produces no egg at all).

Differences of thickness and length of hair are observed according to the different regions of the body where it occurs, and according to the character of the skin on which it grows. For the most part, the stouter the skin, the harder and stouter is the hair, and it is more plentiful and longer on the hollower and moister regions, provided the region is such as to permit the growth of hair. The case is similar with the animals that have fish-scales and horny scales. The hair of soft-haired animals becomes harder if they are well fed; the hair of hard-haired ones becomes softer and less plentiful. Hair also differs according as the region of the body is warm or cold: thus in man the hair is hard in warm parts of the body and soft in cold ones. Furthermore, straight hair is soft, and curly hair hard.

Hair [c] is in its nature fissile: and it exhibits in different animals differences of " the more and less." [d] In some animals the hair gradually increases in hard-

XI

---

pounded by Denys W. Tucker, "A new solution to the Atlantic eel problem," in *Nature*, 183 (1959), pp. 495-501; summary by Maurice Burton in *Ill. Lon. News*, 234 (1959), p. 532. See also *Nature*, 183, pp. 1405-1406.

[c] For a fuller discussion of hair see *G.A.* Book V, chh. 3-6.
[d] See Notes, § 3.

θριξὶν ἐοίκασιν ἀλλ' ἀκάνθαις, οἷον αἱ τῶν ἐχίνων τῶν χερσαίων, παραπλησίως τοῖς ὄνυξιν· καὶ γὰρ τὸ τῶν ὀνύχων γένος ἐν ἐνίοις τῶν ζῴων οὐδὲν διαφέρει τὴν σκληρότητα τῶν ὀστῶν.

Δέρμα δὲ πάντων λεπτότατον ἄνθρωπος ἔχει κατὰ λόγον τοῦ μεγέθους. ἔνεστι δ' ἐν τοῖς δέρμασι πᾶσι γλισχρότης μυξώδης, ἐν μὲν τοῖς ἐλάττων ἐν δὲ τοῖς πλείων, οἷον ἐν τοῖς τῶν βοῶν, ἐξ ἧς ποιοῦσι τὴν κόλλαν· ἐνιαχοῦ δὲ καὶ ἐξ ἰχθύων ποιοῦσι κόλλαν.[1] ἀναίσθητον δὲ τὸ δέρμα τεμνόμενόν ἐστι καθ' αὑτό· μάλιστα δὲ τοιοῦτον τὸ ἐν τῇ κεφαλῇ, διὰ τὸ τὸ μεταξὺ ἀσαρκότατον εἶναι πρὸς τὸ ὀστοῦν. ὅπου δ' ἂν ᾖ καθ' αὑτὸ δέρμα, ἂν διακοπῇ, οὐ συμφύεται, οἷον γνάθου τὸ λεπτὸν καὶ ἀκροποσθία καὶ βλεφαρίς. τῶν συνεχῶν δ' ἐστὶ[2] τὸ δέρμα ἐν ἅπασι τοῖς ζῴοις, καὶ ταύτῃ διαλείπει ᾗ καὶ οἱ κατὰ φύσιν πόροι ἐξικμάζονται, καὶ κατὰ τὸ στόμα καὶ ὄνυχας.

Δέρμα μὲν οὖν ἅπαντα[3] ἔχει τὰ ἔναιμα ζῷα, τρίχας δ' οὐ πάντα, ἀλλ' ὥσπερ εἴρηται πρότερον. μεταβάλλουσι δὲ τὰς χρόας γηρασκόντων καὶ λευκαίνονται ἐν ἀνθρώπῳ· τοῖς δ' ἄλλοις γίγνεται μέν, οὐκ ἐπιδήλως δὲ σφόδρα, πλὴν ἐν ἵππῳ. λευκαίνεται δὲ καὶ ἀπ' ἄκρας ἡ θρίξ. αἱ δὲ πλεῖσται εὐθὺς φύονται λευκαὶ τῶν πολιῶν. ᾗ καὶ δῆλον ὅτι οὐχ αὐότης ἐστὶν ἡ πολιότης, ὥσπερ τινές φασιν· οὐδὲν γὰρ φύεται εὐθὺς αὖον.

Ἐν δὲ τῷ ἐξανθήματι ὃ καλεῖται λεύκη, πᾶσαι πολιαὶ γίγνονται· ἤδη δέ τισι κάμνουσι μὲν πολιαὶ ἐγένοντο, ὑγιασθεῖσι δὲ ἀπορρυεισῶν μέλαιναι ἀν-

---

[1] ἐνιαχοῦ ... κόλλαν secl. A.-W., Dt.
[2] ἐστὶ AC : om. PD, vulg.
[3] ἅπαντα AC : πάντα vulg.

ness until it no longer resembles hair, but rather spines, as in the hedgehog; and the same applies to nails: in some animals the nails are indistinguishable in hardness from bones.

With regard to skin: Man has, for his size, the finest skin of all animals. All varieties of skin contain a mucous sticky substance, some a lesser quantity, some a greater: an example of the latter is the hide of the ox, out of which glue is made. Glue is also made out of fishes in some places. The skin, in itself, has no sensation when cut: an extreme case of this is the skin on the head, because between it and the skull-bone there is an extreme absence of flesh. And wherever the skin is all on its own, it does not grow together again when severed, as is exemplified by the thin part of the jaw, the prepuce, and the eyelid. In all animals the skin belongs to that class of parts which are continuous, and is discontinuous only where the natural passages discharge their fluidity, and at the mouth and the nails.

Now all blooded animals have skin, but not all have hair: the facts are as already stated. The hair changes colour as animals age: in man it becomes white; in other animals its colour certainly changes, but not very noticeably except in the horse. The whitening of the hair begins at the tip. But in most cases of greyness, the hairs are white at the start of their growth, which goes to show that greyness is not a sort of desiccation, as some allege, since nothing is desiccated at the very start of its growth.

In the eruptive disease called the white-sickness all the hairs go grey. Cases have been known where during the sickness the hairs became grey, but after recovery these hairs dropped out and black ones grew

15 ἐφύησαν. [γίγνονταί τε μᾶλλον πολιαὶ σκεπαζομένων τῶν τριχῶν ἢ διαπνεομένων.][1] πρῶτον δὲ πολιοῦνται οἱ κρόταφοι τῶν ἀνθρώπων, καὶ τὰ πρόσθια πρότερα τῶν ὄπισθεν[2]· τελευταῖον δ' ἡ ἥβη.

Εἰσὶ δὲ τῶν τριχῶν αἱ μὲν συγγενεῖς, αἱ δ' ὕστερον κατὰ τὰς ἡλικίας γιγνόμεναι ἐν ἀνθρώπῳ μόνῳ
20 τῶν ζῴων, συγγενεῖς μὲν αἱ ἐν τῇ κεφαλῇ καὶ ταῖς βλεφαρίσι καὶ ταῖς ὀφρύσιν, ὑστερογενεῖς δ' αἱ ἐπὶ τῆς ἥβης πρῶτον, ἔπειτα δ' αἱ ἐπὶ τῆς μασχάλης, τρίται δ' αἱ ἐπὶ τοῦ γενείου· ἴσοι γὰρ οἱ τόποι εἰσὶν ἐν οἷς αἱ τρίχες ἐγγίγνονται αἵ τε συγγενεῖς καὶ αἱ ὑστερογενεῖς. λείπουσι δὲ καὶ ῥέουσι κατὰ τὴν
25 ἡλικίαν αἱ ἐκ τῆς κεφαλῆς καὶ μάλιστα καὶ πρῶται. τούτων δ' αἱ ἔμπροσθεν μόναι· τὰ γὰρ ὄπισθεν οὐδεὶς γίγνεται φαλακρός. ἡ μὲν οὖν κατὰ κορυφὴν λειότης φαλακρότης καλεῖται. ἡ δὲ κατὰ τὰς ὀφρῦς ἀναφαλανθίασις· οὐδέτερον δὲ τούτων συμβαίνει οὐδενὶ πρὶν ἢ ἀφροδισιάζειν ἄρξηται. οὐ
30 γίγνεται δ' οὔτε παῖς φαλακρὸς οὔτε γυνὴ οὔθ' οἱ ἐκτετμημένοι· ἀλλ' ἐὰν μὲν ἐκτμηθῇ πρὸ ἥβης, οὐ φύονται αἱ ὑστερογενεῖς, ἐὰν δ' ὕστερον, αὗται μόναι ἐκρέουσι, πλὴν τῆς ἥβης.

Γυνὴ δὲ τὰς ἐπὶ τῷ γενείῳ οὐ φύει τρίχας· πλὴν ἐνίαις γίγνονται ὀλίγαι, ὅταν τὰ καταμήνια στῇ,
35 καὶ οἷον ἐν Καρίᾳ ταῖς ἱερείαις, ὃ δοκεῖ συμβαίνειν σημεῖον τῶν μελλόντων. αἱ δ' ἄλλαι γίγνονται μέν, ἐλάττους δέ. γίγνονται δὲ καὶ ἄνδρες καὶ γυναῖκες ἐκ γενετῆς ἐνδεεῖς τῶν ὑστερογενῶν τριχῶν· ⟨ὧν⟩[3] ἅμα καὶ ἄγονοι, ὅσοιπερ ἂν καὶ ἥβης στερηθῶσιν.

---
[1] secl. Dt.
[2] ὄπισθεν AC : ὀπισθίων PD, vulg.
[3] ⟨ὧν⟩ Dt.

in their place. [Hair tends to go grey more if it is kept covered than if it is freely exposed to the air.] The first hair to go grey in man is that on the temples; and the forward parts are affected before the back parts. The last hair to go grey is the pubic hair.

Some hair is congenital, some grows later, according to the various stages of life—this applies to man only. The congenital hair is that on the head, on the eyelids, and the eyebrows; the first of the later hair is the pubic, then that under the armpits, and third the hair on the chin. It will thus be seen that the number of places where the congenital hair grows and where the later hair grows are equal. It is the hair on the head which fails and falls out most of all as age advances, and it does so before any other. This applies only to the hair in front; no one goes bald at the back of the head. Smoothness on the top of the head is called baldness, on the eyebrows *anaphalanthiasis*. Neither of these occurs in a man until he has entered upon sexual activity. No boy or woman or castrated man ever goes bald; if castration takes place before puberty, the later growths of hair do not occur; if after, the later growths only (except the pubic) fall away.

Women do not grow the hairs on the chin, except that in some cases a few hairs grow when menstruation ceases; another example is the priestesses in Caria,[a] which when it occurs is considered a sign of events to come. The other later growths occur in women, but on a lesser scale. Cases occur of both men and women who from birth have been lacking in the later growths; of these some are also impotent, *viz.*, those who are destitute of the pubic hairs.

[a] *Cf.* Herodotus i. 175, viii. 104.

ARISTOTLE

518 b

Αἱ μὲν οὖν ἄλλαι τρίχες αὐξάνονται κατὰ λόγον ἢ πλέον ἢ ἔλαττον, μάλιστα μὲν αἱ ἐν τῇ κεφαλῇ, εἶθ' αἱ ἐν[1] πώγωνι, καὶ εἰ[2] λεπτότριχοι, μάλιστα. δασύνονται δέ τισι καὶ αἱ ὀφρύες γιγνομένοις πρεσβυτέροις, οὕτως ὥστ' ἀποκείρεσθαι, διὰ τὸ ἐπὶ συμφύσει ὀστῶν κεῖσθαι, ἃ γηρασκόντων διιστάμενα διίησι πλείω ὑγρότητα. αἱ δ' ἐν ταῖς βλεφαρίσιν οὐκ αὐξάνονται, ῥέουσι δέ, ὅταν ἀφροδισιάζειν ἄρξωνται, καὶ μᾶλλον τοῖς μᾶλλον ἀφροδισιαστικοῖς. πολιοῦνται δὲ βραδύτατα αὗται.

Ἐκτιλλόμεναι δ' αἱ τρίχες μέχρι τῆς ἀκμῆς ἀναφύονται, εἶτα οὐκέτι. ἔχει δὲ πᾶσα θρὶξ ὑγρότητα πρὸς τῇ ῥίζῃ γλίσχραν, καὶ ἕλκει εὐθὺς ἐκτιλθεῖσα τὰ κοῦφα θιγγάνουσα. ὅσα δὲ ποικίλα τῶν ζῴων κατὰ τὰς τρίχας, τούτοις καὶ ἐν τῷ δέρματι προϋπάρχει ἡ ποικιλία καὶ ἐν τῷ τῆς γλώττης δέρματι. περὶ δὲ τὸ γένειον τοῖς μὲν συμβαίνει τὴν ὑπήνην καὶ τὸ γένειον δασὺ ἔχειν, τοῖς δὲ ταῦτα μὲν λεῖα τὰς σιαγόνας δὲ δασείας· ἧττον δὲ γίγνονται φαλακροὶ οἱ μαδιγένειοι. αὔξονται δ' αἱ τρίχες ἔν τε νόσοις τισίν, οἷον ἐν ταῖς φθίσεσι μᾶλλον, καὶ ἐν γήρᾳ καὶ τεθνεώτων, καὶ σκληρότεραι γίγνονται ἀντὶ μαλακῶν· τὰ δ' αὐτὰ ταῦτα συμβαίνει καὶ περὶ τοὺς ὄνυχας. ῥέουσι δὲ μᾶλλον αἱ τρίχες τοῖς ἀφροδισιαστικοῖς αἱ συγγενεῖς· αἱ δ' ὑστερογενεῖς γίγνονται θᾶττον. οἱ δ' ἰξίαν ἔχοντες ἧττον φαλακροῦνται, κἂν ὄντες φαλακροὶ λάβωσιν, ἔνιοι δασύνονται. οὐκ αὐξάνεται δὲ θρὶξ ἀποτμηθεῖσα, ἀλλὰ κάτωθεν ἀναφυομένη γίγνεται μείζων. καὶ αἱ λεπίδες δὲ τοῖς ἰχθύσι σκληρότεραι γίγνονται καὶ

---

[1] sic Dt.: εἶτα vulg.
[2] εἰ Pi.: οἱ vulg.

Hair of course grows in due proportion, whether more or less, wherever it is situated; but it grows most on the head, next most on the chin; and fine hair grows most of all. In some persons as they get older the eyebrows actually become thicker, and have to be cut; the reason for this is that they are situated on a place where bones join, and as age advances these bones draw apart and allow more moisture to exude. The hairs in the eyelashes do not grow, but fall off, when sexual activity begins, and the more the greater this activity is. These hairs take the longest time to go grey.

Hairs plucked out before maturity, but not later, will grow again. Every hair has at its root a sticky fluid, and immediately after being plucked out it can raise light objects by touching them with it. Animals which exhibit diversity of colour in their hair have prior to that a diversity of colour in their skin and in the cuticle of the tongue. In some men the upper lip and the chin are well covered with hair; in others these parts are smooth and the cheeks hairy. In addition, smooth-chinned men are less liable to go bald. In some diseases the hair tends to grow, *e.g.*, in cases of consumption, and in old age and after death, and its softness gives way to hardness. The same phenomena are exhibited by the nails. In those who are given to sexual activity the congenital hair falls sooner, and the later hair grows sooner. Those who suffer from varicose veins go bald less; and sometimes if they are already bald when this malady overtakes them, their hair grows thicker. A hair which has been cut does not grow at the tip, but becomes longer by growing up from below. In fishes the scales become harder and thicker, and in those

30 παχύτεραι, τοῖς δὲ λεπτυνομένοις καὶ τοῖς γηράσκουσι σκληρότεραι. καὶ τῶν τετραπόδων δὲ γιγνομένων πρεσβυτέρων τῶν μὲν αἱ τρίχες τῶν δὲ τὰ ἔρια βαθύτερα μὲν γίγνεται, ἐλάττω δὲ τῷ πλήθει· καὶ τῶν μὲν αἱ ὁπλαὶ τῶν δ' αἱ χηλαὶ γίγνονται γηρασκόντων μείζους, καὶ τὰ ῥύγχη τῶν ὀρνίθων.
35 αὐξάνονται[1] δὲ καὶ αἱ χηλαί, ὥσπερ καὶ οἱ ὄνυχες.

XII Περὶ δὲ τὰ πτερωτὰ τῶν ζῴων, οἷον τοὺς ὄρνιθας, κατὰ μὲν τὰς ἡλικίας οὐδὲν μεταβάλλει, πλὴν γέρανος· αὕτη δ' οὖσα τεφρὰ μελάντερα γηράσκουσα τὰ πτερὰ ἴσχει· διὰ δὲ τὰ πάθη τὰ γιγνόμενα κατὰ τὰς ὥρας, οἷον ὅταν ψύχη γίγνηται μᾶλλον, ἔνια[2]
5 γίγνεται τῶν μονοχρόων ἐκ μελάνων τε καὶ μελαντέρων λευκά, οἷον κόραξ τε καὶ στρουθὸς καὶ χελιδόνες· ἐκ δὲ τῶν λευκῶν γενῶν οὐκ ὦπται εἰς μέλαν μεταβάλλον. [καὶ κατὰ τὰς ὥρας δ' οἱ πολλοὶ τῶν ὀρνίθων μεταβάλλουσι τὰς χρόας, ὥστε λαθεῖν ἂν τὸν μὴ συνήθη.][3]

10 Μεταβάλλουσι δ' ἔνια τῶν ζῴων τὰς χρόας τῶν τριχῶν καὶ κατὰ τὰς τῶν ὑδάτων μεταβολάς· ἔνθα μὲν γὰρ λευκὰ γίγνονται, ἔνθα δὲ μέλανα ταὐτά. καὶ περὶ τὰς ὀχείας δ' ἐστὶν ὕδατα πολλαχοῦ τοιαῦτα, ἃ πιόντα καὶ ὀχεύσαντα μετὰ τὴν πόσιν τὰ πρόβατα μέλανας γεννῶσι τοὺς ἄρνας, οἷον καὶ ἐν
15 τῇ Χαλκιδικῇ τῇ ἐπὶ τῆς Θρᾴκης ἐν τῇ Ἀσσυρίτιδι ἐποίει ὁ καλούμενος ποταμὸς Ψυχρός. καὶ ἐν τῇ Ἀντανδρίᾳ δὲ δύο ποταμοί εἰσιν, ὧν ὁ μὲν λευκὰ ὁ δὲ μέλανα ποιεῖ τὰ πρόβατα. δοκεῖ δὲ καὶ ὁ Σκάμανδρος ποταμὸς ξανθὰ τὰ πρόβατα ποιεῖν·

which are getting thin or old the scales become harder. In quadrupeds, as they grow old, the hair (or fleece as the case may be) becomes deeper, but the component hairs are fewer; the hooves (or claws) grow larger, and birds' beaks do the same. The claws also grow, and so do the nails.

With regard to feathered animals, such as birds, none undergoes a change ⟨of colour⟩ through age except the crane, which begins as ashen-coloured and as it grows old its feathers get blacker. In addition, seasonal conditions, such as unusually sharp frost, cause some birds whose plumage is of a single colour to change. Thus, those with dark or darkish plumage turn white, *e.g.*, the raven, the sparrow, and the swallow, though no instance of a white bird changing to black has been observed. [Further, the majority of birds change their colour according to the seasons, so that anyone unfamiliar with them would not recognize them.]

Some animals change the colour of their hair even on account of a change in the water they drink. In some localities the same animal will be white, in another black. Also, with regard to mating, in some places the quality of the water is such that rams, if they drink it and then have intercourse, beget black lambs: this used to be caused by the river known as Coldstream (*Psychros*) in the district of Assyritis in the Chalcidian peninsula in Thrace. Also, in Antandria there are two rivers, one of which makes sheep white and the other black. It is also generally believed that the Scamander makes them yellow, and

XII Changes of colour.

---

[1] AC: αὔξονται PD, vulg.
[2] ἔνια Sn.: ἐνίοτε vulg.
[3] secl. Dt.

ARISTOTLE

διὸ καὶ τὸν Ὅμηρόν φασιν ἀντὶ Σκαμάνδρου Ξάνθον προσαγορεύειν αὐτόν.

Τὰ μὲν οὖν ἄλλα ζῷα οὔτ' ἐντὸς ἔχει τρίχας, τῶν τ' ἀκρωτηρίων ἐν τοῖς πρανέσιν ἀλλ' οὐκ ἐν τοῖς ὑπτίοις· ὁ δὲ δασύπους μόνος καὶ ἐντὸς ἔχει τῶν γνάθων τρίχας καὶ ὑπὸ τοῖς ποσίν. ἔτι δὲ καὶ τὸ μυστακόκητος[1] ὀδόντας μὲν ἐν τῷ στόματι οὐκ ἔχει, τρίχας δ' ὁμοίας ὑείαις.

Αἱ μὲν οὖν τρίχες αὐξάνονται ἀποτμηθεῖσαι κάτωθεν, ἄνωθεν δ' οὔ· τὰ δὲ πτερὰ οὔτ' ἄνωθεν οὔτε κάτωθεν, ἀλλ' ἐκπίπτει. οὐκ ἀναφύεται δ' ἐκτιλθὲν οὔτε τῶν μελιττῶν τὸ πτερὸν οὔθ' ὅσα ἄλλ' ἔχει ἄσχιστον τὸ πτερόν· οὐδὲ τὸ κέντρον, ὅταν ἀποβάλλῃ ἡ μέλιττα, ἀλλ' ἔκτοτε ἀποθνήσκει.[2]

XIII Εἰσὶ δὲ καὶ ὑμένες ἐν τοῖς ζῴοις ἅπασι τοῖς ἐναίμοις. ὅμοιος δ' ἐστὶν ὁ ὑμὴν δέρματι πυκνῷ καὶ λεπτῷ, ἔστι δὲ τὸ γένος ἕτερον· οὔτε γάρ ἐστι σχιστὸν οὔτε τατόν. περὶ ἕκαστον δὲ τῶν ὀστῶν καὶ περὶ ἕκαστον τῶν σπλάγχνων ὑμήν[3] ἐστι καὶ ἐν τοῖς μείζοσι καὶ ἐν τοῖς ἐλάττοσι ζῴοις· ἀλλ' ἄδηλοι ἐν τοῖς ἐλάττοσι διὰ τὸ πάμπαν εἶναι λεπτοὶ καὶ μικροί. μέγιστοι δὲ τῶν ὑμένων εἰσὶν οἵ τε περὶ τὸν ἐγκέφαλον δύο, ὧν ὁ περὶ τὸ ὀστοῦν ἰσχυρότερος καὶ παχύτερος τοῦ περὶ τὸν ἐγκέφαλον, ἔπειθ' ὁ περὶ τὴν καρδίαν ὑμήν. διακοπεὶς δ' οὐ συμφύεται

---

[1] ita scripsi Th. secutus : μυστόκητος PAD : μυστοκῆτος C : ὁ μῦς τὸ κῆτος vulg.
[2] ἔκτοτε ἀποθνήσκει PD : θνήσκει AC, vulg.
[3] ὑμήν D : ὁ ὑμήν ACP, vulg.

this, they say, is why Homer[a] calls it the Yellow River (*Xanthos*) instead of Scamander.

All animals except one have no hair internally, and on their extremities the hair appears only on the upper surface, never on the lower. The exception is the hare (or " hairyfoot ") : this is the only animal which has hair both inside its mouth and on the underside of its feet. In addition, the moustache-whale[b] lacks teeth in its mouth, and has instead hairs similar to pigs' bristles.

Now hairs that have been cut grow at their base, though not at the tip ; whereas feathers do neither, but fall out. A bee's wing will not grow again if plucked out, nor will the wing of any other creature whose wings are undivided. Nor will a bee's sting grow again if the bee loses it ; from that time the bee begins to die.[c]

XIII. Membranes.

All blooded animals possess membranes. Now a membrane resembles a fine close-textured skin ; but it differs in character, because it admits neither of division nor of extension. Membrane surrounds every one of the bones and every one of the viscera both in the larger and in the smaller animals, though in the smaller ones they are difficult to detect owing to their excessive fineness and smallness. The largest membranes are those which surround the brain—two of them : the one round the bone is stronger and thicker than the one round the brain itself. Next largest is the membrane round the heart. If cut apart

[a] *Cf. Iliad* xx. 74 " Gods call it Xanthos, but men Scamander."

[b] Here, following Th.'s suggestion, I have restored " moustache-whale " for the manuscripts' (and later fabulous) " mouse-whale." See Th.'s note *ad loc.*, and *id.*, *Glossary of Greek Fishes*, p. 168.   [c] *Cf.* 626 a 17.

ARISTOTLE

519 b

ψιλὸς ὑμήν, ψιλούμενά τε τὰ ὀστᾶ τῶν ὑμένων σφακελίζει.

XIV. Ἔστι δὲ καὶ τὸ ἐπίπλοον ὑμήν. ἔχει δ' ἐπίπλοον ἅπαντα τὰ ἔναιμα· ἀλλὰ τοῖς μὲν πῖον τοῖς δ' ἀπίμελόν ἐστιν. ἔχει δὲ καὶ τὴν ἀρχὴν καὶ τὴν ἐξάρτησιν ἐν τοῖς ζῳοτόκοις καὶ ἀμφώδουσιν ἐκ μέσης τῆς κοιλίας, ᾗ ἐστιν οἷον ῥαφή τις αὐτῆς· καὶ τοῖς μὴ ἀμφώδουσι δ' ἐκ τῆς μεγάλης κοιλίας ὡσαύτως.

XV. Ἔστι δὲ καὶ ἡ κύστις ὑμενυειδὴς μέν, ἄλλο δὲ γένος ὑμένος· ἔχει γὰρ τάσιν.[1] ἔχει δὲ κύστιν οὐ πάντα, ἀλλὰ τὰ μὲν ζῳοτόκα πάντα, τῶν δ' ᾠοτόκων ἡ χελώνη μόνον. διακοπεῖσα δ' οὐδ' ἡ κύστις συμφύεται ἀλλ' ἢ παρ' αὐτὴν τὴν ἀρχὴν τοῦ οὐρητῆρος, εἰ μή τι πάμπαν σπάνιον· γέγονε γάρ τι ἤδη τοιοῦτον. τεθνεώτων μὲν οὖν οὐδὲν διήσιν ὑγρόν, ἐν δὲ τοῖς ζῶσι καὶ ξηρὰς συστάσεις, ἐξ ὧν οἱ λίθοι γίγνονται τοῖς κάμνουσιν. ἐνίοις δ' ἤδη καὶ τοιαῦτα συνέστη ἐν τῇ κύστει ὥστε μηδὲν δοκεῖν διαφέρειν κογχυλίων.

Περὶ μὲν οὖν φλεβὸς καὶ νεύρου καὶ δέρματος, καὶ περὶ ἰνῶν καὶ ὑμένων, ἔτι δὲ καὶ περὶ τριχῶν καὶ ὀνύχων καὶ χηλῆς καὶ ὁπλῆς καὶ κεράτων καὶ ὀδόντων καὶ ῥύγχους καὶ χόνδρου καὶ ὀστῶν καὶ τῶν ἀνάλογον τούτοις τοῦτον ἔχει τὸν τρόπον.

XVI. Σὰρξ δὲ καὶ τὸ παραπλησίαν ἔχον τὴν φύσιν τῇ σαρκὶ ἐν τοῖς ἐναίμοις ἐν[2] πᾶσίν ἐστι μεταξὺ τοῦ δέρματος καὶ τοῦ ὀστοῦ καὶ τῶν ἀνάλογον τοῖς ὀστοῖς· ὡς γὰρ ἡ ἄκανθα ἔχει πρὸς τὸ ὀστοῦν, οὕτω καὶ τὸ σαρκῶδες πρὸς τὰς σάρκας ἔχει τῶν ἐχόντων ὀστᾶ καὶ ἄκανθαν.

---

[1] ἔχει γὰρ τάσιν codd.: om. vulg.
[2] ἐν ACPD: om. vulg.

a bared membrane does not grow together again, and bones bared of their membranes mortify.

Another of the membranes is the Omentum. It is present in all blooded animals; and in some it is fatty, in others devoid of fat. In the viviparous animals which have teeth in both jaws its starting-point and its place of attachment is the middle of the stomach, where this has a sort of suture; in those which have not teeth in both jaws it begins at, and is attached to, the main stomach.

XIV
The Omentum.

The bladder, too, is membranous, but in a different way: the membrane here is extensible. Not all animals have a bladder: all Vivipara have one, but the tortoise is the only oviparous animal which has one. Like other membranes, the bladder if cut apart will not grow together again, unless the cut be made just by the beginning of the urethra, except very rarely; such growing together has been observed. In dead bodies the bladder passes no fluid, but in living ones it passes solid matter as well: this gives rise to stones in those who suffer in this way. In some persons objects have actually been formed in the bladder which are indistinguishable from cockle-shells.

XV
The Bladder.

We have now described blood-vessel, sinew, and skin; fibres and membrane; and further, hair, nail, claw, hoof, horn, teeth, beak, cartilage, bones, and their counterparts.

We deal next with Flesh. Flesh, and that which has a similar nature to flesh as found in blooded animals, is in all instances placed between the skin and the bone (or the counterpart of bone): as fish-spine is to bone, so is the flesh-like substance to flesh—some animals have the one pair and some the other.

XVI
Flesh.

ARISTOTLE

Ἔστι δὲ διαιρετὴ ἡ σὰρξ πάντῃ, καὶ οὐχ ὥσπερ τὰ νεῦρα καὶ αἱ φλέβες ἐπὶ μῆκος μόνον. λεπτυνομένων μὲν οὖν τῶν ζῴων ἀφανίζονται, καὶ γίγνονται φλέβια καὶ ἶνες· εὐβοσίᾳ δὲ πλείονι χρωμένων πιμελὴ ἀντὶ σαρκῶν. εἰσὶ δὲ τοῖς μὲν ἔχουσι τὰς σάρκας πολλὰς αἱ φλέβες ἐλάττους καὶ τὸ αἷμα ἐρυθρότερον καὶ[1] σπλάγχνα καὶ κοιλία μικρά· τοῖς δὲ τὰς φλέβας ἔχουσι μεγάλας καὶ τὸ αἷμα μελάντερον καὶ σπλάγχνα μεγάλα καὶ κοιλία μεγάλη, αἱ δὲ σάρκες ἐλάττους. γίγνονται δὲ κατὰ σάρκα πίονα τὰ τὰς κοιλίας ἔχοντα μικράς.

XVII Πιμελὴ δὲ καὶ στέαρ διαφέρουσιν ἀλλήλων. τὸ μὲν γὰρ στέαρ ἐστὶ θραυστὸν πάντῃ καὶ πήγνυται ψυχόμενον, ἡ δὲ πιμελὴ χυτὸν καὶ ἄπηκτον· καὶ οἱ μὲν ζωμοὶ οἱ τῶν πιόνων οὐ πήγνυνται, οἷον ἵππου καὶ ὑός, οἱ δὲ τῶν στέαρ ἐχόντων πήγνυνται, οἷον προβάτου καὶ αἰγός. διαφέρουσι δὲ καὶ τοῖς τόποις· ἡ μὲν γὰρ πιμελὴ γίγνεται μεταξὺ δέρματος καὶ σαρκός, στέαρ δ' οὐ γίγνεται ἀλλ' ἢ ἐπὶ τέλει τῶν σαρκῶν. γίγνεται δὲ καὶ τὸ ἐπίπλοον τοῖς μὲν πιμελώδεσι πιμελῶδες, τοῖς δὲ στεατώδεσι στεατῶδες. ἔχει δὲ τὰ μὲν ἀμφώδοντα πιμελήν, τὰ δὲ μὴ ἀμφώδοντα στέαρ.

Τῶν δὲ σπλάγχνων τὸ ἧπαρ ἐν ἐνίοις τῶν ζῴων γίγνεται πιμελῶδες, οἷον τῶν ἰχθύων ἐν τοῖς σελάχεσιν· ποιοῦσι γὰρ ἔλαιον ἀπ' αὐτῶν, ὃ γίγνεται τηκομένων· αὐτὰ δὲ τὰ σελάχη ἐστὶν ἀπιμελώτατα καὶ κατὰ σάρκα καὶ κατὰ κοιλίαν κεχωρισμένῃ πιμελῇ. ἔστι δὲ καὶ τὸ τῶν ἰχθύων στέαρ πιμελῶδες καὶ οὐ πήγνυται. πάντα δὲ τὰ ζῷα τὰ μὲν κατὰ σάρκα ἐστὶ πίονα· τὰ δ' ἀφωρισμένως. ὅσα δὲ μὴ

---

[1] καὶ CPD : καὶ τὰ A, vulg.

Flesh can be divided in any direction, *i.e.*, not, like sinew and blood-vessel, lengthwise only. When animals become emaciated, the flesh disappears, and they become just blood-vessels and fibres; when they are overfed, fat replaces flesh. In animals which are abundant in flesh, the blood-vessels are smaller and the blood redder, while the viscera and stomach are small; in those that have large blood-vessels the blood is blacker, the viscera and the stomach large, and the flesh smaller in quantity. Animals that have small stomachs tend to become fat in their fleshy parts.

Fat and Suet. These differ from each other as follows: Suet is breakable in all directions and congeals when cooled, whereas fat can be melted but does not congeal. Soups made from fatty animals do not congeal, for example soup made from horse-flesh and pork, whereas those made from animals that contain suet do congeal, for example, mutton and goats' flesh. Also, fat and suet differ in that their place in the body is different: fat is formed between the skin and the flesh, suet only at the extreme of the fleshy parts. Again, in fatty animals the Omentum is fatty, and in animals that have suet it is suety. Also, animals with teeth in both jaws have fat, those without have suet.

The liver, one of the viscera, becomes fatty in some animals, as in the Selachia (to give an instance of a fish): their livers when melted yield an oil. Otherwise these Selachia have no fat round either the flesh or the stomach—that is to say, they have no free fat. Furthermore, the suet in fish is fatty and does not congeal. All animals have fat, some with their flesh, some apart from it. Those whose fat is not free are

ARISTOTLE

ἔχει κεχωρισμένην τὴν πιότητα, ἧττόν ἐστι πίονα κατὰ κοιλίαν καὶ ἐπίπλοον, οἷον ἔγχελυς· ὀλίγον γὰρ στέαρ ἔχουσι περὶ τὸ ἐπίπλοον. τὰ δὲ πλεῖστα γίγνεται πίονα κατὰ τὴν γαστέρα, καὶ μάλιστα τὰ μὴ ἐν κινήσει ὄντα τῶν ζῴων.

Οἱ δ' ἐγκέφαλοι τῶν μὲν πιμελωδῶν λιπαροί, οἷον ὑός, τῶν δὲ στεατωδῶν αὐχμηροί. τῶν δὲ σπλάγχνων περὶ τοὺς νεφροὺς μάλιστα πίονα γίγνεται τὰ ζῷα· ἔστι δ' ἀεὶ ὁ δεξιὸς ἀπιμελώτερος, κἂν σφόδρα πίονες ὦσιν, ἐλλείπει τι ἀεὶ κατὰ τὸ μέσον. περίνεφρα δὲ γίγνεται τὰ στεατώδη μᾶλλον, καὶ μάλιστα τῶν ζῴων πρόβατον· τοῦτο γὰρ ἀποθνήσκει τῶν νεφρῶν πάντῃ καλυφθέντων. γίγνεται δὲ περίνεφρα δι' εὐβοσίαν, οἷον τῆς Σικελίας περὶ Λεοντίνους· διὸ καὶ ἐξελαύνουσιν ὀψὲ τὰ πρόβατα τῆς ἡμέρας, ὅπως ἐλάττω λάβωσι τὴν τροφήν.

XVIII Πάντων δὲ τῶν ζῴων κοινόν ἐστι τὸ περὶ τὴν κόρην ἐν τοῖς ὀφθαλμοῖς ⟨πῖον⟩[1]· ἔχουσι γὰρ τοῦτο τὸ μόριον στεατῶδες πάντα ὅσα ἔχουσι τὸ τοιοῦτον μόριον ἐν τοῖς ὀφθαλμοῖς καὶ μή εἰσι σκληρόφθαλμα. ἔστι δ' ἀγονώτερα πάντα[2] τὰ πιμελώδη καὶ ἄρρενα καὶ θήλεα. πιαίνεται δὲ πάντα πρεσβύτερα μᾶλλον ἢ νεώτερα ὄντα, μάλιστα δ' ὅταν καὶ τὸ πλάτος καὶ τὸ μῆκος ἔχῃ τοῦ μεγέθους καὶ εἰς βάθος αὐξάνηται.

XIX Περὶ δ' αἵματος ὧδε ἔχει· τοῦτο γὰρ πᾶσιν ἀναγκαιότατον καὶ κοινότατον τοῖς ἐναίμοις καὶ οὐκ ἐπίκτητον, ἀλλ' ὑπάρχει πᾶσι τοῖς μὴ φθειρομένοις.

[1] ⟨πῖον⟩ Dt. (πῖον pro κοινόν A.-W.): *quod est prope pupillam oculi in omnibus animalibus est multi sepi* Σ.
[2] πάντα ΑCΣ : om. PD, vulg.

HISTORIA ANIMALIUM, III. xvii–xix

less fat than others with regard to their stomach and their Omentum: an example is the eel, which has very little fat round the Omentum. Most animals grow fat round the belly, and especially those animals which do not move about much.

In fatty animals the brain is oily (*e.g.*, the pig), in suety animals, it is devoid of oil.[a] The viscera in respect of which animals tend to grow fattest are the kidneys. The right kidney is always less fatty than the left; and if both are very fat indeed, there is always a space clear of fat in between them. Animals that have suet tend to have it round the kidneys, and above all the sheep: sheep die through having their kidneys entirely enveloped in suet. Over-feeding brings on this condition, as occurs at Leontini in Sicily; that is why they do not turn the sheep out to feed until late in the evening, to reduce the amount they eat.

A phenomenon common to all animals is fat [b] round XVIII the pupil of the eye: in all animals which have such a part in their eyes (provided these are not hard), it resembles suet. Fat animals, male and female alike, all tend to be poor breeders. Fatness occurs in older animals more than in young, and especially when they have reached their full breadth and length and are beginning to grow in depth.

Now follows a description of the Blood. In all XIX blooded animals this is the most indispensable and Blood. most universal part [c]; it is not something acquired later, but it is present from the outset in all, provided they are not becoming corrupted. Blood is always

[a] Lit., "parched"; *cf.* 522 a 25.
[b] πῖον (fat) is A.-W.'s and Dittmeyer's restoration. It is supported by Michael Scot's translation. *Cf.* 533 a 9.
[c] *Cf.* 511 b 11.

217

ARISTOTLE

520 b

πᾶν δ' αἷμά ἐστιν ἐν ἀγγείῳ, ἐν ταῖς καλουμέναις φλεψίν, ἐν ἄλλῳ δ' οὐδενὶ πλὴν ἐν τῇ καρδίᾳ μόνον. οὐκ ἔχει δ' αἴσθησιν τὸ αἷμα ἁπτομένων ἐν οὐδενὶ τῶν ζῴων, ὥσπερ οὐδ' ἡ περίττωσις ἡ τῆς κοιλίας[1]· οὐδὲ δὴ ὁ ἐγκέφαλος οὐδ' ὁ μυελὸς οὐκ ἔχει αἴσθησιν ἁπτομένων. ὅπου δ' ἄν τις διέλῃ τὴν σάρκα, γίγνεται αἷμα ἐν ζῶντι, ἐὰν μὴ διεφθαρμένη ἡ σὰρξ ᾖ.[2] ἔστι δὲ τὴν φύσιν τὸ αἷμα τόν τε χυμὸν ἔχον γλυκύν, ἐάν περ ὑγιὲς ᾖ, καὶ τὸ χρῶμα ἐρυθρόν· τὸ δὲ χεῖρον ἢ φύσει ἢ νόσῳ μελάντερον. καὶ οὔτε λίαν παχὺ οὔτε λίαν λεπτὸν τὸ βέλτιστον, ἐὰν μὴ χεῖρον ᾖ διὰ φύσιν ἢ διὰ νόσον. καὶ ἐν μὲν τῷ ζῴῳ ὑγρὸν καὶ θερμὸν ἀεί, ἐξιὸν δ' ἔξω πήγνυται πάντων πλὴν ἐλάφου καὶ προκὸς καὶ εἴ τι ἄλλο τοιαύτην ἔχει τὴν φύσιν· τὸ δ' ἄλλο αἷμα πήγνυται, ἐὰν μὴ ἐξαιρεθῶσιν αἱ ἶνες. τάχιστα δὲ πήγνυται τὸ τοῦ ταύρου αἷμα πάντων.

Ἔστι δὲ τῶν ἐναίμων ταῦτα πολυαιμότερα, ἃ[3] καὶ ἐν αὑτοῖς καὶ ἔξω ζῳοτοκεῖ,[4] τῶν ἐναίμων μὲν ᾠοτοκούντων δέ. τὰ δ' εὖ ἔχοντα ἢ φύσει ἢ τῷ ὑγιαίνειν οὔτε πολὺ λίαν ἔχει, ὥσπερ τὰ πεπωκότα πόμα πρόσφατον, οὔτ' ὀλίγον, ὥσπερ τὰ πίονα λίαν· τὰ γὰρ πίονα καθαρὸν μὲν ἔχει ὀλίγον δὲ τὸ αἷμα, καὶ γίγνεται πιότερα γιγνόμενα ἀναιμότερα· ἄναιμον γὰρ τὸ πίον.

521 a

Καὶ τὸ μὲν πίον ἄσηπτον, τὸ δ' αἷμα καὶ τὰ ἔναιμα τάχιστα σήπεται, καὶ τούτων τὰ περὶ τὰ ὀστᾶ. ἔχει δὲ λεπτότατον μὲν αἷμα καὶ καθαρώτατον ἄνθρωπος, παχύτατον δὲ καὶ μελάντατον τῶν ζῳοτό-

---

[1] τῆς κοιλίας AC : ἐν τῇ κοιλίᾳ PD, vulg.
[2] ᾖ hic AC, post μὴ PD, vulg.   [3] ἃ AC : τὰ PD, vulg.
[4] ζῳοτοκεῖ Rhen., Sn., Dt. : ζῳοτόκα [καὶ] vulg.

218

found in a container, *viz.*, in the blood-vessels, and nowhere else except only in the heart. In no animal is the blood susceptible of sensation when touched,[a] no more than the residue from the stomach is : the same is true of the brain and the marrow. If you sever the flesh at any place, blood appears, provided the animal is alive and the flesh is not gangrened. If healthy, blood naturally has a sweet taste, and a red colour ; if in inferior condition, either naturally or through disease, it tends to be black. Blood in its best condition is not excessively thick nor thin, unless it has deteriorated either naturally or through disease. In the living animal it is always fluid and warm, but once it has left the body the blood of all animals congeals—except that of the deer and the roe and other such animals. Otherwise blood always congeals, unless the fibres are taken out. The quickest of all to congeal is bull's blood.

Of blooded animals, those which are both internally and externally viviparous have more blood than the blooded Ovipara. In animals which are in good condition, either naturally or through healthy dieting,[b] the blood is neither unduly abundant (as it is in those which have taken a recent drink), nor scanty, as happens in excessively fat animals : in fat animals, of course, the blood is pure, but small in quantity ; and increase of fat and decrease of blood go together : a fat animal being deficient in blood.

Fat does not putrefy, whereas blood and parts containing blood putrefy rapidly, and of these especially the regions around the bones. Blood is finest and purest in man, thickest and blackest—among Vivi-

[a] *Cf. P.A.* 650 b 4 ff., where the reason is given.
[b] Lit., " in virtue of being healthy."

ARISTOTLE

κων ταῦρος καὶ ὄνος. καὶ ἐν τοῖς κάτω δὲ μορίοις ἢ ἐν τοῖς ἄνω παχύτερον τὸ αἷμα γίγνεται καὶ μελάντερον.

Σφύζει δὲ τὸ αἷμα ἐν ταῖς φλεψὶν[1] ἅπασι πάντη ἅμα τοῖς ζῴοις, καὶ ἔστι τῶν ὑγρῶν μένον καθ᾽ ἅπαν τε τὸ σῶμα τοῖς ζῴοις καὶ ἀεί, ἕως ἂν ζῇ, τὸ αἷμα μόνον. πρῶτον δὲ γίγνεται τὸ αἷμα ἐν τῇ καρδίᾳ τοῖς ζῴοις, καὶ πρὶν ὅλον διηρθρῶσθαι τὸ σῶμα. στερισκόμενα δ᾽ αὐτοῦ καὶ ἀφιεμένου ἔξω πλείονος μὲν ἐκθνήσκουσι, πολλοῦ δ᾽ ἄγαν ἀποθνήσκουσιν. ἐξυγραινομένου δὲ λίαν νοσοῦσιν· γίγνεται γὰρ ἰχωροειδές, καὶ διοροῦται οὕτως ὥστε ἤδη τινὲς ἴδισαν αἱματώδη ἱδρῶτα. καὶ ἐξιὸν ἐνίοις οὐ πήγνυται παντελῶς ἢ διωρισμένως καὶ χωρίς. τοῖς δὲ καθεύδουσιν ἐν τοῖς ἐκτὸς μέρεσιν ἔλαττον γίγνεται τὸ αἷμα, ὥστε καὶ κεντουμένων μὴ ῥεῖν ὁμοίως. γίγνεται δὲ πεττομένων[2] ἐξ ἰχῶρος μὲν αἷμα, ἐξ αἵματος δὲ πιμελή· νενοσηκότος δ᾽ αἵματος αἱμορροῒς ἥ τ᾽ ἐν ταῖς ῥισὶ καὶ ἡ περὶ τὴν ἕδραν, καὶ ἰξία. σηπόμενον δὲ γίγνεται τὸ αἷμα ἐν τῷ σώματι πύον, ἐκ δὲ τοῦ πύου πῶρος.

Τὸ δὲ τῶν θηλειῶν πρὸς τὸ τῶν ἀρρένων διαφέρει· παχύτερόν τε γὰρ καὶ μελάντερόν ἐστιν, ὁμοίως ἐχόντων πρὸς ὑγίειαν καὶ ἡλικίαν, ἐν τοῖς θήλεσιν, καὶ ἐπιπολῆς μὲν ἔλαττον ἐν τοῖς θήλεσιν, ἐντὸς δὲ πολυαιμότερα.[3] μάλιστα δὲ καὶ τῶν θηλέων ζῴων γυνὴ πολύαιμον, καὶ τὰ καλούμενα

---

[1] ἐν add. PD, vulg.
[2] sic PD: πεττόμενον AC, vulg.
[3] corr. Dt.: πολυαιμότερον codd., vulg.

---

[a] *Cf. P.A.* 647 b 34.   [b] See note on 489 a 23.

HISTORIA ANIMALIUM, III. xix

para—in the bull and the ass. In the lower parts of the body the blood is thicker and blacker than in the upper ones.[a]

The blood in animals pulsates in the blood-vessels all over the body at once; and blood is the only fluid which remains throughout the whole body, and throughout life so long as it lasts. Further, blood is formed first in the heart, even before the body as a whole becomes articulated. If they are deprived of blood, or if a considerable quantity is lost, animals fall into a swoon: if too much is lost, they die. If it becomes too fluid, they fall ill, because then the blood becomes ichor-like,[b] and it gets so thin that cases have been known in which a blood-like sweat [c] has been observed. In some persons the blood, when it has left the body, does not congeal at all, or only in separate and isolated places. During sleep the blood becomes less plentiful in the exterior parts, so that if the sleeper is pricked with a sharp object the blood does not flow as copiously as usual. Blood is formed by a process of concoction out of ichor, and similarly fat out of blood. If the blood has become diseased haemorrhoids can be caused in the nostrils or the anus, or varicose veins may be produced. If the blood putrefies in the body it becomes pus, and pus produces *pōrus*.[d]

The blood of females differs from that of males. Thus, if both are on an equality with regard to health and age, the blood is thicker and blacker in females, and while it is less plentiful on the surface it is more plentiful inwardly. Further, of all female animals the human female has the most abundant blood, and

[c] *Cf. P.A.* 668 b 6, and note in Loeb ed. Perhaps a reference to a case of haematoporphyria.

[d] A kind of chalkstone.

καταμήνια γίγνεται πλεῖστα τῶν ζῴων ταῖς γυναιξίν. νενοσηκὸς δὲ τοῦτο τὸ αἷμα καλεῖται ῥοῦς. τῶν δ' ἄλλων τῶν νοσηματικῶν ἧττον μετέχουσιν αἱ γυναῖκες. ὀλίγαις δὲ γίγνεται ἰξία καὶ αἱμορροῖς καὶ ἐκ ῥινῶν ῥύσις· ἐὰν δέ τι συμβαίνῃ τούτων, τὰ καταμήνια χείρω γίγνεται.

Διαφέρει δὲ καὶ κατὰ τὰς ἡλικίας πλήθει καὶ εἴδει τὸ αἷμα· ἐν μὲν γὰρ τοῖς πάμπαν νέοις ἰχωροειδές ἐστι καὶ πλέον, ἐν δὲ τοῖς γέρουσι παχὺ καὶ μέλαν καὶ ὀλίγον, ἐν ἀκμάζουσι δὲ μέσως· καὶ πήγνυται ταχὺ τὸ τῶν γερόντων, κἂν ἐν τῷ σώματι ᾖ ἐπιπολῆς· τοῖς δὲ νέοις οὐ γίγνεται τοῦτο. ἰχώρ δ' ἐστὶν ἄπεπτον αἷμα, ἢ τῷ μήπω πεπέφθαι ἢ τῷ διωρῶσθαι.

XX Περὶ δὲ μυελοῦ· καὶ γὰρ τοῦτο ἓν τῶν ὑγρῶν ἐνίοις τῶν ἐναίμων ὑπάρχει ζῴων. πάντα δ' ὅσα φύσει ὑπάρχει ὑγρὰ ἐν τῷ σώματι, ἐν ἀγγείοις ὑπάρχει, ὥσπερ καὶ αἷμα ἐν φλεψὶ καὶ μυελὸς ἐν ὀστοῖς, [τὰ δ' ἐν ὑμενώδεσι, καὶ δέρμασι καὶ κοιλίαις.][1] γίγνεται δ' ἐν μὲν τοῖς νέοις αἱματώδης πάμπαν ὁ μυελός, πρεσβυτέρων δὲ γενομένων ἐν μὲν τοῖς πιμελώδεσι πιμελώδης, ἐν δὲ τοῖς στεατώδεσι στεατώδης. οὐ πάντα δ' ἔχει τὰ ὀστᾶ μυελόν, ἀλλὰ τὰ κοῖλα, καὶ τούτων ἐνίοις οὐκ ἔνεστιν· τὰ γὰρ τοῦ λέοντος ὀστᾶ τὰ μὲν οὐκ ἔχει πάμπαν, τὰ δ' ἔχει μικρόν,[2] διόπερ ἔνιοι οὔ φασιν ὅλως ἔχειν μυελὸν τοὺς λέοντας, ὥσπερ εἴρηται καὶ[3] πρότερον. καὶ ἐν τοῖς ὑείοις δ' ὀστοῖς ἐλάττων ἐστίν, ἐνίοις δ' αὐτῶν πάμπαν οὐκ ἔνεστιν.

Ταῦτα μὲν οὖν τὰ ὑγρὰ σχεδὸν ἀεὶ σύμφυτά ἐστι

[1] secl. A.-W., Dt.
[2] sic AC : οὐκ ἔχει, τὰ δ' ἔχει πάμπαν μικρόν vulg.

the catamenia are more plentiful in women than in any other animal. This blood if it has become diseased is known as flux (*rhous*). Apart from this one, women are less troubled by maladies ⟨of the blood⟩. Few women suffer from varicose veins, haemorrhoids, or nose-bleeding ; and if any of these occurs, the catamenia deteriorate.

Blood differs both in quantity and in appearance according to age. In very young animals it is ichor-like and plentiful; in the aged, thick, black, and scanty ; in the prime of life, it exhibits intermediate qualities. The blood of aged animals congeals quickly, even if it is at the surface of the body : this does not occur in the young. Ichor is simply unconcocted blood [a]—either because it has not yet been concocted, or because it has become thin again.

The next subject is Marrow. This is another of the fluid parts, which occurs in certain blooded animals. All natural fluids in the body are carried in receptacles ; thus, blood is carried in blood-vessels, and marrow in bones : [others are contained in membranous containers, whether in skin or gut]. In young animals the marrow is quite bloodlike ; when they are older, in the fatty animals it becomes fatty, and in the suety ones suety. Not all the bones contain marrow, but only the hollow ones, and some of these have none. Thus, in the lion some bones have no marrow at all, some only a little ; which explains why, as stated already, some writers assert that lions have no marrow.[b] In the bones of pigs there is a small quantity of marrow, though in some of them there is none at all.

XX
Marrow.

Now the fluids already mentioned are almost always Milk.

---

[a] See note on 489 a 23. [b] *Cf.* 516 b 7.

[3] καὶ om. D, vulg.

ARISTOTLE

521 b

τοῖς ζώοις, ὑστερογενῆ δὲ γάλα τε καὶ γονή. τούτων δὲ τὸ μὲν ἀποκεκριμένον ἅπασιν, ὅταν ἐνῇ,
20 ἐστίν,[1] τὸ γάλα· ἡ δὲ γονὴ οὐ πᾶσιν ἀλλ' ἐνίοις [οἱ καλούμενοι θοροί], οἷον[2] τοῖς ἰχθύσιν.

Ἔχει δ', ὅσα ἔχει τὸ γάλα, ἐν τοῖς μαστοῖς. μαστοὺς δ' ἔχει ὅσα ζῳοτοκεῖ καὶ ἐν αὑτοῖς καὶ ἔξω, οἷον ὅσα τε τρίχας ἔχει, ὥσπερ ἄνθρωπος καὶ ἵππος, καὶ τὰ κήτη, οἷον δελφὶς καὶ φώκαινα[3] καὶ φάλαινα· καὶ γὰρ ταῦτα μαυτοὺς ἔχει καὶ γάλα.
25 ὅσα δ' ἔξω ζῳοτοκεῖ μόνον ἢ ᾠοτοκεῖ, οὐκ ἔχει οὔτε μαστοὺς οὔτε γάλα, οἷον ἰχθὺς καὶ ὄρνις.

Πᾶν δὲ γάλα ἔχει ἰχῶρα ὑδατώδη, ὃς καλεῖται ὀρός, καὶ σωματῶδες, ὃ[4] καλεῖται τυρός· ἔχει δὲ πλείω τυρὸν τὸ παχύτερον τῶν γαλάκτων. τὸ μὲν οὖν τῶν μὴ ἀμφωδόντων γάλα πήγνυται (διὸ καὶ
30 τυρεύεται τὸ[5] τῶν ἡμέρων), τῶν δ' ἀμφωδόντων οὐ πήγνυται, ὥσπερ οὐδ' ἡ πιμελή, καὶ ἔστι λεπτὸν καὶ γλυκύ. ἔστι δὲ λεπτότατον μὲν γάλα καμήλου, δεύτερον δ' ἵππου, τρίτον δ' ὄνου· παχύτατον δὲ τὸ βόειον. ὑπὸ μὲν οὖν τοῦ ψυχροῦ οὐ πήγνυται τὸ
522 a γάλα, ἀλλὰ διοροῦται μᾶλλον· ὑπὸ δὲ τοῦ πυρὸς πήγνυται καὶ παχύνεται. οὐ γίγνεται δὲ γάλα, πρὶν ἢ ἔγκυον γένηται, οὐδενὶ τῶν ζῴων ὡς ἐπὶ τὸ πολύ. ὅταν δ' ἔγκυον ᾖ, γίγνεται μέν, ἄχρηστον δὲ τὸ πρῶτον καὶ ὕστερον. μὴ ἐγκύοις δ' οὔσαις
5 ὀλίγον μὲν ἀπ' ἐδεσμάτων τινῶν, οὐ μὴν ἀλλὰ καὶ βδαλλομέναις ἤδη πρεσβυτέραις προῆλθε, καὶ τοσ-

---

[1] ἐστίν AC : ἔνεστι PD, vulg.
[2] οἷον om. vulg. : ἐνίοις οἷον οἱ καλ. θ. ἐν ἰχθύσιν PD : οἱ καλ. θ. seclusi ut ex adnot. orta, om. Σ ; quid θοροί pluraliter scriptum significet non liquet.
[3] φώκαινα Karsch : φώκη codd., vulg. ἵππος καὶ φώκη, καὶ τὰ κήτη, οἷον δ. καὶ φάλαινα Meyer, p. 150.

HISTORIA ANIMALIUM, III. xx

connate in animals; milk and semen, however, come to be formed later. Milk, whenever it is present, is in all cases present in an already secreted and separated state; the semen, however, is not so produced in all animals, but only in some, as in fishes.[a]

As many animals as have milk have it in their breasts; and breasts are present in all animals that are both internally and externally viviparous, *e.g.*, all animals that have hair, such as man and horse, and the Cetacea (such as the dolphin, the porpoise, and the whale): these have breasts and milk. Those which are viviparous externally only, or which are oviparous, have neither breasts nor milk: examples are fishes and birds.

All milk contains a watery ichor known as whey, and a consistent stuff known as curd or cheese: the thicker sorts of milk contain more of the latter. Now the milk of the non-ambidentates curdles (hence the milk of the domesticated ones is made into cheese), but that of the ambidentates does not (nor indeed does their fat), and it is thin and sweet. The thinnest of all milk is the camel's, next the mare's, third the ass's; cow's milk is the thickest. Cold does not cause milk to congeal, but rather to turn to whey; heat makes it congeal and thicken. Generally speaking, milk is not produced in any animal until the time of pregnancy; then it is produced, but the first milk is unfit for use, as also is the latest. In non-pregnant animals a small quantity of milk has been produced by the use of certain foods; indeed, it has been known for quite elderly females to produce some

---

[a] *Cf. G.A.* 717 b 23 ff.

[4] ὅς Dt., Sn. secutus.
[5] ⟨τὸ⟩ Richards.

ARISTOTLE

οὗτον ἤδη τισὶν ὥστ' ἐκτιτθεῦσαι παιδίον. καὶ οἱ περὶ τὴν Οἴτην δέ, ὅσαι ἂν μὴ ὑπομένωσι τὴν ὀχείαν τῶν αἰγῶν, λαμβάνοντες κνίδην τρίβουσι τὰ οὔθατα βίᾳ διὰ τὸ ἀλγεινὸν εἶναι· τὸ μὲν οὖν πρῶτον αἱματῶδες ἀμέλγονται, εἶθ' ὑπόπυον, τὸ δὲ τελευταῖον γάλα ἤδη οὐδὲν ἔλαττον τῶν ὀχευομένων.

Τῶν δ' ἀρρένων ἔν τε τοῖς ἄλλοις ζῴοις καὶ ἐν ἀνθρώπῳ ἐν οὐδενὶ μὲν ὡς ἐπὶ τὸ πολὺ γίγνεται γάλα, ὅμως δὲ γίγνεται ἔν τισιν, ἐπεὶ καὶ ἐν Λήμνῳ αἴξ ἐκ τῶν μαστῶν, οὓς ἔχει δύο ὁ ἄρρην παρὰ τὸ αἰδοῖον, γάλα ἠμέλγετο τοσοῦτον ὥστε γίγνεσθαι τροφαλίδα, καὶ πάλιν ὀχεύσαντος τῷ ἐκ τούτου γενομένῳ συνέβαινε ταὐτόν. ἀλλὰ τὰ μὲν τοιαῦτα ὡς σημεῖα ὑπολαμβάνουσιν, ἐπεὶ καὶ τῷ ἐν Λήμνῳ ἀνεῖλεν ὁ θεὸς μαντευομένῳ ἐπίκτησιν ἔσεσθαι κτημάτων.[1] ἐν δὲ τοῖς ἀνδράσι μεθ' ἥβην ἐνίοις ἐκθλίβεται ὀλίγον· βδαλλομένοις δὲ καὶ πολὺ ἤδη τισὶ προῆλθεν.

Ὑπάρχει δ' ἐν τῷ γάλακτι λιπαρότης, ἣ καὶ ἐν τοῖς πεπηγόσι γίγνεται ἐλαιώδης. εἰς δὲ τὸ προβάτειον ἐν Σικελίᾳ, καὶ ὅπου πῖον,[2] αἴγειον μιγνύουσιν. πήγνυται δὲ μάλιστα οὐ μόνον τὸ τυρὸν ἔχον πλεῖστον, ἀλλὰ καὶ τὸ αὐχμηρότερον ἔχον.

Τὰ μὲν οὖν πλέον ἔχει γάλα ἢ ὅσον εἰς τὴν ἐκτροφὴν τῶν τέκνων, καὶ χρήσιμον εἰς τύρευσιν καὶ ἀπόθεσιν, μάλιστα μὲν τὸ προβάτειον καὶ τὸ αἴγειον, ἔπειτα τὸ βόειον· τὸ δ' ἵππειον καὶ ὄνειον

[1] μαντευσαμένῳ et χρημάτων PD, vulg.
[2] πῖον A¹C, Dt.: πλεῖον PDA², vulg.: *ubi coagulum caprinum miscent* Gul., unde ὀπὸν καὶ πυὸν αἴγειον coni. Th.

when milked, and some have produced sufficient to suckle a youngster. Furthermore, the inhabitants of the region round Mount Oeta take she-goats which refuse to submit to the male and rub their udders forcibly with nettles enough to cause pain; the fluid they obtain to begin with is bloodlike, then it is mixed with pus, and finally it is true milk, no less than that which is obtained from those which submit to the male.

Generally speaking, milk is not produced in any male animals, man included, but it is in some individuals. For example, at one time there was in Lemnos a he-goat which used to be milked by its dugs (the male has two near the penis), and it produced so much that cheese was made out of it; and the same occurred with a male of which it was the sire. Nevertheless, phenomena of this sort are regarded as signs from heaven. When the owner of the goat in Lemnos consulted the oracle, the god's reply was that he would come into possession of a fortune. There are men from whom, after reaching puberty, a small amount can be squeezed out; some when milked have actually produced large quantities.

Milk contains a fatty ingredient, and in congealed milk this becomes oily. In Sicily, and wherever sheep's milk is fat, they mix goats' milk with it. The milk which congeals best is not only that which contains most cheese, but also that in which the cheese is less fatty.[a]

Now some animals produce more milk than is required to nourish their young, and this is useful for cheese-making and for storage. First in this respect is sheep's milk, next goats', then cows'. Mares' milk

[a] Lit., " parched "; *cf.* 520 a 27.

μίγνυται εἰς τὸν Φρύγιον τυρόν. ἔνεστι δὲ τυρὸς πλείων ἐν τῷ βοείῳ ἢ ἐν τῷ αἰγείῳ· γίγνεσθαι γάρ φασιν οἱ νομεῖς ἐκ μὲν ἀμφορέως αἰγείου γάλακτος τροφαλίδας ὀβολιαίας μιᾶς δεούσης εἴκοσιν, ἐκ δὲ βοείου τριάκοντα. τὰ δ' ὅσον τοῖς τέκνοις ἱκανόν, πλῆθος δ' οὐδὲν οὐδὲ[1] χρήσιμον εἰς τύρευσιν, οἷον πάντα τὰ πλείους ἔχοντα μαστοὺς δυοῖν· οὐδενὸς γὰρ τούτων οὔτε πλῆθός ἐστι γάλακτος οὔτε τυρεύεται τὸ γάλα.

Πήγνυσι δὲ τὸ γάλα ὀπός τε συκῆς καὶ πυετία. ὁ μὲν οὖν ὀπὸς εἰς ἔριον ἐξοπισθείς, ὅταν ἐκπλυθῇ πάλιν τὸ ἔριον εἰς γάλα ὀλίγον· τοῦτο γὰρ κεραννύμενον πήγνυσιν. ἡ δὲ πυετία γάλα ἐστίν· τῶν γὰρ ἔτι θηλαζόντων γίγνεται ἐν τῇ κοιλίᾳ. ἔστιν οὖν[2] ἡ πυετία γάλα ἔχον ἐν ἑαυτῷ πῦρ ὃ ἐκ τῆς[3] τοῦ ζῴου θερμότητος πεττομένου τοῦ γάλακτος γίγνεται. ἔχει δὲ πυετίαν τὰ μὲν μηρυκάζοντα πάντα, τῶν δ' ἀμφωδόντων δασύπους. βελτίων δ' ἐστὶν ἡ πυετία ὅσῳ ἂν ᾖ παλαιοτέρα· συμφέρει γὰρ πρὸς τὰς διαρροίας ἡ τοιαύτη μάλιστα καὶ ἡ τοῦ δασύποδος· ἀρίστη δὲ πυετία[4] νεβροῦ.

Διαφέρει δὲ τῷ πλέον ἱμᾶσθαι γάλα ἢ ἔλαττον τῶν ἐχόντων γάλα ζῴων κατά τε τὰ μεγέθη τῶν σωμάτων καὶ τὰς τῶν ἐδεσμάτων διαφοράς, οἷον ἐν Φάσει μέν ἐστι βοΐδια μικρὰ ὧν ἕκαστον βδάλ-

---

[1] οὐδὲ Richards : οὔτε vulg.
[2] θ. ἐν τῇ κοιλίᾳ γίγνεται. ἔστιν οὖν A.-W. : θ. ἐστὶν ἐν τῇ κοιλίᾳ. γίγνεται οὖν vulg.
[3] sic PD : τυρόν· ἐκ δὲ τῆς C, A.-W., Dt.
[4] add. ἡ τοῦ PD, vulg.

---

[a] See add. note, p. 239.
[b] Some mss. and edd. here read τυρόν (cheese) ; see critical

HISTORIA ANIMALIUM, III. xx-xxi

and asses' milk is included in the blend for making Phrygian cheese. There is more cheese in cows' milk than in goats': dairy-farmers tell us that out of one *amphoreus* of goats' milk they can get nineteen *obol*-cheeses,[a] and from the same amount of cows' milk, thirty. Other animals produce only enough milk to feed their young, and none over, none fit for cheese-making: this is true of all animals which have more than two dugs; none of these animals produces a superabundance of milk nor is their milk made into cheese.

Milk is curdled by fig-juice aud rennet. The fig-juice is squeezed out into some wool, and then the wool is rinsed out into a small quantity of milk: this milk when mixed with other milk curdles it. Rennet XXI is a sort of milk; it is formed in the stomach of young animals while still being suckled. Rennet is thus milk which contains fire,[b] which comes from the heat of the animal while the milk is undergoing concoction. All ruminants contain rennet; the hare is the only ambidentate which does so. The older rennet is, the better it is: this sort is a very good remedy for looseness of the bowels; so is hare's rennet, and young deer's rennet is the best of all.

The milk-producing animals yield a greater or less amount of milk in relation to their various sizes and the various pastures they feed on. Thus, in Phasis there is a breed of small cattle every one of which

note. *Cf. G.A.* 739 b 23 " Rennet is milk containing vital heat, which integrates the homogeneous substance and makes it set " (εἰς ἓν ἄγει καὶ συνίστησι); *cf.* 772 a 25. The action of rennet is compared with that of semen, though the " setting " brought about by the latter involves quality as well as quantity, whereas that brought about by rennet involves quantity only.

ARISTOTLE

λεται γάλα πολύ, αἱ δ' Ἠπειρωτικαὶ βόες αἱ μεγάλαι βδάλλονται ἑκάστη ἀμφορέα καὶ τούτου τὸ ἥμισυ κατὰ τοὺς δύο μαστούς[1]· ὁ δὲ βδάλλων ὀρθὸς ἕστηκεν, μικρὸν ἐπικύπτων, διὰ τὸ μὴ δύνασθαι ἂν ἐφικνεῖσθαι[2] καθήμενος. γίγνεται δ' ἔξω ὄνου καὶ τἆλλα μεγάλα ἐν τῇ Ἠπείρῳ τετράποδα, μέγιστοι δ' οἱ βόες καὶ οἱ κύνες. νομῆς δὲ δέονται τὰ μεγάλα πλείονος· ἀλλ' ἔχει πολλὴν ἡ χώρα τοιαύτην εὐβοσίαν, καὶ καθ' ἑκάστην ὥραν ἐπιτηδείους τόπους. μέγιστοι δ' οἵ τε βόες εἰσὶ καὶ τὰ πρόβατα τὰ καλούμενα Πυρρικά, τὴν ἐπωνυμίαν ἔχοντα ταύτην ἀπὸ Πύρρου τοῦ βασιλέως.[3]

Τῆς δὲ τροφῆς ἡ μὲν σβέννυσι τὸ γάλα, οἷον ἡ Μηδικὴ πόα, καὶ μάλιστα τοῖς μηρυκάζουσιν· ποιεῖ δὲ πολὺ ἑτέρα,[4] οἷον κύτισος καὶ ὄροβοι, πλὴν κύτισος μὲν ὁ ἀνθῶν οὐ συμφέρει (πίμπρησι γάρ), οἱ δ' ὄροβοι ταῖς κυούσαις οὐ συμφέρουσι (τίκτουσι γὰρ χαλεπώτερον). ὅλως δὲ τὰ φαγεῖν ⟨πολὺ⟩[5] δυνάμενα τῶν τετραπόδων, ὥσπερ καὶ πρὸς τὴν κύησιν

[1] μαστεύσεις coni. Th.
[2] AC : ἐφικέσθαι PD, vulg.
[3] haec de Pyrrho rege suspic. Dittm.; cf. 595 b 18, ubi πυρρίχας βοῦς codd.; fortasse hic etiam πυρρίχα scribendum, et verba quae ad regem spectant delenda.
[4] ἑτέρα Dt. : ἕτερα APD, vulg.
[5] ⟨πολὺ⟩ Dt.

---

[a] Large cows in Epeirus are mentioned again at 595 b 18 ff., where they are described as πυρρίχαι (not πυρρικαί); see note on l. 25 below.

[b] See add. note, p. 239.

[c] In this passage the only animals in Epeirus to which the adjective "Pyrrhic" is actually applied are the sheep, although the large cows in Epeirus are mentioned a few lines

yields a copious supply of milk. In Epeirus [a] there are large cows which yield an *amphoreus* [b] and half of this from the two teats. The milker has to stand upright, bending slightly forward, because he would not be able to reach far enough if seated. In addition to the cattle, all the quadrupeds (except the ass) in Epeirus are large in size, and in proportion the cattle and the dogs are the largest. Large animals of course require more extensive pasturage than small ones—which is supplied by this district: it provides just the rich feed which they need, and suitable regions for every season of the year. The cattle are very large indeed, and so are the Pyrrhic sheep as they are called (they are named after King Pyrrhus.) [c]

Some pasture, *e.g.*, lucerne,[d] causes a failure of milk, especially in ruminants; other pasture increases the milk, *e.g.*, cytisus and vetches—though cytisus when in bloom is not wholesome, because it causes burning, and vetch is not wholesome for pregnant animals, because it makes parturition more difficult. Generally speaking, quadrupeds that can eat plentifully not only benefit in regard to pregnancy, but also

above, and at 595 b 18 ff. we find the large cows in Epeirus described as πυρρίχαι. It seems doubtful whether the explanation of the epithet given in the present passage can be genuine, since the famous King Pyrrhus of Epeirus was not born (*c.* 318 B.C.) until after Aristotle's death. Pyrrhus the son of Achilles, however, was said to have settled in Epeirus after the Trojan war and become the founder of the race of Molossian kings (*cf.* Pindar, *Nem.* iv. 82, vii. 54, etc., and Vergil, *Aen.* iii. 333). It seems possible that the name may be accounted for in this way. At Theocritus iv. 20 ὁ ταῦρος ὁ πυρρίχος occurs, where πυρρίχος is usually translated " ruddy " simply. I suspect that the original reading in the present passage was πυρρίχα (at 595 b 18 the mss. give πυρρίχας), and that it was altered to suit the inserted explanation.

[d] Lit., " the Medic grass." *Medicago sativa* L.

ARISTOTLE

συμφέρει, καὶ βδάλλεται πολὺ τροφὴν ἔχοντα. ποιε͂, δὲ γάλα καὶ τῶν φυσ⟨κ⟩ωδῶν[1] ἔνια προσφερόμεναῖ οἷον καὶ κυάμων πλῆθος οἳ καὶ αἰγὶ καὶ βοῒ καὶ χιμαίρᾳ· ποιεῖ γὰρ καθιέναι τὸ οὖθαρ. σημεῖον δὲ τοῦ γάλα πλέον ἱμήσεσθαι,[2] ὅταν πρὸ τοῦ τόκου τὸ οὖθαρ βλέπῃ κάτω.

Γίγνεται δὲ πολὺν χρόνον γάλα πᾶσι τοῖς ἔχουσιν, ἂν ἀνόχευτα διατελῇ καὶ τὰ ἐπιτήδεια ἔχωσι, μάλιστα δὲ τῶν τετραπόδων πρόβατον· ἀμέλγεται γὰρ μῆνας ὀκτώ. ὅλως δὲ τὰ μηρυκάζοντα γάλα πολὺ καὶ χρήσιμον εἰς τυρείαν ἀμέλγεται. περὶ δὲ Τορώνην αἱ βόες ὀλίγας ἡμέρας πρὸ τοῦ τόκου διαλείπουσι, τὸν δ' ἄλλον χρόνον πάντα ἔχουσι γάλα. τῶν δὲ γυναικῶν τὸ πελιώτερον[3] γάλα βέλτιον τοῦ λευκοῦ τοῖς τιτθευομένοις· καὶ αἱ μέλαιναι τῶν λευκῶν ὑγιεινότερον ἔχουσιν. τροφιμώτατον μὲν οὖν τὸ πλεῖστον ἔχον τυρόν, ὑγιεινότερον δὲ τοῖς παιδίοις τὸ ἐλάττονα.[4]

XXII Σπέρμα δὲ προΐενται πάντα τὰ ἔχοντα αἷμα. τί δὲ συμβάλλεται εἰς τὴν γένεσιν καὶ πῶς, ἐν ἄλλοις λεχθήσεται. πλεῖστον δὲ κατὰ τὸ σῶμα ἄνθρωπος προΐεται. ἔστι δὲ τῶν μὲν ἐχόντων τρίχας γλίσχρον, τῶν δ' ἄλλων ζῴων οὐκ ἔχει γλισχρότητα.

---

[1] φυσκωδῶν scripsi : tale desideravit Th., cf. 595 b 6 : φυσωδῶν codd., edd. φασκωλοειδῶν coni. Warmington.
[2] ἱμήσεσθαι Buss., Pi., A.-W., Dt. : ἱμήσασθαι A¹C : γενήσεσθαι PDA², vulg.
[3] πελιώτερον A¹C : πελιδνότερον PDA², vulg.
[4] ἐλάττονα Sylb. : τὸ ἔχον ἐλάττονα Pi. : ἔλαττον codd., vulg.

---

[a] The mss. here read φυσωδῶν, "causing flatulence"; the same adjective occurs at 588 a 9; and at 595 b 6 the adjective φυσητικοῖς, where two mss. read φυσικοῖς. In the first

yield much milk because they are well nourished. Some of the podded [a] plants if used as feed are milk-producing; a good diet of beans will do this for the ewe, the she-goat, the cow, and the young she-goat: it causes them to drop their udders. A sign that there will be a good yield of milk is when the udder faces downwards before parturition.

All milk-producing animals go on producing it for a long time if they are kept away from the male and are given suitable food: so far as quadrupeds are concerned this applies especially to the ewe, which can be milked for eight months continuously. Generally, the ruminants yield an abundance of milk, suitable for cheese-making. In the district round Toronē cows run dry for a few days before calving, but are in milk all the rest of the time. In women, milk which is rather livid in colour is better than white milk for suckling; and dark women's milk is healthier than fair ones'. The most nourishing milk is that which contains most cheese, but milk containing less cheese is more healthy for children.

All blooded animals emit semen. We shall state elsewhere what is its contribution to generation, and the method of it. For his size, man emits more than any animal. The semen of hairy animals is sticky, but that of others is not. In all animals it is white.

XXII Semen.

and third of these passages the adjective is applied to beans and such plants. In his note on 595 b 6 Th. suggests that φασήλοις, or some word meaning "podded," is required; and in the present passage he in fact translates by "leguminous." Following his suggestion I have conjectured φυσκωδῶν, from φύσκη, a large intestine, especially applied to one stuffed with meat to make a sausage or black pudding. φύσκη is also used by Dioscorides (*Alex.* 22) to denote the gall-bag on a plant; it thus has some botanical connexion. The passage at Herodotus iv. 2 perhaps tells in favour of keeping φυσωδῶν.

λευκὸν δὲ πάντων· ἀλλ' Ἡρόδοτος διέψευσται γράψας τοὺς Αἰθίοπας προΐεσθαι μέλαιναν τὴν γονήν.

Τὸ δὲ σπέρμα ἐξέρχεται μὲν λευκὸν καὶ παχύ, ἂν ᾖ ὑγιεινόν, θύραζε δ' ἐλθὸν λεπτὸν γίγνεται καὶ μέλαν. ἐν δὲ τοῖς πάγοις οὐ πήγνυται, ἀλλὰ γίγνεται πάμπαν λεπτὸν καὶ ὑδατῶδες καὶ τὸ χρῶμα καὶ τὸ πάχος· ὑπὸ δὲ τοῦ θερμοῦ πήγνυται καὶ παχύνεται. καὶ ὅταν ἐξίῃ χρονίσαν ἐν τῇ ὑστέρᾳ, παχύτερον ἐξέρχεται, ἐνίοτε δὲ ξηρὸν καὶ συνεστραμμένον. καὶ τὸ μὲν γόνιμον ἐν τῷ ὕδατι χωρεῖ κάτω, τὸ δ' ἄγονον διαχεῖται. ψευδὲς δ' ἐστὶ καὶ ὃ Κτησίας γέγραφε περὶ τῆς γονῆς τῶν ἐλεφάντων.

HISTORIA ANIMALIUM, III. xxii

Herodotus [a] is mistaken when he writes that the Ethiopians emit black semen.

Semen when it issues from the body is white and thick, if it is healthy; afterwards it becomes thin and black. In frosty weather it does not congeal but becomes quite thin and watery both in colour and consistency; whereas heat makes it congeal and thicken. If it remains in the uterus a long time before coming out, it is thicker when it does so, and sometimes when it comes out it is dry and compact. Fertile semen sinks in water, infertile dissolves. Ktesias's statement [b] about the semen of elephants is untrue.

[a] Herodotus iii. 101; *cf. G.A.* 736 a 10.
[b] See *G.A.* 736 a 4 ff., and note there, in Loeb ed., on Ktesias.

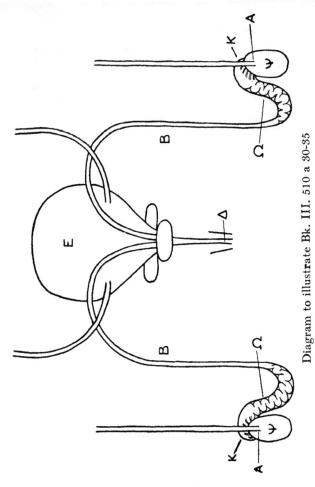

Diagram to illustrate Bk. III. 510 a 30-35

ADDITIONAL NOTES

Page 34, note *b*. See also *G.A.* 748 b 34, and note there in Loeb edition. P. Louis (*Revue de Philologie*, XXI (1957), pp. 63-65) argues that properly *ginnos* is simply a small ass or horse.

Page 88, note *a* (*cf.* page 84, note *a*). W. H. Riddell " Concerning Unicorns," *Antiquity*, XIX (1945), pp. 194 ff., records that he has seen oryx with only one horn as the result of fighting. The prototype of the unicorn he thinks may have been the nylghau (see 498 b 32). Wiegmann too (*ad loc.*), writing in 1826, quotes Prof. H. Lichtenstein as reporting oryx which had lost a horn : this he supposed had given rise to tales about the unicorn. Photius (*Bibl.* cod. 72, 48 b 19 ff. Bk.) quotes Ktesias as describing wild asses in India, equal in size to horses or larger, with a horn on their foreheads, a cubit in length, white at the base, red or purple at the tip, and black in the centre. Since Aristotle appears to be speaking of wild animals, artificially produced unicorns (*cf.* W. Franklin Dove, " Artificial production of the fabulous unicorn," *The Scientific Monthly*, XLII (1936), pp. 431-436) probably need not be taken into account. Nevertheless, Dr. Dove's operation on a male Ayrshire day-old calf confirms in a remarkable way Ktesias's description, as he himself points out (*loc. cit.*, p. 435) : " the horn sheath is white or greyish-white at the base, and is tipped with black. Had the unicorn been a female, the horn would be tipped with red, since the color appears as a sex-limited factor in this particular breed." (The animal at the time of writing was $2\frac{1}{2}$ years old). Cuvier had argued that a cloven-footed ruminant with a single horn would be impossible, because its frontal bone would be divided, and no horn could grow above the division. Dr. Dove's experiment showed (*J. of Experim. Zoology*, 69 (1935), pp. 347-405) that this argument was unfounded, since the horn spike grows not *from* the skull but *upon* the skull, first as an epiphysis, later to fuse to the frontal bones over the suture as a horn spike solidly attached to the skull ; a single united sheath covers the horn spike. The bony horn cores of ruminants are not outgrowths of the

skull (frontal) bones, as comparative anatomists (including Cuvier) had generally considered them to be, but are rather the products of separate ossification centres with their anlagen residing in tissues above the frontal bones ; and these anlagen or horn buds may be transplanted in whole or in part to other regions of the head, where they take root and develop as true horns or parts of horns either solidly or loosely attached to the skull, according to the method of transplantation. Dr. Dove also refers to a statement in Pliny (*N.H.* xi. 127), which indicates that the transplantation of horn-buds in young oxen to produce four-horned animals was practised in ancient times. (I am greatly indebted to my friend Mr. William F. Dove for sending me an offprint of his father's article in *The Scientific Monthly*, from which these excerpts are taken.) It is interesting to note that Aristotle seems to have known something of the way in which ruminants' horns grow, since at *H.A.* 500 a 8 he says that " the hollow part grows out rather from the skin," and at 517 a 27 that " the horns grow attached to the skin rather than to the bone," although in both passages he speaks of the central core as growing out of the bone of the skull. It may also be of interest to note that at *H.A.* 595 a 13 ff. Aristotle remarks that the horns of young cattle if smeared with hot wax can be turned in any direction one wishes.

Page 89, note *b*. Aristotle probably had in mind the half-astragal which appears in high relief as a ready means of identification on the top of the one-stater weights of Athens. For an example of a one-stater bronze weight of the fifth century found in the agora, see *Hesperia*, VII (1938), p. 360, fig. 51 ; fourth-century examples are similar. The labyrinth design is found on frescoes at Knossos, and on fourth-century coins of Knossos (see, *e.g.*, C. T. Seltman, *Greek Coins*, pl. xxxvii). For this information I am indebted to the kindness of Professor Homer A. Thompson.

Page 99, note *b*. The passage about the martichoras, however, appears in Michael Scot's translation. Photius' version (*Bibl.* cod. 72, 45 b 31 ff. Bk.) of Ktesias' account agrees closely with that given in the text of *H.A.*

Page 113, note *a*. Regenbogen thinks the account of the chamaeleon is probably taken from Theophrastus' treatise on animals that change their colour (see Photius, *Bibl.* cod.

HISTORIA ANIMALIUM

278, 525 a 31 ff. Bk.). This, however, seems unlikely, since the present passage is concerned with much else beside colour-changes.

Page 228, note *a* (p. 230, n. *b*). An *amphoreus* for wine is usually taken as equal to about nine gallons, but many other sizes are known of jars so called. At Eurip. *Cyclops*, 327 the Cyclops tells Odysseus that he drinks off an *amphoreus* of milk, but deductions from this about the actual quantity might be unreliable. The obol here probably refers to price, since weight and shape both seem inappropriate. At *Odyssey*, ix. 246 ff. we read that the Cyclops, when he milked his animals, put half the milk into vessels for drinking later (as above), and the other half he curdled (θρέψας) and put into wicker baskets. Professor Homer Thompson tells me that we may have an indication of the kind of basket used for making cheeses, in a clay imitation of a small bowl-shaped basket, with almost straight sides, made by pressing clay into a finely-woven basket: the size of this vessel is 2¾" high by nearly 4½" diameter at the rim (see Eva T. H. Brann, *The Athenian Agora* VIII, *Late Geometric and Protoattic Pottery*, 1962, p. 62, no. 271; illustrated on plate 16). Professor Thompson has also drawn my attention to the word ὀβολιαῖος or ὀβελιαῖος in Philo, *De Spec. Leg.* ii, § 20 (273 M.) used of loaves, with cheese in the context: ὅμως ἐφ' ἃ καὶ οἱ πένητες ἡμεῖς ἐστιν ὅτε τρέπονται, κεραμιαίας κύλικας καὶ ὀβελιαίους ἄρτους καὶ ἐλαίας ἢ τυρὸν ἢ λάχανα προσόψημα, where ὀβελιαίους is the reading of the oldest MS. (R), ὀβολιαίους of F; editors change it to ὀβελίας (ὀβελέας is the reading of M); the MSS. show similar variants for κεραμιαίας, which editors change to κεραμεᾶς. Colson here translates (reading ὀβελίας) "spit-baked loaves," and Cohn "geröstetes Brot." The word ὀβελιαῖος, in the sense of "sagittal," is used by Galen of the suture of a skull, and it has occurred to me that, since the bowl mentioned above has three bands of thin red glaze on the inside of the flat bottom, crossing at the centre, the baskets themselves might have had similar cross-bands, which would be impressed on the cheeses made in them, and this pattern could give rise to the description ὀβολιαῖος. The word τροφαλίς is no doubt connected with τρέφω (*cf.* θρέψας above).